Reconceptualizing th

DIBNER
INSTITUTE
FOR THE HISTORY
OF SCIENCE AND
TECHNOLOGY

Dibner Institute Studies in the History of Science and Technology
George Smith, general editor

Isaac Newton's Natural Philosophy
Jed Z. Buchwald and I. Bernard Cohen, editors

Histories of the Electron: The Birth of Microphysics
Jed Z. Buchwald and Andrew Warwick, editors

Science Serialized: Representations of the Sciences in Nineteenth-Century Periodicals
Geoffrey Cantor and Sally Shuttleworth, editors

The Kantian Legacy in Nineteenth-Century Science
Michael Friedman and Alfred Nordmann, editors

Natural Particulars: Nature and the Disciplines in Renaissance Europe
Anthony Grafton and Nancy Siraisi, editors

The Enterprise of Science in Islam: New Perspectives
J. P. Hogendijk and A. I. Sabra, editors

Instruments and Experimentation in the History of Chemistry
Frederic L. Holmes and Trevor H. Levere, editors

Systems, Experts, and Computers: The Systems Approach in Management and Engineering, World War II and After
Agatha C. Hughes and Thomas P. Hughes, editors

From Embryology to Evo-Devo
Manfred D. Laubichler and Jane Maienschein, editors

The Heirs of Archimedes: Science and the Art of War through the Age of Enlightenment
Brett D. Steele and Tamera Dorland, editors

Ancient Astronomy and Celestial Divination
N. L. Swerdlow, editor

The Artificial and the Natural: An Evolving Polarity
Bernadette Bensaude-Vincent and William R. Newman, editors

Communications Under the Seas: The Evolving Cable Network and its Implications
Bernard S. Finn and Daqing Yang, editors

Reconceptualizing the Industrial Revolution
Jeff Horn, Leonard N. Rosenband, Merritt Roe Smith, editors

Reconceptualizing the Industrial Revolution

---

edited by Jeff Horn, Leonard N. Rosenband, and Merritt Roe Smith

The MIT Press
Cambridge, Massachusetts
London, England

For information about special quantity discounts, please e-mail special_sales@mitpress.mit.edu.

This book was set in Bembo by Toppan Best-set Premedia Limited.

Printed and bound in the United States of America.

Library of Congress Cataloging-in-Publication Data

Reconceptualizing the Industrial Revolution / edited by Jeff Horn, Leonard N. Rosenband, Merritt Roe Smith.
p. cm.
Includes bibliographical references and index.
ISBN 978-0-262-51562-7 (pbk. : alk. paper)
1. Industrial revolution 2. Industrialization. 3. Economic history. 4. Technological innovations—History. I. Horn, Jeff, Ph.D. II. Rosenband, Leonard N. III. Smith, Merritt Roe, 1940–
HD2329.R43 2010
330.9′034—dc22

2010000386

10 9 8 7 6 5 4 3 2

Contents

Contributors

Maxine Berg Professor of History, University of Warwick

Eric Dorn Brose Professor of History, Drexel University

Kristine Bruland Professor of Economic History, University of Oslo, and Professor of History, University of Geneva

Peter Gatrell Professor of Economic History, University of Manchester

Anne G. Hanley Associate Professor of History, Northern Illinois University

Jeff Horn Professor of History, Manhattan College

Ian Inkster Professor of Global History, Wenzao Ursuline College of Languages, Kaohsiung, Taiwan

Robert Martello Associate Professor of the History of Science and Technology, Franklin W. Olin College of Engineering

Joel Mokyr Robert H. Strotz Professor of Arts and Sciences and Professor of Economics and History, Northwestern University

Patrick K. O'Brien Professor of Global Economic History, London School of Economics

Prasannan Parthasarathi Associate Professor of History, Boston College

Peter C. Perdue Professor of History, Yale University

Leonard N. Rosenband Professor of History, Utah State University

Merritt Roe Smith Leverett and William Cutten Professor of the History of Technology, Massachusetts Institute of Technology

Marta V. Vicente Associate Professor of History and Women's Studies, University of Kansas

Acknowledgments

This book is the outcome of a conference on April 1–2, 2005, at the Dibner Institute for the History of Science and Technology, which was then on the campus of the Massachusetts Institute of Technology. Above all, the editors of this book, who were also the co-conveners of the meeting, are grateful to the Dibner family and the Institute for their generosity. George R. Smith, the acting director of the Dibner Institute, supported both the conference and this book, and served as a wonderful host at the event itself. Bonnie Edwards, the executive director of the Institute, graced our efforts with her organizational skills and enthusiasm. The administrative staff of the institute—Carla Chrisfield, Rita Dempsey, Trudy Kontoff, and Dawn Davis Loring—took expert care of matters large and small. David McGee of the Burndy Library established a Web portal that enabled us to share the papers effectively. And Judy Spitzer, of MIT's Science, Technology, and Society Program, eased our concerns on a number of occasions. Finally, Reva Rosenband did extraordinary work in the final preparation of the manuscript for this book. This volume is dedicated to our parents: Eleanor Smith Fox and Wilson N. Smith, Regina and Maurice Rosenband, and Carole and Richard Horn.

1

Introduction

Jeff Horn, Leonard N. Rosenband, and Merritt Roe Smith

Johann Wolfgang von Goethe claimed in 1825 that "wealth and speed are what the world admires, and what all are bent on." He took particular note of "railways, express mail-coaches, steamboats," all products and symbols of the early stirrings of large-scale industrialization.[1] This book considers the Industrial Revolution in a broad range of national settings. Each chapter explores a distinctive production ecology—a complex blend of natural resources, demographic pressures, cultural impulses, technological assets, commercial practices, and a host of other features. Yet these studies also reveal the portability of skilled workers and technologies, as well as the porosity of political borders. Taken together, the chapters emphasize the acceleration of wealth making and the spread of "useful knowledge" from Goethe's day onward. But this book is tempered as well by the historians' attention to the inefficiencies of every production regime, whether at the level of the shop floor or of state policies. Moreover, these chapters reveal the layers of cultural and institutional sediment that gathered around the manufacture of both steam turbines and porcelain plates.

Not long ago, many of the fundamental issues concerning the origins and nature of the Industrial Revolution seemed settled. Most scholars accepted some version of the idea that a clear boundary, or at least a set of visible markers, separated traditional and modern society. Perhaps W. W. Rostow's work was the most prominent example of this approach. He detected a linear pathway to modern industrial production that was punctuated by "stages," discrete toeholds on an ever-ascending journey.[2] But matters grew more complicated than the binary oppositions of modernization theory could contain. Capitalist practices had deep roots in traditional societies, and nonmarket customs and motives survived alongside industrial capitalism. Family firms continued to flourish, especially in continental Europe, among the impersonal managerial behemoths.[3] Batch production prospered beside continuous output.[4] Even the machine itself, the source and the symbol of the displacement of fingertip skills, often left gaps between promise and performance. As Raphael Samuel observed, "In many cases the machines . . . failed to execute their appointed tasks. Even if brought 'nearly . . . to perfection' by its inventor, a machine would often prove difficult to operate."[5] Maxine Berg has rightly reminded us that "the journeyman in a *large-scale* [our italics] brass finishing works was designer, supervisor, tool maker, tool setter, and all-around workman."[6] Where, then, was the precise tipping point between traditional and modern production?

Despite a series of scholarly challenges, economic historians for the most part still regard England as the primary site of the First Industrial Revolution. In a path-breaking essay, Peter Mathias posed a famous query: Was English industrial development first because it was unique—or unique because it was first? He replied in the affirmative to both dimensions of the problem. After all, he reasoned, England had engineered the original industrial track. Followers might seek to duplicate the journey or craft their own route, but the territory would never again be uncharted.[7] According to Eric Hobsbawm, "Subsequent [industrial] revolutions could use the British experience, example and resources. Britain could use those of other countries only to a very limited and minor extent."[8] From Birmingham and like-minded precincts, English migrants, machines, and goods moved abroad, confronting moss-covered manufacture wherever they landed. What returned, orthodox opinion claimed, had little import. Thus, David Landes deemed the industrial transformation of Belgium, France, and Germany "Continental Emulation," and there was no doubt about the location of the prime mover.[9]

If a certain English exceptionalism lingers in learned debates about the era of Wedgwood and Watt, convergence has taken on fresh meaning. As Patrick K. O'Brien contends in chapter 2, "techniques used to manufacture, bleach, dye, and print cotton cloth are no longer acclaimed as peculiarly 'English.'" Berg points out in chapter 3 that even commodities that had acquired a British "brand" mimicked Asian designs and European luxuries. Finally, Joel Mokyr draws attention in chapter 4 to a "European Enlightenment" that brought "the full force of human knowledge to bear on technology." The arrow of industrial espionage in the early years of industrialization doubtless pointed from continental Europe to England. But the Dutch, Huguenot, German, and Flemish craftsmen who crossed the Channel influenced the taste and technique of British production, not to mention British finance. It is tempting, then, to construe the "first industrial nation" as a hybrid—a hybrid whose perch did not constitute a neatly demarcated stage.

While discussion swirled around the place of Britain in the coming of industrialization, ferocious exchanges characterized assessments of the pace of change. Summing up years of controversy, C. Knick Harley, an ardent participant in these disputes, concluded in 1990 that "it seems impossible to sustain the view that British growth was revolutionized in a generation by cotton spinning innovation."[10] In fact, after a generation of intellectual bloodletting, a rough consensus was reached about early British industrial growth and its measurement. The best entry point to the wrangle is to consider the limits of the numbers themselves. Reliable figures for British national income and output before the twilight of the nineteenth century are simply beyond reach or, as Charles Feinstein put it, "guesstimates."[11] Even then, these numbers likely conceal more than they reveal. For instance, both aggregate production statistics and loyalty to familiar devices masked the impact of innovative devices. And O'Brien correctly notes that time and money passed quickly in the slow process of moving from blueprint to learning by doing to a commercially successful process or machine. So radical shifts in the numerical indicators of British growth before 1850 were unlikely, and that is the tale told by the figures. But as Landes contends, trends often tell more than tallies. The anxieties of Alexander Hamilton, Friedrich List, and Jean-Antoine Chaptal about new practices in Manchester and Prestonpans may well express more than Feinstein's "insecure guesses."[12]

Yet large questions remain: How fast is fast, and how long must it last, in order to label industrial growth revolutionary? This book offers a variety of responses to these issues. In chapter 7, Kristine Bruland describes a deeply entrenched culture of growth in Scandinavia that has enabled the region repeatedly to create and adapt cutting-edge advances. Marta Vicente explains in chapter 8 that enduring family structures of work and accumulation underpinned the periodic moments of Spanish industrial expansion. And Ian Inkster speaks in chapter 14 of a Meiji policy of "cultural engineering" that "induced the institutionalization of technological progress," an approach to growth beyond direct measurement but surely of ceaseless significance.

Whether gradual or sudden, economic growth inevitably provokes questions of distribution. Hardened by Cold War convictions and the durable problem of development in "backward" states, the "standard of living" debate quickly became another fierce dispute.[13] At its heart was the issue of whether English workers suffered or profited from the complex changes of early English industrialization. Once again, convincing, wide-ranging statistics proved hard to locate. Still, the material standards of

England's workers apparently varied little, when measured quantitatively, from 1760 to 1820, and then improved rapidly during the next three decades.[14] But what of the cultural dislocations endured by the laboring poor as they left the cottage for the mill and close communities for congested cities? Sharply etched images of the pain of Robert Blincoe and his counterparts in the "dark Satanic mills" also opened lingering doubts about gains among England's first factory laborers.[15]

Daniel Roche's evocative depiction of the diffusion of mirrors, razors, and hairbrushes among the *menu peuple* of eighteenth-century Paris helped to renew attention to consumption patterns beyond the *haut monde*.[16] Jan de Vries's elegant concept of "the industrious revolution" aggressively linked the study of popular demand to production. He contended that pre-Revolutionary Europe's laboring men and women toiled more regularly, turned out more for the market and less for their personal use, and put their children more frequently into waged labor than their predecessors had. They did all this in order to consume more. Here was a major pivot in the history of industrialization: a "moral economy" that took comfort in sufficiency, leisure, and community ritual gradually gave way to personal enrichment and the accumulation of goods. Even among the lower ranks, tastes were changing as old wants became new needs.[17]

In chapter 3, Berg considers a fresh array of middle-class products that bespoke "high design," modernity, and novelty. These stylish, successful wares, she concludes, advertised the materials they were made of and the machines that made them. Not surprisingly, the social relations in the workshops that fashioned these goods grew more unsettled. With the arrival of machines and the continued use of perishable raw materials, time increasingly became money. Josiah Wedgwood dreamed of making "*Machines* of the *Men*."[18] But the skilled craftsmen of Wedgwood's potbank and elsewhere labored to keep this know-how scarce. They honeycombed labor markets with custom, negotiated for the last pence, and decamped when opportunity beckoned or their brothers needed numbers to swell a strike. Mechanized production may have altered this balance, but not always to the full satisfaction of entrepreneurs. Writing about Harpers Ferry Armory and Rockdale, a textile manufacturing hamlet in Pennsylvania, Merritt Roe Smith and Anthony F. C. Wallace respectively affirmed that "the big question was 'who *should* control the machines of the Industrial Revolution?'"[19]

Often the men and women who sweated in cottage industry and early factories emerge as irreducibly hostile to mechanical change. But the unskilled and the craftsmen were not inevitable Luddites. Many adapted to new devices so long as their stake in production—a job for a son or a daily afternoon dram—remained intact.[20] Moreover, workers themselves engaged in a great deal of anonymous invention, creating machines and instruments in a broad range of sectors. Still, the importance of machine-breaking—in fact, even the rumor of machine-breaking—sometimes had enormous consequences. In chapter 5, Jeff Horn argues that French Luddism and its intimate links to popular political revolution in 1789 frightened manufacturers sufficiently to delay industrial mechanization for a generation.

E. P. Thompson slyly noted that the rational system of labor discipline pioneered by the pugnacious Josiah Wedgwood collapsed with his death.[21] Nevertheless, Wedgwood successfully imposed an innovative, enlightened system of flow production at his pottery works. Historians long ago abandoned distinctions between the head (science) and the hand (technology) as sources of large-scale industrialization. Today, most favor such formulas as the "inventive intersection" of science, technology, and the shop floor, or the stimulating effects of the diffusion of openly accessible, objective knowledge.[22] Production guided by science excited the *philosophes*, and a long line of scholarly luminaries, spanning A. P. Usher, T. S. Ashton, and Margaret Jacob, have shared this enthusiasm. Here, these historians proclaim, is the essential explanation for the progressive industrial destiny of the West and the delay of the rest.

Currently, the historian most closely identified with the central place of science in Western industrial distinction is Joel Mokyr, author of the eloquent aphorism "the industrial Enlightenment" and the equally formidable notion of "the knowledge economy."[23] Mokyr insists in chapter 4 that early industrial experiments in Flanders and the Italian states "fizzled" because they were not buttressed by a systematic, scientific approach to technological development. The era of the Enlightenment, however, finally broke ranks with its forerunners and embraced the Baconian program of production flavored by scientific principles. Thus, the road was paved to self-generating, sustained growth. Britain, of course, was first, since there, free minds enjoyed the widest play and free markets ensured that intellectual liberty was uniquely profitable.[24] The island's "doctrine of *economic reasonableness*" was a far cry from the guild-ridden, constricted manufacturing environment of Enlightenment France.[25] The "good guys," Mokyr affirms, had won.

Inarguably, this is forceful stuff. Plantations, the slave trade, abundant coal, a powerful navy, and painfully long hours of child labor all take a backseat to the influence of science. Mokyr is too discerning a historian to miss the importance of these institutions and practices; still, he insists that scientifically informed production was the *telling* difference. It is therefore worth inquiring if Mokyr's tightly wrapped case is too confining. First, has he paid too little attention to the shop floor empiricism of multitudes of artisans? After all, A. Rupert Hall noted that "craftsmen have always experimented and talked of experiments: but not scientifically."[26] Equally, Charles Gillispie, the distinguished historian of Enlightened science, observed that "one searches" the history of the cotton manufacture in industrializing England "in vain for any trace of scientific influence, except in the bleaching or the dyeing of the finished product." Across the Channel, Gillispie knew, the makers of lightweight textiles "were shown the way, not by scientific research, but by Englishmen and Scotsmen."[27]

In this book, Mokyr continues to refine his most daring claims. The "European Enlightenment" he describes blurs the boundary between Britain and the Continent. He depicts the influence of science on production before 1850 as primarily "foundational," as more of a broadly dispersed method and model than a frequent, hands-on contributor. But even Mokyr's amendments remind us that Enlightened Europe was, in Lorraine Daston's lovely phrase, "a great echo chamber."[28] Experiments and ideas, discoveries and delusions reverberated from coffeehouses to learned societies, from knock-off versions of the *Encyclopédie* through the huge crowds at balloon flights, and from the celebrity of the charlatan Franz Mesmer to the use of hydrometers to pursue excise frauds. Industrial production could hardly remain immune from all this noise for very long. And that is why Mokyr concentrated on an era of reform that helped unleash industrial revolution, however gradual it was.

Eighteenth-century British statesmen used every mercantilist trick to build and buttress competitive trades. Their mid-nineteenth-century heirs pressed for unfettered markets and an end to tariffs. Britain's position as the "workshop of the world" accounted for this shift. Yet the decades-long struggle by England's manufacturers to overturn the Corn Laws, which raised their labor costs, revealed an imperfect fit between the appetites of industry and state policy. Political and production institutions rarely evolved in lockstep, which is why distinctive paths were the immediate experience of all of Britain's "followers." "Emulation" ensnared every borrowed technology in the recipient nation's economic assumptions, work and entrepreneurial cultures, and political designs and possibilities. It was never simply emulation.

In Stalin's Russia, Peter Gatrell reports in chapter 11, "His Excellency, the Harvest" had lost ground to "His Excellency, the Plan" in the economic system. Illuminating as these aphorisms are, they do not offer a full definition of industrial revolution, even in the midst of a Five-Year Plan. Alone among the authors in this collection, Smith and Robert Martello take a stab at this task in chapter 9. Theirs is a technology-centered statement, which combines "the replacement of craft methods with mechanized methods of production, the organization of work into larger and more specialized units, the more rigorous and pervasive management of labor, the construction of a national transportation system, and the growth of markets." It is surely essential to add the active presence of a variety of organizations engaged in the diffusion of "useful and reliable knowledge." More controversial, but likely no less necessary, was the willingness of states to facilitate the naturalization of the new and to eliminate roadblocks, including those of their own making.

What, then, were the causes and nature of the industrial revolutions in the states examined in this book? How much did free minds and free markets really matter? Ha-Joon Chang maintains that all this openness arrived after the first industrial nations found their places in the sun. Consequently, he insists that the advanced industrial world is not prescribing what it did but what it admires now.[29] Finally, what of the enduring question of technological determinism? In recent years, this issue has been retooled. Systems of labor discipline, accounting practices, and workshop adaptations are now understood as intrinsic elements of every technology. Is it still possible, then, to speak of technological determinism as *the* force that drives history? Or are we well advised to construe technologies as lodged in a delicate, reciprocal balance—the shapers of much, but endlessly reshaped themselves by a bevy of forces?[30]

By 1974, modernization theory had seen its best days. Moreover, orchestrating industrial revolutions on several continents had proven frustratingly elusive. Now "failure" was increasingly the order of the scholarly day. But Carlo Cipolla wondered if the whole notion of failed economies turned on "the (arbitrary) assumption that all proto-capitalistic economies could or should lead to an industrial revolution." He condemned the claim that certain states were on the verge of industrial transformation because they possessed "traits" that scholars construed as typical of the "industrial world." Cipolla's analysis recognized that the presence of a particular mix of assets does not inevitably yield industrialization.[31] Every industrial course was divergent; long before the concept appeared, Cipolla understood that there was no single, "great divergence."[32] Rather, as Peter

Perdue explains in chapter 15, timing, ambitions, factor endowments, and market features joined in unpredictable, combustible ways to spark industrial revolutions.

Patrick K. O'Brien's wide-ranging essay (chapter 2) downplays the claim that Britain's industrial primacy stemmed largely from its distinctive arsenal of "useful knowledge." Instead, he emphasizes a British "maritime strategy for security and development." A cost-effective Royal Navy protected Britain's international commerce and defended the island's shores. With external threats minimized, English assets were freed to maintain domestic order and take advantage of natural endowments. Smith's invisible hand mattered particularly because it was guarded by the visible iron hand of a formidable navy.

In chapter 3, Maxine Berg challenges the familiar formula that early British industrial goods were little more than cheap wares destined for the plebs. Instead, she locates a bounty of English manufactures for the middle classes, "fashion leaders" intended to compete with French goods on the terrain of tastefulness. Most notably, she uses these quality wares to reconceptualize the precocity of Britain's Industrial Revolution. "Product innovation," she insists, "brought in its wake productivity gains that we have not yet even tried to estimate." For Berg, changing tastes played a prominent part in driving technological shifts in industrializing England. Her work provides a valuable complement to Mokyr's cultural case by explaining not just how things were made but why they were made. The "useful knowledge" of desire and display that drove commerce, she concludes, counted too.

Mokyr's chapter, of course, focuses on the values of the Enlightenment, such as reform, utility, and efficiency, that created an apt setting for industrial change. By linking "the sphere of learning and the sphere of production," Europe finally escaped the Malthusian trap. In addition, the broadsides of the Enlightenment took aim at "bad" institutions, that is, monopolies and habits that permitted states to inhibit productive visions and ventures. Yet Colbertian planners and Ming bureaucrats were attracted by the Enlightened virtues of accountability, standardization, and transparency long before Condorcet and Joseph Black were.[33] Their ardent pursuit of revenues and order permitted nothing less.

Jeff Horn calls both cultural and technological determinism into question in chapter 5. During the eighteenth century, Horn suggests, the French state's improving officials, seconded by a substantial flock of entrepreneurs, scientists, and technical experts, sought to nurture a homegrown version of British industrial expansion. But the "threat from below" unleashed by the French Revolution derailed their plans: machine-breaking trumped mechanization. It took Napoleon's strong regime and the vision of his tough-minded minister of the interior, Jean-Antoine Chaptal, to usher in decisive technological change. Chaptal installed *dirigisme* with a liberal agenda; he put the resources of the state behind technical competitiveness and markets that functioned "more efficiently." Over time, French manufacturers carved a distinctive (and successful) industrial pathway within the confines of these policies and institutions. Finely grained French habits of work and consumption, the skill and training of the nation's engineers, and recurrent problems of access to tools and machines surely influenced French production. But perhaps even more, so did the rise of a powerful, centralized French state, the imperatives of France's all-embracing competition with Britain, and the barely buried legacy of the machine-breakers of 1789.

Industrial *revolutions* certainly fits the German experience. Eric Brose opens the issue of regional economic development in this book in chapter 6. More precisely, he considers Prussia's nascent industrial stirrings within a wider framework of "German Europe," including Austria. There he finds a complex tapestry of commercial and production institutions, as well as diverse ideas about the worth and dangers of factories and railroads. Emma Rothschild claims that an English *Kopf*, the triumph of the political economists over the moral economists, took shape in the eighteenth century.[34] But the many German states and their heterogeneous policies and institutions left little room for the rapid configuration of a German *Kopf*. Brose therefore offers an intriguing response to the query, "How fast is fast?" The answer, he suggests, depends on the internal balance of powers and ideas in the regions that sustained industrialization.

Accordingly, Brose's discussion of three turning points of early German industrialization is at once a consideration of the particularities of place and pace. Whether debating more licenses for factories, a customs union (*Zollverein*) among the German governments, or state concessions to the railroads, "tradition-based obstinacy" figured in every dispute. Social conservatives sought to protect guilds, maintain tariff barriers between the German states, and avoid the squalor and pollution of England's industrial cities. They also feared the rise of a concentrated working class. Consequently, the challenge of British mills went largely unanswered, and the smuggling of British goods into the German states rose. Still, Bavaria, for instance, vacillated ceaselessly about whether to license factories, guaranteeing their delay. Württemberg and Saxony settled for a middle way between Bavaria and Birmingham. Even the establishment of the railroads was not inevitable; for many years, they were a matter of choice, and the answer was no. Opponents insisted that railroads would divert capital from textile manufacturing, deplete coal mines, and provide rapid transport for invading French armies. The recalcitrants championed waterways and believed that railroads would aid neighboring German states more than their own. Hence Prussia did not turn the corner on railroads until 1838. Britain's rivals, David Landes has often written, dreaded its precocious industrial practices and desperately wanted to copy or steal them.[35] But Brose clearly demonstrates that many in "German Europe" did not believe that the British model of production was desirable, much less inevitable, even while Britain had become the "workshop of the world."

Kristine Bruland treats Scandinavia as a single region in chapter 7. After all, she reasons, the Scandinavian nations share formidable challenges of climate and geography, and in the nineteenth century, each of these states lost a substantial portion of its working-age population to emigration. So the persistent competitiveness of Scandinavian manufacturers in a broad array of high-income industries beat the odds. What accounted for this achievement? Bruland's compelling answer constitutes social history from above and below. Less the result of a single big push or heroic efforts by strong nation-states, Bruland credits this success to "a broad social process of change in Scandinavia that supported the building of capabilities and the creation of an industrial culture." She traces this pattern to the Enlightenment, in its Scandinavian embodiment, with its mosaic of formal and informal means of monitoring foreign developments. Learned societies, technical journals, and newspapers emerged and flourished. The state subsidized journeys abroad to ensure information flows home. Here Bruland echoes Mokyr and shares some notes with Berg, particularly her emphasis on the internationalization of Scandinavian standards and tastes. Nevertheless, Bruland's case is thoroughly original: it shows how Enlightened values and forms served as the essential background music for the maturation of a set of industrial latecomers.

The title of Marta Vicente's chapter, "Crafting the Industrial Revolution," is both precise and suggestive. Her study centers on the role of artisanal families in Catalonia's cotton calico industry. Sweating in small workshops, these skilled laborers were the indispensable collaborators of the large-scale manufacturers in and around Barcelona. It has been many years since historians depicted handicraft production and factory-based industry locked in an inexorable, life-and-death struggle. But the exact links among mill, cottage, and garret too often remain obscure. In the tough times of Revolutionary war and Napoleonic occupation, Catalonia's calico producers survived through "outsourcing"—the labor of swarms of artisans in upper-story flats. Like strands of industrial DNA, large manufacturers and small men combined, separated, and recombined their efforts. This reciprocity was made possible by the continuing transfer of social capital and its practices, such as family-based and family-like labor discipline, from backstreet shops to factory workrooms.[36] Although "a controlling bureaucracy . . . fossilized economic growth" in Spain, Vicente concludes in chapter 8 that there was surprising flexibility and vitality in Catalonia's actual production settings.

By the late nineteenth century, Americans routinely celebrated their native technological genius. They connected this distinction with their industrial progress and the exceptional qualities of their democracy, and were satisfied. (Some commentators did worry about inequities in the distribution of wealth and class conflict, but the master narrative rolled along.) Smith and Martello consider generations of debate among historians about the transformation of American industrial production from the early republic to the Civil War. They are particularly attentive to the organization of this production and the experience of increasingly rapid technological change. They also recognize the shackles imposed on Southern industry by slavery and the enduring struggle between free laborers and their bosses for shop floor control. At bottom, then, their chapter is about the political economy of American industrial production, warts and all.

Perhaps the most intense challenges in chapter 9 are reserved for Charles Sellers's important book, *The Market Revolution*.[37] Whereas previous interpretations of Jacksonian economic change emphasized plentiful natural resources, technological enthusiasm, infrastructural reform, and the reorganization of production by merchants, Sellers turned our attention to demand and its prompting of "capitalist transformation." As migrants from the farm and from overseas entered the labor market, the communal, face-to-face world of the "moral economy" declined. Dislocation and a never-ending thirst for new goods yielded "capitalist hegemony," social turmoil, and feverish politics, with Jacksonian democracy as a counterweight to capitalism. Managers and laborers occupied the periphery in this depiction of America's early industrial era, as did the machines themselves. All are staging a comeback—to Smith and Martello's obvious relief—but the terrain remains hotly contested.

Meanwhile, the American state has reentered the debate in an intriguing guise. Drawing on the scholarship of Carter Goodrich, Smith and Martello point out that both federal and state governments were frequently in the business of financing risky technological ventures until they demonstrated their market worthiness. Then the government retreated, and the enterprises gathered steam "under private ownership." Smith and Martello insist that "the industrial revolution in America was more incremental than revolutionary in nature and followed many paths." Certainly their resurrection of Goodrich's "state in, state out" process, with its complexity and inevitable delays, helps explain why fast was not terribly fast, even in America's Industrial Revolution.

Leonard Rosenband's contribution is a microhistorical consideration of many of the macrohistorical issues raised in this collection. His portrait of the papermaker Ebenezer Stedman in chapter 10 illuminates the transatlantic nature of his trade, the complex circumstances of international technological convergence, and the customary loyalties and calculating practices that drove this worker-turned-manufacturer. Born in 1808, Stedman spent most of his productive career, which lasted deep into the nineteenth century, in Kentucky. His was a category-defying experience, or at least a tale of trespass of historians' categories. He produced on both sides of the modernization divide, making paper by hand and by machine. He employed the custom of his craft, much of which was rooted in England and France, and exploited every opportunity to the last cent. He labored beside slave paper workers, respected their skills, and added slaves to his holdings. His downfall reminds us that "useful knowledge" was never enough; chance and bravado tripped him up. He lived, as Fernand Braudel might put it, waist-deep in the local, national, and transnational ways of his trade, but he chose among them to suit his needs. He was a man of the newly forged industrial Atlantic.

Peter Gatrell examines "two dramatic periods" in recent Russian economic change, the late Czarist years (1885–1913) and the early Stalinist era (1928–1941). Both regimes faced enormous challenges, Gatrell explains in chapter 11, and both stepped in to provide critical fiscal and entrepreneurial resources. Here Gatrell reaches back to the work of Alexander Gerschenkron, who emphasized the close connection between the political and the economic in both Russian backwardness and the measures used to overcome it.[38] Like Gerschenkron, Gatrell concludes that "the Russian state simultaneously constrained and fostered economic activity."

Russia's Industrial Revolution came late, and it hardly fit the classic Smithian model. During the twilight of czarist rule, the government encouraged a mix of public and privately owned businesses. Russia was increasingly integrated in the world economy, which stimulated the inflow of foreign investment along with mining and metallurgical technology. But problems persisted. Patent law was inadequate, reducing the opportunity for innovators to capitalize on their work. State investment in research was low, and bureaucratic meddling in engineering reflected "the power of politics to constrain creative endeavor." Meanwhile, Sergei Witte, the key figure in czarist industrialization, was in a hurry, so his policies curtailed consumption in favor of heavy industry and featured "mass coercion" first into and then on the shop floor.

Stalin's program of rapid industrialization rested on central economic planning, the vast mobilization of internal resources (including labor) for new and established industries, and the absence of external capital as well as limited recourse to foreign technical expertise. But this extensive structure had thin underpinnings. Russian Taylorism failed, accounting practices were immature and contaminated, and secondary needs, such as storage facilities, were often absent. The ratio of workers to machines tripled during the first Five-Year Plan. Stakhanovism, with its rewards for quota busters, served largely to recast the principal-agent problem into a contest between bureaucratic overlords and indifferent subordinates. And, as ever, there was a shortage of consumer goods. Stalin answered with the empty shout of sacrifice on behalf of future generations.

Gatrell insists that other similarities linked the two eras of rapid Russian industrial growth. Both regimes intended to place "the state's security on firmer economic foundations." Bureaucratic wrangling about backwardness and its remedies inevitably influenced (and damaged) reform in both periods. Czarist and Stalinist planners relied heavily on massive labor inputs as a substitute for capital equipment. Finally, Witte's officials and the "commissarocracy" shared a "'romance' of technology." Perhaps this attraction to the "technological sublime" combined with the routine, hard use of huge reserve armies of labor accounts for the infamous episode of the construction of the Belomor Canal.[39] The building of this waterway, which linked the Baltic and the White seas, cost the lives of an estimated 200,000 "prisoners."

BRIC is the contemporary acronym for a quartet of states with rapidly growing industrial sectors: Brazil, Russia, India, and China. Yet, Anne Hanley insists in chapter 12, Brazil did not experience an industrial revolution. Instead, Brazil's gradual industrial ascension is intriguing due to its distinctive origins: Brazil's turn to industrial production was rooted in the export of an agricultural product, coffee. But the turning point in Brazil's manufacturing history was institutional. In 1890, Hanley writes, a revised legal code "permitted the formation of joint-stock corporations and created the mechanisms to trade their shares." Previously, family partnerships and retained earnings were the sole sources of Brazilian industrial capital. It took the coffee boom, a political coup, and innovations in corporate law to transform the scale of Brazilian industry and, ultimately, to gain the nation a spot among the BRIC producers.

Nineteenth-century Brazil offered forlorn soil for large-scale industry. Mercantilist Portugal inhibited the growth of Brazilian manufacturing, as did England's treaty-based, preferential access to the country's markets. Moreover, these markets were spare and isolated. Then the coffee boom exploded through every sphere of Brazil's economy. It attracted more than a million migrants to Brazil's shores, creating a vast market for home-produced textiles and other goods. It prompted planters to finance railroad development. And it was accompanied by the political events of 1889, which swept away a centralized, conservative empire and installed a federalist, republican regime sympathetic to economic development and diversification. Within a year, Brazil's rulers accepted the limited liability of shareholders. Of course, Brazil's economy has endured its speculative bubbles and periods of quiet despair since. Such straitened circumstances are the inexorable partners of industrial growth. Still, in 1895, a period of instability led Brazilian policymakers to argue for a return to the economy's "true calling, commodity exporting." A similar demand today would be unthinkable.

India, a second member of the BRIC quartet, also suffered economically as a colonial subject. In the middle of the eighteenth century, Prasannan Parthasarathi explains in chapter 13, the Indian region accounted for a quarter of the world's manufacturing output. By 1900, the subcontinent's share of global industrial output amounted to only 2 percent. Why had such a drastic shift occurred? Parthasarathi quotes Niall Ferguson, who claimed that India's colonial masters were not at fault: "Victorian India . . . was booming. Immense sums of British capital were being invested in a range of new industries: cotton and jute spinning, coal mining and steel production."

Parthasarathi dismisses Ferguson's case as "imperial revision." Instead, he contends that Britain's industrial and commercial policy in India served the interests of British industry. He begins his story in the eighteenth century when the subcontinent was experiencing a sort of "industrial Enlightenment." This attention to "useful knowledge," like Mokyr's European "industrial Enlightenment," had the potential to act as a foundation for later mechanization and industrial science. But Britain ruled its

Indian holdings "on the cheap" and "took little interest in patronage for technical change" or the diffusion of "useful knowledge." Meanwhile, Britain's highly centralized rule in the subcontinent sapped the dynamic competition between the previous kingdoms—a competition that had provided "enormous incentives to adopt new techniques, most critically in the metal industries, in order to produce better armaments." And long before Britain submitted its industries in general to free trade in the 1840s, Indian manufacturers received no protection from British producers. Finally, the purchasing activities of the British Indian state, which privileged metropolitan wares, also damaged the development of the subcontinent's industry. So the indigenous technical skills and entrepreneurship of nineteenth-century India went largely to waste. "What India lacked" in the nineteenth century, Parthasarathi concludes, "was the power to make economic policy." Consequently, when the subcontinent's industry began to take off in the interwar years of the twentieth century—in part due to tariff insulation from foreign and especially British goods—India's manufacturers lacked sufficient capital. It took time for the subcontinent's industrial production to recover from Ferguson's supposed Victorian boom.

In the sixteenth century, the ruler of Japan remarked that the "printing and diffusion of books is the most important task of a benevolent government." In chapter 14, Ian Inkster treats the Meiji watershed in Japanese industrial history as an episode of "cultural engineering" by the state. The Tokugawa regime, the predecessors of the Meiji, did little to aid the transfer of foreign "useful knowledge" into Japan's workshops. The Meiji were determined to take advantage of such knowledge, the "Best of the West." Yet the Meiji ran a "very cheap state." The gathering of foreign technical experts as well as the creation of model factories and experimental stations had to be done at low cost. Worse yet, the private borrowing of capital from abroad was negligible. So the securing of advanced technology fell on the meager means of the Meiji state—small wonder then that "cultural suasion and momentum" bulked large in their plans.

Inkster explores the seeming paradox of limited resources and large-scale aspirations by taking us into both the villages and the ministries. At the local level, the relatively seamless (and hence inexpensive) absorption of new techniques and devices was eased by such features of Japanese culture as powerful group identity and a strong work ethos. Borrowing had deep roots in Japan, smoothing the importation of tea, silk, and sugarcane production from China. Accordingly, political breakdown (and its high price) did not accompany product and process revolution in Japan. The islands' past was at once invaluable and, in Inkster's elegant phrase, "revocable without revolution."

The Meiji state sponsored trade associations and internal exhibitions to diffuse "useful and reliable knowledge." From 1867 to 1910, Japanese delegates attended thirty-eight major international trade expositions. Through translations and cheap editions, the dispatch of students to innovative locations abroad, and model enterprises, the regime spread the word. When necessary, as in the United States, the government also served as a "loss leader," absorbing costs until a technology could turn a private profit. Free minds and free markets mattered in Japan's late rise as an industrial power. But Japan's Industrial Revolution certainly would not have happened as rapidly without the state's "cultural engineering" and the solid social traditions on which it built. The Japanese state drove the society's technological maturation.

Peter Perdue begins chapter 15 with two queries: Was imperial China a contender for the title of first industrial nation; and, why was China's industrial ascension so late in light of the advanced state of its economy? A generation of scholarship on the Qing economy during its "flourishing age" (c.1670–1760) has prompted these questions. Moreover, the "great burst" of Chinese industrial expansion since the reforms of the 1980s has led scholars to search for both the long and the short roots of the "Chinese miracle."

Perdue fells a forest of orthodoxies about the imperial Chinese economy with an unsparing ax. Chinese life expectancies and standards of living around 1800 were roughly comparable to those of Europe; there was a Chinese agricultural surplus and hence no Malthusian trap that hamstrung industry. China's internal capital flows and labor mobility resembled those of the Netherlands and England, the most advanced economies of Europe, rather than Russia and the German states. China enjoyed a peace dividend while continental Europe's wars destroyed people and capital, even if the conflicts stimulated technological change. Remarkably, in the era of Mokyr's "industrial Enlightenment," Perdue notes that the physiocrats, France's liberal political economists, idealized China. Indeed, Perdue sees the same foundational cultural work in the Qing's "flourishing age" that Mokyr spotted in the European Enlightenment.

These comparisons lead Perdue to claim that "long-term structural processes do not inevitably determine outcomes." Put precisely, Perdue does not construe England's Industrial Revolution as an "accident."

Instead, he contends that it was "late and sudden," the result of delicate timing, effective public and private policies, and a suitable production ecology. Of course, England's imperial profits also contributed to its industrial precocity. But Chinese imperialism had the opposite consequence: the cost of defending the Qing Empire ate away at the prospects of Chinese industry. Meanwhile, China's economic success on its "normal" path of Smithian commercial growth turned China away from Schumpeterian (technological) profit making. Familiar patterns of investment and frontier pressures, Perdue concludes, created an economic environment with little space for industrial transformation. Thus, China's Industrial Revolution was even later and more sudden than England's.

Exceptionalism is a loaded term in industrial history. At one time, it served as a shorthand for distinctive English (or Anglo-American) virtues. With the recent rise of Japanese and then Chinese industry, this sense of exceptionalism has receded, often to be replaced by the unfortunate label *miracle*. This book generally eschews both *exceptionalism* and *miracles*; instead, its authors consider a wide range of industrial transformations in all their diversity. Intervention by states, for good and ill, played a substantial role in industrialization everywhere. But so did access to consumer goods, the timing of efforts to mechanize, and the military impulses and needs of many societies. Put simply, one size did not fit all. Motivations were many and complicated: states drove technology, and technologies developed their own momentum. As for the "great divergence," it may serve best as a description of England's industrial collapse, while American, German, and Japanese industry, however wounded, have held on.

The intricate industrial pathways of the BRIC quartet, the Pacific tigers, and other "latecomers" have drawn—and deserve—scholarly attention in their own right. These trajectories have also stimulated new questions and revived old lines of inquiry about Europe's past (and Eurocentrism's present). For example, Anne Hanley's and Peter Gatrell's concern with finance, coupled with the current banking crisis, likely foreshadows renewed interest in capital itself in the history of European industrial capitalism. Consider too Hanley's study of the São Paulo region and Eric Brose's yeoman work on the various German states. Perhaps this work anticipates fresh attention to the industrial region and its place in economic development across the globe.[40]

Looking back at the chapters as a whole, the editors' most deeply felt recommendation for future laborers on industrial history would be to reconsider the shop floor. We still have far more work from the top down than the bottom up. We need to know much more about what craftsmen and workers contributed to the development of both hand and mechanized technologies. How often were entrepreneurial artisans eager to expand their shops and their purses? In sum, we must reconceptualize industrial revolutions as lived experiences, rich in exchanges and conflicts among masters and men as well as countries and cultures.

Notes

1. Goethe to Carl F. Zelter, June 6, 1825, *Letters from Goethe*, trans. M. von Herzfeld and C. Melvil Sym, ed. M. von Herzfeld (Edinburgh: Edinburgh University Press, 1957), 463.

2. Walt W. Rostow, *The Stages of Economic Growth* (Cambridge: Cambridge University Press, 1960).

3. Harold James, *Family Capitalism: Wendels, Haniels, Falcks, and the Continental European Model* (Cambridge, MA: Belknap/Harvard University Press, 2006); David S. Landes, *Dynasties: Fortunes and Misfortunes of the World's Great Family Businesses* (New York: Viking, 2006).

4. Philip Scranton, *Proprietary Capitalism: The Textile Manufacture at Philadelphia, 1800–1885* (Cambridge: Cambridge University Press, 1983).

5. Raphael Samuel, "Workshop of the World: Steam Power and Hand Technology in Mid-Victorian Britain," *History Workshop Journal* 3 (1977), 51.

6. Maxine Berg, "Factories, Workshops, and Industrial Organisation," in *The Economic History of Britain since 1700,* vol. 1: *1700–1860*, ed. Roderick Floud and Donald McCloskey, 2nd ed. (Cambridge: Cambridge University Press, 1994), 137.

7. Peter Mathias, *The Transformation of England: Essays in the Economic and Social History of England in the Eighteenth Century* (New York: Columbia University Press, 1979), 3–20, esp. 3, 20.

8. Eric Hobsbawm, *Industry and Empire: From 1750 to the Present Day*, revised and updated with Chris Wrigley (New York: New Press, 1999), 13.

9. The quoted phrase is the title of chapter 3 of David S. Landes, *The Unbound Prometheus: Technological Change and Industrial Development in Western Europe from 1750 to the Present* (Cambridge: Cambridge University Press, 1969), 124.

10. Quoted in David S. Landes, "The Fable of the Dead Horse; or, the Industrial Revolution Revisited," in *The British Industrial Revolution: An Economic Perspective*, ed. Joel Mokyr, 2nd ed. (Boulder, CO: Westview, 1999), 140. For important discussions of British economic growth, see, among many others, Phyllis Deane and William A. Cole, *British Economic Growth, 1688–1959*, 2nd ed. (Cambridge: Cambridge University Press, 1969), and Nicholas Crafts, *British Economic Growth during the Industrial Revolution* (Oxford: Clarendon Press, 1985).

11. Quoted in Landes, "Fable," 145.

12. On Hamilton, Chaptal, and List, see ibid., 153–54; for Feinstein, see ibid., 145.

13. Arthur Taylor, ed., *The Standard of Living in Britain in the Industrial Revolution* (London: Methuen, 1975).

14. Mokyr, "Editor's Introduction: The New Economic History and the Industrial Revolution," in Mokyr, ed., *British Industrial Revolution*, 116.

15. John Brown, *A Memoir of Robert Blincoe* (Manchester: J. Doherty, 1832).

16. Daniel Roche, *The People of Paris: An Essay in Popular Culture in the Eighteenth Century*, trans. Marie Evans in association with Gwynne Lewis (Berkeley: University of California Press, 1987).

17. Jan de Vries, *The Industrious Revolution: Consumer Behavior and the Household Economy, 1650 to the Present* (Cambridge: Cambridge University Press, 2008), 1–185.

18. Wedgwood in 1769, quoted in Jenny Uglow, *The Lunar Men: Five Friends Whose Curiosity Changed the World* (New York: Farrar, Straus and Giroux, 2002), 213.

19. Merritt Roe Smith, *Harpers Ferry Armory and the New Technology* (Ithaca, NY: Cornell University Press, 1977); Anthony Wallace, *Rockdale* (New York: Knopf, 1978).

20. Catharina Lis and Hugo Soly, "'An Irresistible Phalanx': Journeymen Associations in Western Europe, 1300–1800," in *International Review of Social History*, suppl. 2, *Before the Unions*, 39 (1994), 42, 51.

21. Edward P. Thompson, *Customs in Common: Studies in Traditional Popular Culture* (New York: New Press, 1991), 386.

22. Lissa Roberts, Simon Schaffer, and Peter Dear, eds., *The Mindful Hand: Inquiry and Invention from the Late Renaissance to Early Industrialisation* (Amsterdam: Royal Netherlands Academy of Arts and Sciences, 2007).

23. Joel Mokyr, *The Gifts of Athena: Historical Origins of the Knowledge Economy* (Princeton, NJ: Princeton University Press, 2002).

24. William J. Ashworth, in a private communication, is responsible for the elegant linkage of free minds and free markets.

25. Joel Mokyr, "The Intellectual Origins of Modern Economic Growth," *Journal of Economic History* 65 (2005), 336.

26. A. Rupert Hall, "Engineering and the Scientific Revolution," *Technology and Culture* 4 (1961), 338.

27. Charles Coulston Gillispie, *Essays and Reviews in History and History of Science* (Philadelphia: American Philosophical Society, 2007), 97.

28. Lorraine Daston, "Afterword: The Ethos of Enlightenment," in *The Sciences in Enlightened Europe*, ed. William Clark, Jan Golinski, and Simon Schaffer (Chicago: University of Chicago Press, 1999), 498.

29. Ha-Joon Chang, *Kicking Away the Ladder: Development Strategy in Historical Perspective* (London: Anthem Press, 2002), 2–3.

30. Merritt Roe Smith and Leo Marx, eds., *Does Technology Drive History? The Dilemma of Technological Determinism* (Cambridge, MA: MIT Press, 1994).

31. Carlo Cipolla, "The Italian 'Failure,'" in *Failed Transitions to Modern Industrial Society: Renaissance Italy and Seventeenth Century Holland*, ed. Frederick Krantz and Paul Hohenberg (Montreal: Interuniversity Centre for European Studies, 1975), 8.

32. The quoted phrase is from Kenneth Pomeranz, *The Great Divergence: China, Europe, and the Making of the Modern World Economy* (Princeton, NJ: Princeton University Press, 2000).

33. On Colbertian France, see *Un Nouveau Colbert* (Paris: SEDES/CDA, 1985); on Ming China, see Francesca Bray, *Technology and Society in Ming China (1368—1644)* (Washington, DC: American Historical Association, 2000).

34. Emma Rothschild, "The English *Kopf*," in *The Political Economy of British Historical Experience, 1688–1914*, ed. Donald Winch and Patrick O'Brien (Oxford: Oxford University Press, 2002), 31–59.

35. Landes, *Unbound Prometheus*, chap. 3.

36. For social and cultural ties like those explored by Vicente in a variety of Western settings and eras, see Charles Sabel and Jonathan Zeitlin, eds., *Worlds of Possibilities: Flexibility and Mass Production in Western Industrialization* (Cambridge: Cambridge University Press, 1997).

37. Charles Sellers, *The Market Revolution: Jacksonian America, 1815–1846* (New York: Oxford University Press, 1991).

38. Alexander Gerschenkron, *Economic Backwardness in Historical Perspective* (Cambridge, MA: Harvard University Press, 1962).

39. David Nye, *American Technological Sublime* (Cambridge, MA: MIT Press, 1994).

40. On the industrial region, see the classic work of Sidney Pollard, *Peaceful Conquest: The Industrialization of Europe, 1760–1970* (Oxford: Oxford University Press, 1981).

2

Deconstructing the British Industrial Revolution as a Conjuncture and Paradigm for Global Economic History

Patrick K. O'Brien

Representations of the First Industrial Revolution

Industrialization is an important historical process, drawn out or truncated in time and occurring in local, regional, national, continental, and global contexts. While it involves social, cultural, political, and geopolitical forces, its outcome can be parsimoniously encapsulated in statistical form as a conjuncture of economic transformation from an agrarian to an industrial economy.[1] In quantitative terms, what economic historians have observed and measured is structural change, proceeding more or less rapidly until the majority of a national workforce ceases to be engaged with the production and servicing of primary products and becomes employed either directly or indirectly with the production and servicing of manufactured goods. Statistically, the trend toward an industrial market economy can be tracked with reference to data displaying shares of the workforce employed in industry and related services and in imperfect tabulations spanning long periods of time that display shares of gross national products labeled as industrial outputs.[2]

Although claims have been made for the Netherlands to be recognized as "the First Modern Economy," nobody disputes that Great Britain became the first national economy to complete a transition to an industrial economy.[3] For more than two centuries, that realm's famous transformation has been narrated and explained under such labels as *the First* Industrial Revolution, *the First* Industrial Nation, or simply as *The* Industrial Revolution. Anglo-American historians have analyzed the cycles of rapid development in British economic history to delineate subperiods running from the mid-eighteenth until the mid-nineteenth centuries that have been represented metaphorically with terms such as *watershed, great divergence, turning point,* and *take-off.* Others claim that the British Industrial Revolution was a more pervasive and universal achievement than the Florentine Renaissance or the French Revolution.[4] Thus, the Industrial Revolution has been represented not only as a profound discontinuity for the history of the Hanoverian kingdom, but also as a conjuncture of transnational significance for the future of the world economy. It positions and periodizes European, American, Asian, and African histories into a "before" and "after" the Industrial Revolution.[5]

Although nothing approximating a paradigm for industrialization was either initially established or fully developed in Great Britain from 1760 to 1830, there is no need to denigrate the precocious range of innovative economic achievements that came onstream after the country's decisive victory in the Seven Years War of 1756–1763. Comprehended historically as *the* century that marked discernible and irreversible accelerations in the rates of increase of real income per head, in shares of the increment both to rates of growth in income per capita and labor productivity emanating from technical and structural changes including urbanization, it seems solely polemical to engage in semantic attempts designed to purge the label "Industrial Revolution" from academic discourse and public consciousness.[6]

Considered within world history, according to the indicators constructed since the publication of T. S. Ashton's classic *The Industrial Revolution, 1760–1830* in 1948, economic transformation (although discernibly slow by subsequent standards) became rapid enough to achieve the position of competitive superiority that the kingdom enjoyed relative to all other European, American, and Asian economies during the Victorian boom (1846–1873).[7]

Britain's naval and commercial hegemony, as well as the efficiency of its agriculture, was recognized by the second half of the eighteenth century.[8] Thereafter, and as its industries matured, the rest of the world paid deference to clear comparative advantages in several sectors of its industry while retaining strong reservations about the social and political consequences of the British pattern of urbanization and structural change. Thus, a plethora of well-calibrated data complemented by a bibliography of impressions recorded by visitors from the mainland and the United States justifies the representation of the accelerated transformations that occurred after the Seven Years War as the First Industrial Revolution.[9] That century witnessed the development of novel techniques of production, the construction of engines to harness a new and potentially hegemonic source of energy (steam), the extension of improved modes of internal transportation (canals, turnpikes, and railways), the diffusion of efficient forms of business and commercial organization, the spread of responsive systems of financial intermediation and distribution, and the closer integration of commodity and factor markets, all at a pace and on a scale that ex post facto looks extraordinary, if not revolutionary.[10]

As their outlook has become more global and cosmopolitan, historians of the First Industrial Revolution have become less inclined to ignore not only its European but its Chinese, Indian, and African antecedents. Modern interpretations are now unlikely to exaggerate elements in British political institutions, social structure, and culture that not long ago formed the foundations of explanations for that nation's precocious, relative, and short-lived economic success. Only a few Whig historians and economists continue to reify core features and factors behind Britain's peculiar transition toward the first industrial market economy into a paradigm that could be readily transferred to rival but "retarded" economies on the mainland, that became rational enough to adopt best practice (i.e., British) technologies of production and modes of economic organization.[11]

Modern historical scholarship has become aware of the imperial, European, and more recently the African and Asian dimensions of the British Industrial Revolution and has struggled to understand the rather rapid convergence of Western economies to comparable levels of per capita income and labor productivity in terms of the peculiarities of each national case and theories of path dependency. Diffusion models that effectively elevated the status of Britain's precocious transition to a paradigmatic case are no longer regarded as an illuminating way to comprehend the industrialization of mainland Europe, let alone the United States, East Asia, and South America. Such models have been degraded into consoling but simplistic narratives purveyed by nationalistic communicators of British exceptionalism.[12]

Narrated, interpreted, and contextualized as a conjuncture formed by the ebb and flow of global history, the historicized status and heuristic potential for the First Industrial Revolution breaks down into a range of innovations of world significance (e.g., the steam engines of Newcomen and Watt, Cort's path-breaking technique for puddling iron, and the weaving machines of Kay and Cartwright), which can be represented as more or less novel and as indigenous to the British Isles. Other achievements of the period, such as the invention of roller spinning, Wedgwood's "china," or the techniques used to manufacture, bleach, dye, and print cotton cloth are no longer acclaimed as peculiarly "English."[13]

Economic history no longer focuses on separating out indigenous from exogenous components from among the myriad manufactured goods produced in England in the reign of George III.[14] Thirty years of research has allowed us to escape from nationalism to assign conjectural, but plausible, weights to major forces behind the accelerated growth of Britain's per capita output and labor productivity from 1763 to 1860.[15] The significant causes or origins of the First Industrial Revolution now include the kingdom's highly productive and responsive agriculture; its abundant and accessible supplies of minerals, particularly coal; foreign trade, sustained by massive and cost-effective state investment in naval power; and, last but not least, technological discovery and innovation. As usual, emphases accorded to forces behind any large-scale, complex conjuncture in history never settle into a consensus, but these factors, if not their ordering, let alone their weights, are widely accepted as major causes by economic historians.[16] Indeed, it may now prove possible to sum up Britain's famous transition as a "conjuncture" in the global history of material progress that occurred when and where it did largely as a result of the Island state's favorable national endowments and massive investments in naval power. The First Industrial Revolution can be perceived and conceived of as both a case of precocious and exceptional industrialization and as an island story explicable in geographical and geopolitical terms.

NATURAL ENDOWMENTS AND THE INSTITUTIONS FOR THEIR EXPLOITATION

For centuries before 1756, the British Isles had been blessed with a geography and an agricultural sector exemplified by very good (but not extraordinary) yields per arable hectare cultivated and, above all, when compared with other parts of Europe and particularly with India and China, high levels of output per worker.[17] But apart from its favorable soils and climates, from where did its prior but basic advantages in agriculture emanate? Supporters of the traditional Anglocentric view insist that a rather distinctive set of property rights and tenurial arrangements for access to land had appeared earlier on the British Isles than on the Continent or in Eurasia. Over time, the evolution of this English system of property rights promoted the formation of large-scale units of production, flexible markets for tenure, a concentration of rents from the ownership of natural resources, and a steady reduction in the extent and control by peasant families over both land and labor, which became available to capitalist farmers and, later, to proto-industry and the towns.[18] Among those following Arthur Young, who represented the kingdom's aristocracy and gentry as distinctively entrepreneurial, there has been a celebration of unequal landownership as a benign outcome of market forces.[19]

Markets are recommended by economists as rational institutions for the transfer of property rights to land, forests, and minerals into the private ownership and/or control of those who can manage their productive use most effectively. The system of agrarian property rights (already in place well before the First Industrial Revolution) embodied advantages for the realm's precocious transition to an industrial economy, including the outstanding capacities of British agriculture to release ("expel") labor to other sectors of the economy. Nevertheless, there can be no presumption that the emergence in medieval times or the linear evolution thereafter of markets for the sale and purchase of land and of contractual rules governing access to land were solely (or even mainly) an efficient outcome of English individualism or market forces.[20] Political and legal histories of the frameworks surrounding property and tenurial rights suggest that they also emanated from far less "benign" historical forces, which included conquest, internal colonization, the violent expropriation of ecclesiastical and common land, and the systematic accumulation of power by closed aristocratic elites.[21] For centuries before urban industry demanded a rapidly increasing share of the workforce, a "push" from above coupled with an intensifying "pull" from high wages potentially available to migrants from the countryside in London and other maritime cities provided Britain with exceptionally flexible markets for labor.[22]

No matter how they represent the long-term evolution toward a distinctive and inegalitarian system of property rights, most economic historians are now inclined to agree that over time, powerful elites pushed agriculture in directions conducive to the attainment of higher levels of labor productivity and away from the disadvantages for rapid industrialization and urbanization associated with peasant proprietors and household units for production that survived on the mainland and remained omnipresent in South and East Asian societies.[23]

Physiocratic improvers who visited England in the eighteenth century advanced more reductionist accounts of the advantages of the British Isles for a precocious transition. Although they lauded its distinctive set of tenurial institutions, coupled with concentrated landownership and aristocratic management of large estates, most observers insisted on the primacy of geography. They perceived that Britain's favorable environmental endowments (particularly grass) had encouraged the steady accumulation of sheep, cattle, pigs, and, above all, horses, a perception that is now commonplace in agrarian history.[24] By the English Civil War, the kingdom's large population of animals provided high-value outputs, extra supplies of energy, and flows of organic fertilizer that carried English agriculture to a plateau where the primary sector could (with increasing help from Ireland) lend support to accelerated population growth, proto-industrialization, and extensive urbanization. Geography not only mattered more than institutions, it also goes a long way toward explaining their form and evolution.

E. Anthony Wrigley recently brought back into the foreground of the First Industrial Revolution another of Britain's natural advantages: easy access by water to abundant supplies of cheap inorganic energy: coal.[25] True, its European competitors, particularly Belgium and Germany (and even France and China), also possessed "subterranean forests," but not of the same quality and not nearly as cheap to transport to coastal cities. Britain began and completed the transition from organic to inorganic (mineral) sources of energy several decades before the rest of Europe.[26] By the early nineteenth century, households and firms consumed around 15 million tons of coal a year compared to 3 million tons for Europe as a whole.[27]

Mainland European and East Asian economies and cities found substitutes such as peat, wood, water, wind, and human energy, but the advantages for industrialization of using cheaper and more efficient thermal energy were substantial. For example, wind power and waterpower are less reliable and predictable. Coal replaced the land used to feed horses and oxen, as well as the manpower employed in forestry. As a substitute for wood fuel, coal allowed more land and other resources to be devoted to growing food and agrarian raw materials. Given that the energy from a ton of coal equals the energy from 2 tons of timber and that an acre of land produces 2 tons of dry wood, Britain's coal output for 1815 implies that 15 million acres (equivalent to 88 percent of the arable area) could counterfactually be released from forestry to grow grains, vegetables, animal products, and industrial raw materials.[28]

Heat-intensive industrial processes in metallurgy, glassmaking, brewing, refining sugar and salt, chemistry, in baking food and bricks, and so on could all be conducted more efficiently with cheap coal. The feedbacks and technological spin-offs from these industries to metallurgy and to the making of kilns, pots, vats, and containers were important for industrial development. Cheaper fuel kept workers warmer, diminishing their need for calories and allowing greater efforts in production. Lower-cost bricks and metals saved capital that could be invested in social overhead facilities and in industry itself.

Energy accounts constitute a heuristic and illuminating complement to national income accounts for the analysis of transitions to modern systems of production requiring modern sources of energy. At a time when technological progress that augmented labor productivity remained slow and confined to a few industrial sectors, economies favorably endowed with fertile land, minerals, natural waterways, and, above all, cheaper fuel linked to a maturing network technology (steam power), enjoyed a head start in the "leap forward" to become industrial market economies.[29]

The Nature and Economic Significance of Britain's Maritime Strategy for Security with Development

Debates about the precise nature and significance of foreign trade for the British Industrial Revolution remain unresolved.[30] Contemporary perceptions and histories that maintained that overseas commerce was a major component in British industrialization are being restored as valid even though the range and significance of mechanisms involved are not captured within a modern and statistical framework based on national accounts. Over the eighteenth century, the volume of goods sold overseas multiplied four times compared to a multiplier of over just two from 1500 to 1700. Ratios of exports to gross national product increased from a little over 4 percent in the reign of Elizabeth, to 6 percent after the Restoration, to 8 percent at the Glorious Revolution, and reached 12 percent under George III. At least half of the increase in industrial production during the long eighteenth century (1688–1815) was sold overseas.

Shares of the outputs of the most rapidly growing and technically progressive of British export industries (cottons, woolens, metals, and shipbuilding) became exceptional. For a British economy led by modernizing industries, the nation's multifaceted involvement in the world economy is now understood as an essential precondition for growth accompanied by structural change and diversification that took place both before and during the Industrial Revolution. By the close of the Seven Years War, something like half of the nation's nonagricultural workforce depended directly or indirectly on markets overseas for its livelihood. As *pôles de croissance*, London, Bristol, Hull, Glasgow, Newcastle, Liverpool, and other maritime cities provided infrastructures, skilled workforces, and transportation and distribution networks to service internal as well as overseas trade. Their high wages attracted labor from the countryside. Their hinterlands matured into productive fiscal bases for the state's rapacious demand for customs and excise duties. No estimates for the total values of commodities and services exchanged across the world's frontiers between 1660 and 1860 exist, but few historians would disagree that Britain (not France, Portugal, Spain, the Netherlands, let alone China or

Japan) reaped a lion's share of the gains from expanding international trade and commerce over that period.[31]

Was that, as Whig historians maintain, because the country's institutions (particularly its parliamentary constitution, legal system, and embedded cultures of enterprise) had become more hospitable to private investment and innovation than institutions in rival mainland economies or the maritime provinces of China and Tokugawa Japan?[32] Research into continental European economies has left historians more agnostic about the superiorities of the Hanoverian realm's institutions.[33] Recently rediscovered economic worlds of "surprising resemblances" across a range of advanced regions of Eurasia, also undergoing Smithian growth for centuries before the First Industrial Revolution, has effectively degraded both Marxian and Weberian perceptions that only certain countries and regions of northwestern Europe (particularly England, but also Holland) were proceeding along trajectories leading to modern economic growth.[34] Most might argue, however, that both societies appropriated growing shares of the gains to be reaped from mercantilistic engagements in global trade and commence.

One potentially significant contrast between Britain and all other premodern candidates (including Holland) for a First Industrial Revolution has, however, become clearer: the country's geographically conditioned but sustained commitment to a naval strategy for the defense of the realm carried unintended but important consequences for the development of a leading maritime public-cum-private sector of the British economy over time.[35] Not long after the Hundred Years War (1337–1453), England's kings, aristocrats, and merchants began to conceive of naval power, funded and sustained by the state, as the first line of defense against external threats and as the force required to back conquest and commerce with continents outside Europe.[36]

For reasons that cannot be expanded here, that conception took a long time to mature. Only after the restoration of the monarchy and aristocracy in 1660 did Britain's elite sustain the political consensus required to form a highly effective fiscal naval state.[37] Despite vicissitudes following the Dutch coup d'état of 1688 and the loss of sovereignty over thirteen American colonies in 1783, the restored British state became outstandingly successful in raising the taxes and loans required for external security and for the survival of an essentially ancien régime that protected an established and inegalitarian system of property rights.[38] The rights to own and use natural resources and capital located within a unifying kingdom; merchant shipping and merchandise on the high seas; and bases, plantations, mines, and colonies in an expanding empire all became better protected for Britons than for any other propertied elite in Western Europe, the Americas, Africa, or Asia at the time.

This quite exceptional level of protection, stability, and good order supplied by the state to its wealthier citizens rested on an expanding fiscal and financial base.[39] Between 1670 and 1815, total revenues from taxes rose by a factor of around seventeen, while national income increased by a multiplier of three. Most of these appropriations were allocated by the central government to service a national debt incurred to fund no fewer than eleven wars against other European powers and economic rivals–mainly France and Spain, but including four naval wars against the Netherlands.

From a nominal capital of less than 2 million pounds sterling in the reign of James II, Britain's national debt reached the astronomical sum of 854 million pounds sterling, or 2.7 times the national income, in 1819. The tax share devoted to servicing this public debt jumped from modal ratios of 2 to 3 percent before the Glorious Revolution to 60 percent after the Napoleonic Wars.[40] When Castlereagh signed the Treaty of Vienna in 1815, all Europeans were acutely aware of the costs of geopolitical strife. Yet the recently formed United Kingdom of England, Wales, Scotland, and Ireland enjoyed virtually complete security from external aggression, possessed the largest occidental empire since Rome, and enjoyed extraordinary shares of world trade and income from servicing global commerce.[41]

For a European economy to thrive in a mercantilist economic order riven by dynastic and imperial rivalries, an island state needed to allocate considerable resources to preclude invasion, preserve internal stability, and retain advantages over equally violent competitors in armed struggles related to global commerce and colonization. Geopolitical conditions formed inescapable parameters within which state formation, institution building, and macroeconomic growth occurred.[42] For the age of mercantilism, post hoc analysis of taxation based on counterfactual scenarios concerned with distortions from competitive equilibria look like interesting but anachronistic exercises in applied economics.[43] These analyses are surely irrelevant to questions of whether the state successfully raised and allocated the resources that carried the kingdom and its economy to a plateau of safety, political stability, and potential for future development envied by the rest of Europe. Since nobody then (or historians later) elaborated alternative strategies that combined security for the realm and internal order with growth for the economy, the comparison of the maritime strategy for security and development pursued by the English state with strategies pursued by other European and Asian powers could lead only to a Panglossian conclusion that virtually everything that was done looks unavoidable, was done for the best in the worst of all possible worlds, and paid off.[44]

Inaugurated under the republic, the essence of England's strategy for geopolitical security with economic power can be found in the persistent and relatively high levels of expenditure on the Royal Navy.[45] That sustained commitment provided the kingdom with the world's largest fleet manned by a largely coerced workforce of able seamen, under the command of a well-motivated and well-rewarded corps of professional officers.[46] The fleet was constructed and maintained in readiness for multiple missions at sea by an onshore workforce of skilled shipwrights, carpenters, and other artisans and serviced by an infrastructure of ports, harbors, dockyards, stores, ordnance depots, and other facilities under collaborative and coordinated public/private ownership and control.[47]

This huge fleet and massive onshore infrastructure of human and physical capital operated primarily to keep ships of the line strategically placed at sea as the first bastion of defense. Secondarily, but at falling average cost, these well-armed ships performed mercantilist missions to protect British trade and colonies while preying on hostile and potentially hostile merchant marines and threatening enemy coastal cities and colonies. Thus, Britain's evolving maritime strategy for defense with trade and growth included all kinds of attendant spin-offs for internal stability, for the protection of property rights, and for the extension of domestic as well as colonial markets.[48]

The nation's fleet of durable, strategically placed, and proficient ships provided external security at a relatively high level of efficiency compared to the logistical costs per joule of force delivered by large armies; recruited, mobilized, equipped, and supplied with food and forage; and moved overland to battlegrounds and vulnerable borders to repel enemy attacks.[49] An economically efficient offshore strategy for defense also allowed the British state to allocate greater proportions of revenues provided by an elastic fiscal and financial system not only to complementary mercantilist and imperial missions pursued at sea, but also to sustain surprisingly high levels of military expenditure.[50] Paradoxically and throughout the period 1688–1815, expenditures on armies by the Eurasian state most committed to naval power amounted to a modal 60 percent.[51]

Part of that allocation included hiring mercenary regiments of Hanoverians, Hessians, and other soldiers for combat outside the kingdom;

part consisted of subsidies and subventions to European allies willing to field troops to contain and thwart the continental and colonial designs of France and its allies; and finally, part consisted of the commitment of British troops to continental warfare, notably in 1702–1712 and 1808–1815. Strategic expenditures on the military forces of Britain's clients and allies prevented the Bourbon states of France and Spain, as well as other antagonists, from allocating funds to construct fleets capable of seriously challenging the Royal Navy.[52]

But a considerable proportion of state revenue was allocated to British regiments, militias, volunteers, and yeomanry on station in the realm. They served as a less-than-credible second line of defense against foreign invasion, but were used consistently, during a period of population growth, industrialization, and urbanization, to preserve the stability of the regime against subversion or disruption of internal trade on its Celtic fringes and to protect hierarchy and property rights from challenges to law and order.[53]

With external security taken for granted, other public goods, such as stability, good order, respect for traditional property rights, and the maintenance of hierarchy over potentially unruly employees, became the key political-cum-economic interest for landowners, merchants, farmers, industrialists, and other businessmen of Hanoverian Britain. On the whole, a monarchical and aristocratic state met such concerns for the protection of property and the maintenance of authority over their workforces.[54] When necessary, the state also redefined legal rights for new forms of wealth by promulgating statutes that superseded custom and common law that counterfactually could have been used to protect the welfare of the majority of the nation's workforce threatened by market forces associated with industrialization.[55]

Parliament's antipathies to large standing armies in times of peace look like Whig rhetoric because the actual numbers of troops, mobilized militiamen, and patriotic volunteers on station in Britain and Ireland year after year (and particularly in wartime) were more than adequate to repress disturbances to the peace. For the purposes of political stability, maintaining internal order, the protection of property, and upholding hierarchies of all kinds, it is not at all obvious that on a per capita basis, the political and legal authorities of constitutional Britain commanded a smaller or less coercive force than the despotisms of mainland Europe, who deployed soldiers (and not capital-intensive navies) to defend their more vulnerable frontiers. Indeed, in 1812, the number of soldiers mobilized to combat Luddites in the Midlands and North of England exceeded Wellington's expeditionary force in the Peninsular War.[56] The navy allowed the political authorities of Hanoverian Britain to allocate less revenue to external security while providing an effective military presence and the exemplary displays of armed force required to maintain order, protect property, and preserve authority among a potentially ungovernable society that was becoming ever more urban, industrial, and "dangerous."

THE DISCOVERY, TAKE-UP, AND DIFFUSION OF "ENGLISH" TECHNOLOGY

For several reasons, the invention and diffusion of a familiar list of machines, energy converters, and industrial processes long represented as "English" and regarded as prime movers in the national economy's precocious transition has moved into contexts where their importance has become problematic. That has occurred not only by way of significance testing by cliometricians, but also because the Industrial Revolution is no longer Anglocentrically conceived as a short, sharp discontinuity based on fundamental breakthroughs in industrial technologies emanating from and developing within a singularly progressive set of Anglo-Saxon institutions and culture.[57]

Several major inventions certainly emerged and matured in Britain after the Seven Years War, but their effects were confined to particular sectors of industry, like cotton textiles, metallurgy, shipbuilding, transportation, and the generation of energy from steam.[58] Furthermore, these technologies that became first the wonders and eventually the marks of a modern economy, like machines, steam power, processes for making and shaping metals, chemicals, and factories, appeared early but matured rather slowly over that century of "revolutionary transition" after 1750. Tabulations purporting to account in quantitative terms for the sources of British economic growth—derived from exercises that "fit" production functions to extant but imperfect data for national output and inputs of land, labor, and capital—expose the persistence of an entirely traditional, extensive form of aggregated economic growth, emanating mainly from faster rates of capital accumulation and upswings in the size and hours worked by the workforce rather than from innovations or even new sources of energy.[59] These essentially taxonomic exercises provide a nationwide perspective derivable from cliometric models designed to measure proximate sources of British economic growth. Nevertheless, the contribution of technological change (which had proceeded slowly over the centuries in many regions of a connected but not integrated world economy) is properly measured and defined by two widely recognized hallmarks of modern economic growth: accelerated and sustained rates of growth in output per worker and per capita income.[60] For the British case, after protracted debate over the models and the statistics, cliometricians now take into account the tentative quality of the data at their disposal and the reciprocal interactions between profitable opportunities provided by the appearance of new process and product innovations on the one hand, and higher rates of investment, on the other. In this light, technological progress evolved over time to reach a point around the mid-nineteenth century when its outcome can be retrospectively perceived and heuristically represented as highly significant—if not overwhelming. In this macroeconomic context, without the discovery, development, and diffusion of technologies and improved modes of organization that augmented the average productivity of its workforce, the British economy would never have been designated as the locus of the First Industrial Revolution.[61]

Nevertheless, the role for new technology coming onstream over that time can be relegated to a chapter in a longer and more complex historical narrative, which recognizes its confined scale, scope for transformation, and potential across all sectors, not only of the national economy but of manufacturing itself. Economic histories of a range of industries, other than that paradigm case of revolutionary change, cotton textiles, have made us aware of the decades taken and costs incurred to move from a blueprint, through several stages of development and protracted periods of learning by using, until original and promising designs became marketable prototype machines, processes, or artifacts.[62]

We now realize that forward planning and investment are required to embody a backlog of known product and process innovations in firms that were connected to markets for commodities, labor, and capital and must be networked to suppliers of raw materials and to transportation and distribution services so that entrepreneurs exploiting new knowledge could realize external economies of scale and agglomeration by locating in industrial towns and maritime cities. The costs of systemwide investments to develop, embody, and relocate production in factories and towns were large multipliers of the original outlays required to come up with potentially useful and commercially viable knowledge.[63]

As pioneers, British investors and entrepreneurs lacked examples of prior experiments and experience from elsewhere, as well as an extensive and reliable base of systemic scientific knowledge that, later in the nineteenth century, could expose the problems, ramifications, and potential of untried knowledge more rapidly and at lower cost.[64] Subsequent industrializers had advantages unavailable to Britain.[65]

Although British investors and businessmen lacked references to prior practice and to science to inspire the confidence to undertake risky investments in new technologies, their direct support for research and development and a more rapid and extensive diffusion of the potentially useful knowledge already available in the eighteenth century does not appear to have been particularly entrepreneurial. Considered as a national group, British businessmen promoted and managed one of the slowest and, for the working classes, more miserable transitions to an industrial economy in world history.[66]

Subsequent faster and often more socially benign industrial revolutions are marked by higher rates of saving and investment and by a more rapid take-up of advanced technology.[67] For example, in the British case, the ratio of gross investment to national income took more than a century to double from a rather low base point of around 6 percent in 1760.[68] In comparison to "follower" nations, this looks like unimpressive average and marginal propensities to save or to invest in the social overhead and industrial capital required to promote urbanization.[69]

The gradual rise in domestic capital accumulation required to exploit new technology has, however, been attributed to the massive sums of otherwise surplus investment funds borrowed by the state to fund three wars (1756–1763, 1775–1783, and 1793–1815) against France, other European rivals, and the United States.[70] Counterfactually, government borrowing for the purposes of waging war might have "crowded out" the potential for higher rates of private capital formation, but the overall effect could well have been trivial. First, the observed variations between years of war and interludes of peace in real rates of interest for investors in low-risk government securities floated and sold on the London capital market do not suggest that Britain was an economy constrained by incapacity to save. On the contrary, during all three wars, the overall supply of investible funds responded elastically to additional demands from a state that offered both domestic and international capital markets attractive and secure paper assets. Government borrowing also promoted the development of financial intermediation in London and the integration of a national capital market that raised the elasticity of supply and improved the allocation of investible funds.[71]

Furthermore, and to return to the analysis of strategic expenditures, models of crowding out that neglect the benefits (and incentives for investment) provided by high rates of state expenditure on external security, the protection of overseas commerce and colonization, and a repressive but effective system of internal order are seriously underspecified.

Balance sheets for these indispensable public goods would be difficult to model and impossible to add up. Given that rather high levels of expenditure on the army and navy were necessary for state formation and the preservation of British institutions, particularly in wartime, the crowding-out hypothesis needs to be reformulated as a historical problem of ascertaining and measuring the proportions of taxes and loans devoted to security and stability that might conceivably be defined as wasteful allocations by the Hanoverian state. Few mercantilists of the period suggested that the depressing effects on private savings and investment flowing from the operations of the fiscal and financial system exceeded the benign effects of "crowding in" that depended on the effective provision of external security, mercantilism, stability, and internal order.[72] Adam Smith certainly appreciated that defense came before opulence and that unilateral withdrawal from the prevailing geopolitical order by the rulers of the British Isles was never an option.[73]

Once expenditures by the state are reconfigured as positive or at least unavoidable, then rates of development and take-up of advanced technologies as urban systems of production during an ostensibly revolutionary period in British economic history cease to appear as entrepreneurial and historically remarkable as Anglo-American historiography has maintained for too long. Indeed, the way back into a properly conceptualized and contextualized historical analysis of the Industrial Revolution is already underway in the program of the Cambridge school in the history of political economy. They have reconstructed the discourses of the day to indicate that classical economists recognized there was nothing particularly "progressive" about the majority of the country's economic elite.[74] The owners and controllers of property reinvested rather low proportions of their rentier-type gains that accrued from industrialization.

Generations of the national history profession researching Britain's agriculture, commerce, and industry have published what aggregates to a library of case studies of British landowners, farmers, merchants, industrialists, bankers, professional experts, and others with surpluses to save and invest in the new technologies and urban systems of production. Numerous well-documented examples of commendable foresight, perseverance, risk taking, innovation, and entrepreneurship, particularly for leading industries, can be drawn from the rich historiography of the First Industrial Revolution.[75] But did British capitalists manifest a national *Geist* or *Kopf* for risk taking and improvement that was exceptional?[76]

Nevertheless, these questions must be located with the findings of a generation of quantitative research that has constructed a statistical base to engage with potentialities derived from macroeconomic modeling. This program in economic history has seriously qualified (if not degraded) the notion that an insular "culture" ordering economic behavior on the British Isles could be represented as exceptionally enterprising.[77] Looking at the Industrial Revolution as a macroeconomic event, connected to, if not embedded in, a wider world economy, several statistically validated reasons suggest that (within an environment of incomparable security provided and sustained by the Hanoverian state for the nation's businessmen and wealthy elites) the take-up of new technology, the construction of urban agglomerations, and the formation of social overhead capital required to realize the full potential of technologies that appeared after the Seven Years War seem anything but impressive. On the contrary, macroeconomic trends as currently measured all look favorable for the promotion of higher rates of savings investment and innovation. For example, after falling below the 10 percent mark during the recession that surrounded the crisis with England's thirteen North American colonies, average rates of return on all forms of capital other than agricultural land fluctuated cyclically, but doubled by the mid-nineteenth century. By then, even rents from farmed land (the sector in relative decline) had risen by nearly 50 percent.

Over the century that succeeded the Seven Years War, average real wages passed through three cycles or phases: slow improvement (c. 1761–1800), virtual stasis (1801–1820), and upswing (1821–1851) to reach a point around midcentury some 45 percent above their initial level.[78] Labor productivity followed a different trajectory and enjoyed a faster rate of increase to arrive at a level 87 percent above its baseline average. Classical features of all industrial revolutions—higher rates of growth in labor productivity emanating from advanced technologies and externalities derived from the agglomeration of production in towns—became more evident during the First Industrial Revolution than during the Italian Renaissance or the Dutch Golden Age.[79] The British case was also marked by a uniquely gradual rate of change, a slow take-up of new technology, and "deplorably" low rates of investment in housing and the infrastructure of towns required to support a more rapid transition to industrial society.[80]

This feature of the First Industrial Revolution, rather than machinery and factories as such, attracted the condemnations of visitors from the mainland, as well as generations of British reformers.[81] Amelioration and a jack-up in investment rates took a long time to achieve, partly because the fiscally emasculated state that emerged from the Napoleonic Wars could not raise the taxes required to do much more than continue to protect the realm's commerce and expanding overseas empire, partly because average real wages (and aggregate demand) increased very slowly, but partly because British economic elites reinvested such small proportions of the rising share of the rentier-type income that they obtained from their secure property rights.[82] Of course, commendable examples of entrepreneurship supporting innovation and invention in this period demonstrate the activities of enterprising Britons. But these laudable achievements need to be contexualized within the macroeconomic frameworks constructed by Allen, Crafts, Harley, Mokyr, Clark, and Voth, and other cliometricians that, taken together, reconfigure the Industrial Revolution as a precocious, unremarkable, and rather predictable transition in a global history of slow but accelerating technological change. Furthermore, very few economic historians now regard this conjuncture in British economic history as a paradigm for comparable changes that followed elsewhere or believe that standards of living or labor productivity in the world's industrial market economies would look very different today without the transformation that occurred in Britain between 1750 and 1850.[83]

Insofar as the discovery and development of new technologies for industry, transportation, and agriculture that appeared during this period can be linked to an evolving base of systemic knowledge, the scale, scope, and utilitarian relevance of that kind of knowledge can be realistically depicted as European rather than British in origin. Britain's advantages resided more in the development, improvement, and diffusion of technology than in discovery itself.[84] Yet some historians, notably Margaret Jacob and Ian Inkster, argue that British "culture" became more receptive to an intermingling of science with business, religion, and politics than was the case elsewhere in Eurasia.[85] Studies of several contexts for the advance and diffusion of useful and reliable knowledge in France, Italy, and even Spain have, however, made it more difficult to accept Anglocentric assertions that mainland European monarchs, aristocracies, ecclesiastical and political elites, and especially the military were somehow less "rational" or less open to the potentialities of new knowledge than their offshore counterparts.[86] That debate seems to be something of a hangover from religious controversies over the Reformation, including memorable, but unproven, theories about the positive connections between Protestantism and entrepreneurship, Protestantism and hard work, as well as Protestantism and science, lifted uncritically from Max Weber and Robert Merton.[87] The urban and commercial cultures of Europe's or Asia's maritime cities cannot be singled out as looking discernibly less rational, calculating, and utilitarian than cultures operating in British towns, among British educational institutions, or in peculiarly British publishing and information flows.[88] Roy Porter has also made claims for the exceptionalism of an English Enlightenment that have been challenged by a controversial interpretation of the "long eighteenth century" in British history that characterized the years 1660 to 1832 as a period marked by the persistence of an ancien régime presided over by an autocratic, aristocratic, and confessional state. Cultural turns by nations or cities toward progress are difficult to expose, let alone measure.[89]

Early in the eighteenth century, European visitors did recognize, however, that British industry was moving ahead in certain spheres of industrial technology. Indeed, several states engaged in espionage in order to close any gaps, particularly for technologies with military potential.[90] The appearance of British machines on the mainland even in such seemingly conservative cultures as Catalonia occurred rather rapidly before the outbreak of Revolutionary warfare, 1793–1815, arrested diffusion to much of Europe. Across Europe, technological advances tended to appear, moreover, in branches of industrial production that had reached a certain scale and diversity in production. In some well-known British cases like cotton and bar iron, this occurred after processes of import substitution. Foreign products obtained and pioneered access to their home market and tempted British businessmen to press for protection and to engage in a search for indigenous ways to satisfy first domestic, then imperial, and eventually foreign demand. The process involved the creation, by a sympathetic mercantilist state, of helpful matrices of legislation and fiscal incentives surrounding commodity and labor markets for Britain and its imperial possessions.[91]

Technological progress depended above all on the prior and persistent accumulation of a skilled and mobile industrial workforce of artisans and craftsmen. To explain how, when, and why the British economy managed to build up the range of skills required to facilitate breakthroughs and improvements in technological knowledge to survive a necessary stage of development to the point of commercial viability has not been easy.[92] Economic theory is not particularly helpful in explaining the formation of human capital, but economic history is generating promising research into the records of Europe's guilds, and their connections to the rise, embodiment, and maintenance of skills among European workforces.[93]

For Eurasia, the relevant contexts were invariably urban. In the United Kingdom, London, Bristol, Nottingham, Birmingham, and even Dublin all became important locations for the development of skilled workforces. Immigrant German, Flemish, Dutch, and Huguenot craftsmen, merchants, and financiers clearly played important roles in starting and sustaining the process of human capital formation in Britain. They could be attracted from the mainland to a kingdom that promised security from external aggression, religious toleration, and from time to time offered them royal protection and subsidies. If and when they developed interests in trade with the Americas, Africa, and Asia, they could be assured of protection by the Royal Navy. Europeans settled and, as part of extended families and mini-diasporas, maintained links with communities of knowledge and skill on the mainland. In an age in which the diffusion and adaptation of technology occurred through the migration of skilled and professional manpower, the obvious attractions of a shorter or longer domicile in English towns were reinforced by warfare and religious persecution on the mainland.[94]

CONCLUSION: DECONSTRUCTING AND RECONFIGURING THE FIRST INDUSTRIAL REVOLUTION

After the Seven Years War, the British economy moved onto a century-long trajectory of accelerated growth with structural change that merits the appellation of the First Industrial Revolution. This long cycle, together with the wars against Revolutionary and Napoleonic France, carried the United Kingdom to the clear position of competitive advantage it enjoyed over the economies of continental Europe and the rest of the world between 1846 and 1873. That "moment" of economic dominance took centuries to mature, looks brief, and was based to a significant degree on natural advantages and naval power. Britain's technological hegemony was, it seems, European (Eurasian) in origin; confined to textiles, metallurgy, and engineering; and destined to pass away through the traditional and familiar workings of diffusion, adaptation, and convergence.[95]

In order to explain the First Industrial Revolution and the rather rapid convergence of Western Europe into an interrelated and ultimately integrated set of highly successful industrial market economies, it is necessary to explore longer time spans and wider geographical frames that include Africa, the Americas, and East Asia, as well as the mainland.[96] In this long stream of time and recently revealed premodern "world of surprising resemblances," the Industrial Revolution can be recontextualized as a precocious but not that remarkable conjuncture in humanity's escape from the diminishing returns endemic to organic economies. Real growth (efflorescences) in labor productivity and incomes per capita occurred in other places and in other times prior to the Seven Years War, but in these instances, natural disasters, geopolitical shocks, and Malthusian checks returned organic economies to stasis or very slow growth. Geography ensured that the British Isles were predestined to avoid the first. In the wake of an interregnum of civil war and republican rule, a properly funded Royal Navy was developed to protect the economy from the second. A less than impressive diffusion of new technologies and inorganic sources of energy turned out to be sufficient to confound Malthus by producing a First Industrial Revolution. Britain escaped first, and mainland Europe and its European offshoots overseas soon followed. High and rising standards of living can now be observed in many regions of an integrating world economy. In this frame of historical reference, being first matters a lot less than the North–South divide and the persistence of mass poverty. For solutions to that problem, there is no British model and no need for patriotic histories of a First Industrial Revolution, proclaiming Britain, Holland, or any other nationally constructed location or culture as the original locus, let alone the paradigm for modern economic growth. As modern Chinese and Japanese scholars now correctly observe, neither English nor European history represented global destiny.[97]

Notes

1. Patrick K. O'Brien, ed., *Industrialization: Critical Perspectives on the World Economy*, 4 vols. (London: Routledge, 1998).

2. Simon Kuznets, *Modern Economic Growth* (New Haven, CT: Yale University Press, 1996).

3. Jan De Vries and Ad Van Der Woude, *The First Modern Economy: Success, Failure and Perseverance of the Dutch Economy, 1500–1815* (Cambridge: Cambridge University Press, 1997).

4. Peter Mathias and John Davis, eds., *The First Industrial Revolutions* (Oxford: Blackwell, 1990), 1–24.

5. Jack Goldstone, "Efflorescences and Economic Growth in World History: Rethinking the 'Rise of the West' and the Industrial Revolution," *Journal of World History* 13 (2002), 323–392.

6. Rondo Cameron, "The Industrial Revolution: Fact or Fiction" in *Leading the World Economically*, ed. François Crouzet and Armand Clesse (Amsterdam: Dutch University Press, 2003), 169–184.

7. Nicholas Crafts and C. Knick Harley, "Output Growth and the British Industrial Revolution: A Restatement of the Crafts-Harley View," *Economic History Review* 45 (1992), 703–730; T. S. Ashton, *The Industrial Revolution, 1760–1830* (London: Oxford University Press, 1948).

8. Paul Langford, "The English as Reformers: Foreign Visitors' Impressions 1750–1850," in *Proceedings of the British Academy: 100 Reforms in Great Britain and Germany, 1750–1850*, ed. T. C. W. Blanning and Peter Wende (Oxford: Oxford University Press, 1999), 101–119.

9. Giorgio Riello and Patrick K. O'Brien, "Reconstructing the Industrial Revolution: Analyses, Perceptions and Conceptions of Britain's Precocious Transition to Europe's First Industrial Society," *Economic History* 84 (2004), 1–41.

10. Roderick Floud and Paul Johnson, eds., *The Cambridge Economic History of Modern Britain*, vol. 1, *Industrialization, 1700–1860* (Cambridge: Cambridge University Press, 2004).

11. David S. Landes, *The Wealth and Poverty of Nations: Why Some Are So Rich and Some So Poor* (New York: Little Brown, 1998); Douglass North, *Institutions, Institutional Change and Economic Performance* (Cambridge: Cambridge University Press, 1990).

12. Christine Rider and Michael Thompson, eds., *The Industrial Revolution in Comparative Perspective* (Malabar, FL: Krieger, 1999).

13. Ian Inkster, *Technology and Industrialisation: Historical Case Studies and International Perspectives* (London: Variorium Press, 1998), 40–58.

14. Maxine Berg, *Luxury and Pleasure in Eighteenth-Century Britain* (Oxford: Oxford University Press, 2005).

15. Nicholas Crafts, "Productivity Growth in the Industrial Revolution: A New Growth Accounting Perspective," *Journal of Economic History* 64 (2004), 521–535.

16. Martin Daunton, *Progress and Poverty: An Economic and Social History of Britain, 1750–1850* (Oxford: Oxford University Press, 1995).

17. Bart Van Bavel and Eric Thoen, eds., *Land Productivity and Agro Systems in the North Sea Area, Middle Ages—20th Century: Elements for Comparison* (Turnhout: Brepols Publications, 1999).

18. Maarten Prak, ed., *Early Modern Capitalism: Economic and Social Change in Europe* (London: Routledge, 2001).

19. Robert C. Allen, *Enclosure and the Yeoman* (Oxford: Oxford University Press, 1992).

20. Alan Macfarlane, *The Origins of English Individualism: The Family, Property, and Social Transition* (Oxford: Oxford University Press, 1979); Richard Britnell, *The Commercialisation of English Society, 1000–1500* (Cambridge: Cambridge University Press, 1993).

21. Tom Scott, ed., *The Peasantries of Europe from the Fourteenth to the Eighteenth Centuries* (London: Longman, 1998).

22. Robert C. Allen, "The Great Divergence in European Wages from the Middle Ages to the First World War," *Explorations in Economic History* 38 (2001), 411–447.

23. Kenneth Pomeranz, "Beyond the East-West Binary: Resituating Development Paths in the Eighteenth-Century World," *Journal of Asian Studies* 61–62 (2002), 539–590.

24. Patrick K. O'Brien and Daniel Heath, "English and French Landowners 1688–1789," in *Landowners, Capitalists and Entrepreneurs: Essays for Sir John Habakkuk*, ed. Michael Thompson (Oxford: Oxford University Press, 1994), 23–62.

25. E. Anthony Wrigley, *Continuity, Chance and Change: The Character of the Industrial Revolution in England* (Cambridge: Cambridge University Press, 1988).

26. E. Anthony Wrigley, "The Divergence of England: The Growth of the English Economy in the Seventeenth and Eighteenth Centuries," *Transactions of the Royal Historical Society* 10 (2000), 117–141.

27. Ian Inkster and Patrick K. O'Brien, eds., "The Global History of the Steam Engine," *History of Technology* 25, special issue (2004).

28. Rolf Sieferle, *The Subterranean Forest: Energy Systems and the Industrial Revolution* (Cambridge: Cambridge University Press, 2001).

29. Vaclav Smil, *Energy in World History* (Boulder, CO: Westview Press, 1994).

30. Joel Mokyr, ed., *The British Industrial Revolution: An Economic Perspective* (Boulder, CO: Westview, 1993).

31. Javier Cuenca Esterban, "Comparative Patterns of Colonial Trade: Britain and Its Rivals," in *Exceptionalism and Industrialization: Britain and Its European Rivals, 1688–1815*, ed. Leandro Prados de la Escosura, 35–69.

32. Charles Kindleberger, *World Economic Primacy, 1500–1990* (Oxford: Oxford University Press, 1996).

33. Richard Sylla and Gianni Toniolo, *Patterns of European Industrialisation: The Nineteenth Century* (London: Routledge, 1991).

34. Kenneth Pomeranz, *The Great Divergence: China, Europe, and the Making of the Modern World Economy* (Princeton, NJ: Princeton University Press, 2000).

35. Patrick K. O'Brien, "Mercantilism and Imperialism in the Rise and Decline of the Dutch and British Economies," *De Economist* 148 (2000), 469–501.

36. Nicholas Rodger, *The Safeguard of the Sea: A Naval History of Britain,* vol. 1, *660–1649* (London: Allen Lane, 1997).

37. Henry Roseveare, *Financial Revolution, 1660–1760* (London: Longman, 1991).

38. John Brewer, *The Sinews of Power: War, Money and the English State, 1688–1783* (London: Unwin Hyman, 1991).

39. Prados de la Escosura, ed., *Exceptionalism and Industrialisation.*

40. Patrick K. O'Brien, "The Political Economy of British Taxation, 1660–1815," *Economic History Review* 41 (1988), 1–32.

41. Patrick K. O'Brien, "Fiscal Exceptionalism: Great Britain and Its European Rivals from Civil War to Triumph at Trafalgar and Waterloo," in *The Political Economy of British Historical Experience, 1688–1914*, ed. Donald Winch and Patrick K. O'Brien (Oxford: Oxford University Press, 2002), 246–265.

42. Kenneth Morgan, "Mercantilism and the British Empire, 1698–1815," in *The Political Economy of British Historical Experience*, ed. Winch and O'Brien, 165–192.

43. Jan Glete, *War and the State in Early Modern Europe: Spain, the Dutch Republic and Sweden as Fiscal-Military States, 1500–1660* (London: Routledge, 2002).

44. Leonard Gomes, *Foreign Trade and the National Economy: Mercantilist and Classical Perspectives* (Basingstoke: Macmillan, 1987).

45. *Parliamentary Papers 1868–69* (XXXV); C. D. Chandaman, *English Public Revenue, 1660–1688* (Oxford: Oxford University Press, 1975); Frederick Dietz, *English Government Finance, 1458–1641* (New York: Frank Cass, 1964).

46. Nicholas Rodger, *The Command of the Ocean: A Naval History of Britain,* vol. 2, *1649–1815* (London: Allen-Lane, 2004).

47. Roger Morris, *Naval Power and British Culture: Public Trust and Government Ideology* (Aldershot: Ashgate, 2004).

48. Daniel Baugh, "The Eighteenth-Century Navy as a National Institution," in *The Oxford Illustrated History of the Royal Navy*, ed. John Hill (Oxford: Oxford University Press, 1995), 120–160.

49. Richard Harding, *The Evolution of the Sailing Navy, 1509–1815* (Basingstoke: Macmillan, 1995); John Landers, *The Field and the Forge: Population, Production and Power in the Pre-Industrial West* (New York: Oxford University Press, 2003).

50. Patrick K. O'Brien and Philip Hunt, "England 1485–1815," in *The Rise of the Fiscal State in Europe, c.1200–1815*, ed. Richard Bonney (Oxford: Oxford University Press, 1999), 53–100.

51. *Parliamentary Papers 1868–69* (XXXV).

52. Daniel Baugh, "Great Britain's Blue Water Policy, 1689–1815," *International History Review* 10 (1988), 33–58.

53. Patrick K., O'Brien, "The State and the Economy, 1688–1815," in *The Economic History of Britain since 1700,* vol. 1, ed. Roderick Floud and Donald McCloskey (Cambridge: Cambridge University Press, 1994), 205–241; Louis Cullen, *An Economic History of Ireland since 1660* (London: Batsford, 1987).

54. John Brewer and John Styles, eds., *An Ungovernable People: The English and Their Law in the Seventeenth and Eighteenth Centuries* (London: Hutchinson, 1980).

55. John Rule, *Albion's People: English Society, 1714–1815* (London: Longman, 1992).

56. Clive Emsley, *Crime and Society in England, 1750–1900* (London: Longman, 1987).

57. Arnold Pacey, *The Maze of Ingenuity: Ideas and Idealism in the Development of Technology* (Cambridge, MA: MIT Press, 1994).

58. Maxine Berg and Pat Hudson, "Rehabilitating the Industrial Revolution." *Economic History Review* 45 (1992), 269–335; Peter Temin, "Two Views of the British Industrial Revolution," *Journal of Economic History* 57 (1997), 63–82; Crafts and Harley, "Output Growth and the British Industrial Revolution," 703–730; Nicholas Crafts and C. Knick Harley, "Simulating the Two Views of the Industrial Revolution," *Journal of Economic History* 60 (2000), 819–841.

59. Crafts, "Productivity Growth in the Industrial Revolution," 521–535.

60. Joel Mokyr, "Accounting for the Industrial Revolution," in *Industrialisation, 1700–1860,* ed. Floud and Johnson, 1–27.

61. Nicholas Crafts, "The First Industrial Revolution: Resolving the Slow Growth/Rapid Industrialization Paradox?" *Journal of the European Economic Association* 3 (2005), 525–534.

62. Roy Church and E. Anthony Wrigley, eds., *The Industrial Revolutions,* 11 vols. (Oxford: Blackwell, 1994), vols. 8–10.

63. Vernon Ruttan, *Technology, Growth and Development: An Induced Innovation Perspective* (New York: Oxford University Press, 2001), pt. 2.

64. Joel Mokyr, *The Gifts of Athena: Historical Origins of the Knowledge Economy* (Princeton, NJ: Princeton University Press, 2002).

65. Alice Amsden, *The Rise of the Rest: Challenges to the West from Late Industrializing Economies* (Oxford: Oxford University Press, 2001).

66. Riello and O'Brien, "Reconstructing the Industrial Revolution."

67. Angus Maddison, *The World Economy, a Millennium Perspective* (Paris: OECD, 2001).

68. Charles Feinstein and Sidney Pollard, eds., *Studies in Capital Formation in the United Kingdom, 1750–1820* (Oxford: Oxford University Press, 1988).

69. Peter Mathias and Munia M. Poston, eds., *The Cambridge Economic History of Europe,* vol. 7, parts 1 and 2, *Capital, Labour and Enterprise* (Cambridge: Cambridge University Press, 1978).

70. Ann Digby, Charles Feinstein, and David Jenkins., *New Directions in Economic and Social History,* vol. 2 (Basingstoke: Macmillan, 1992), 37–48.

71. Prados de la Escosura, ed., *Exceptionalism and Industrialisation,* 35–69.

72. Terence Hutchison, *Before Adam Smith: The Emergence of Political Economy* (Oxford: Oxford University Press, 1988).

73. Keith Tribe, "Mercantilism and Economics of State Formation," in Lars Magnusson, ed., *Mercantilist Economics* (Norwell, MA: Kluwer, 1993).

74. Gareth Stedman-Jones, *An End to Poverty: A Historical Debate* (London: Profile Books, 2004); Donald Winch, *Riches and Poverty: An Intellectual History of Political Economy in Britain, 1750–1934* (Cambridge: Cambridge University Press, 1996).

75. Daunton, *Progress and Poverty*.

76. Emma Rothschild, "The English Kopf," in *The Political Economy of British Historical Experience, 1688–1914*, ed. Winch and O'Brien, 31–60.

77. Robert C. Allen, "Capital Accumulation, Technological Change and the Distribution of Income during the British Industrial Revolution," unpublished manuscript, Nuffield College, Oxford (2005).

78. Charles Feinstein, "Pessimism Perpetuated: Real Wages and the Standard of Living in Britain during and after the Industrial Revolution," *Journal of Economic History* 38 (1998), 625–658.

79. Jan Luiten Van Zanden, "Wages and Standards of Living in Europe, 1500–1800," *European Review of Economic History* 3 (1999), 175–198.

80. Nicholas Crafts, "British Industrialization in an International Context," *Journal of Interdisciplinary History* 19 (1989), 415–428.

81. Riello and O'Brien, "Reconstructing the Industrial Revolution."

82. Patrick K. O'Brien, "Aristocracies and Economic Progress under the Ancien Regime," in *European Aristocracies and Colonial Elites: Patrimonial Management Strategies and Economic Development, 15th–18th Centuries*, ed. Paul Janssens and Bartolomé Yun-Casalilla (Aldershot: Ashgate, 2005).

83. Landes, *The Wealth and Poverty of Nations*, and *The Unbound Prometheus: Technological Change and Industrial Development in Western Europe from 1750 to the Present Day*, 2nd ed. (Cambridge: Cambridge University Press, 2004).

84. Joel Mokyr, *The Lever of Riches: Technological Creativity and Economic Progress* (Oxford: Oxford University Press, 1990).

85. Margaret Jacob, *Scientific Culture and the Making of the Industrial Revolution* (Oxford: Oxford University Press, 1997); Ian Inkster, "Potential Global. A Story of Useful and Reliable Knowledge and Material Progress in Europe 1474–1914," *International History Review* 18 (2006), 237–286.

86. William Clark, Jan Golinski, and Simon Schaffer, eds., *The Sciences in Enlightened Europe* (Chicago: University of Chicago Press, 1999).

87. John Brooke, *Science and Religion: Some Historical Perspectives* (Cambridge: Cambridge University Press, 1991).

88. Patrick K. O'Brien et al., eds., *Urban Achievements in Early Modern Europe: Golden Ages in Antwerp, Amsterdam and London* (Cambridge: Cambridge University Press, 2001).

89. Roy Porter, "The English Enlightenment," in *English Society, 1688–1832*, ed. Jonathan Clark (Cambridge: Cambridge University Press, 2002).

90. John R. Harris, *Industrial Espionage and Technology Transfer: Britain and France in the Eighteenth Century* (Aldershot: Ashgate, 1998).

91. Joseph Inikori, *Africans and the Industrial Revolution in England: A Study in International Trade and Development* (Cambridge: Cambridge University Press, 2002).

92. Lilliane Hilaire-Pérez, *L'invention technique au siècle des Lumières* (Paris: Albin-Michel, 2000).

93. Maarten Prak and Stephan R. Epstein, eds., *Guilds, Innovation and the European Economy, 1400–1800* (New York: Cambridge University Press, 2008).

94. Derek Keene and Stephan R. Epstein, eds., *The Rise of a Skilled Workforce in London, 1500–1800.* Forthcoming.

95. Kristine Bruland, ed., *Technology Transfer and Scandinavian Industrialization* (Oxford: Berg, 1991).

96. Mikulás Teich and Roy Porter, eds., *The Industrial Revolution in National Context* (Cambridge: Cambridge University Press, 1996).

97. Roy Bin Wong, "The Political Economy of Agrarian Empire and Its Modern Legacy," in *China and Historical Capitalism: Genealogies of Sinological Knowledge*, ed. Tim Brook and Gregory Blue (Cambridge: Cambridge University Press, 1999), 210–245; Kaoru Sugihara, "The East Asian Path to Economic Development: A Long Term Perspective," in *The Resurgence of East Asia: 500, 150, and 50 Year Perspectives*, ed. Giovanni Arrighi (London: Routledge, 2003).

3

The British Product Revolution of the Eighteenth Century

Maxine Berg

New Products of the Eighteenth Century

Historians seeking to explain Britain's industrial ascendancy in the eighteenth century now focus on its distinctive energy sources and, above all, on its technological precocity. Where once they identified mechanization and steam power as the keys to Britain's success, they now look more broadly to a peculiar concentration of "useful knowledge" deployed to invention and to productivity gains.[1] While "useful knowledge" has contributed to a wider concept of invention, invention must also connect outward from process to product, from production to consumption. To what extent was Britain's Industrial Revolution about products as much as it was about processes? What can we learn about that crest of products that rose on T. S. Ashton's "wave of gadgets"?

In 1792, Joseph Priestley, that Lunar man whose chemical experiments and radical dissenting values were so important to Britain's "industrial enlightenment," submitted to the Warwickshire assizes an inventory of his losses during the Birmingham riots of 1791.[2] The inventory, valued at over 4,000 pounds sterling, included meticulous descriptions and values of his possessions.[3] These were all commodities new to the eighteenth century, most were produced in Britain, and all were described with what were then recognizable British attributes. There were "willow" and "Scotch" carpets, Manchester and calico curtains, plated silver tea urns and cutlery, plated buckles and medallions, cut glass, Nankeen and Wedgwood chinaware, patent candlesticks, and mahogany tea tables. What strikes any reader of customs accounts, shipping lists, or advertisements of the time is the enormous range and variety of these goods. This was no cornucopia of the world's commodities, but a dense description of new goods with "British" attributes. Our industrial histories, however, tell us little about them.

We have, perhaps, focused too narrowly on technologies that tell us of process innovation, and too little on the product innovation that went with this. Product innovation brought in its wake productivity gains that we have not yet even tried to estimate. Economists now claim that estimates of productivity growth in the last half of the twentieth century are biased downward because these do not adjust for quality improvements; nor do they count new products. New products, and improvements in the quality of existing ones, do not show up in the output statistics by which productivity growth is measured; such statistics require long runs of data for the same goods over time. Failure to account for new products and for improved quality introduces a bias against output growth in earlier periods. Cotton quality improvements and new varieties in the last third of the eighteenth century, and improvements in oil lamps in the late eighteenth century, along with the introduction of gas lighting at the beginning of the nineteenth century, are important examples. Joel Mokyr adds the part played by new services and especially new medical interventions, such as the discovery of smallpox inoculation in 1796. What economist would want to deny their significance to economic growth? Yet they are not included in measures of economic growth between 1760 and 1830.[4] Running alongside the "industrial revolution," there was a "product revolution" of equal significance.

Over the course of the eighteenth century, this product revolution achieved the invention and global dominance of a category of identifiably British consumer goods that conveyed modernity and novelty, quality and good value. The goods in this category were high design products, goods that invoked "art and industry." Their quality went with price competitiveness, but consumers bought them not because they were cheap, and certainly not because they were mass products. They bought them because they were fashion leaders. They were quality goods, and they worked. Such products claimed attributes of convenience, utility, ingenuity, and novelty. In the discourse deployed at the time, they were "modern luxuries." How did manufacturers, merchants, retailers, and consumers transform these new products—Birmingham buckles on French shoes, Sheffield forks and knives in Connecticut and Barbados, Staffordshire chinaware on Anglo-Indian tables—into global commodities?

This product revolution was not a domestic event, for it found its context in a global economy. It marked out a British pathway to providing quality consumer goods for rapidly expanding middling-class markets at home and abroad. Philosophers and pundits at the time proclaimed a "new luxury," and statesmen and economists debated the impact of world commerce in such goods on their economies and national identities. The British developed new products in response to European luxuries, and they built on the success story of Asian export ware. They then branded these goods with British identity markers during their great period of expansion in world markets in the period just before and after the Seven Years War of 1756–1763.

INDUSTRIALIZATION AND PRODUCT LEADERSHIP

Our theories of industrialization focus on process innovation; most of them treat the final products of these processes as generic manufactured output. Debates on British and French, or indeed wider continental industrialization, rarely contain the word *consumption*. They concentrate on technological leadership, productivity growth, and rates of economic growth. In many cases, economic historians estimate productivity on data gathered on spun cotton or woolen yarn, with no reference to yarn counts or quality; the output of pig iron and consumption of coal are endlessly reassessed with no reference to the things made from the iron or processed with the coal.[5] The comparative data we work with as economic historians are full of inputs, not real outputs, of generic categories, not commodities. Certainly the categories of the economist are simplifications necessary for analysis, but with this, there is an elision that cuts out the essential characteristics of the product, including its varieties and its qualities. In assuming that one output suffices to cover all outputs, we lose the significance of the products.

Economic theories of industrialization owe something of this simplification to the sharp dichotomies drawn between the industrial histories of Britain and France that we have come to accept. The economic historians of the 1960s typecast Britain as a place of mechanical process innovation and standardized consumer goods, in contrast to France, which they associated with taste, style, and variety of goods. These accounts replicated the distinction drawn by Alfred Marshall at the beginning of the twentieth century between French and British social structures, and with this a French consumer market based in luxury and fashion, and a British one founded on substantial simple goods and solid comforts. The sharp divide between the developmental paths of Britain and France, as set out by Marshall, continues to pervade the approach to the history of luxury and consumer goods production in both countries. France's elite luxuries, its court, and its patronage of the arts contrast with Britain's ordinary commodities for middling and lower-class consumers.[6] Although historians of France are now challenging this comparison, Britain's historians have been slow to confront their favorite technological and economic theories.[7] These leave by the wayside the appearance of new and various commodities. The British Industrial Revolution, as we know it, is about technology rather narrowly defined; it is not about the role of the arts or aesthetic aspects of the history of goods.

Those who wrote comparative histories of industrialization accepted these polarities. François Crouzet, summing up the assessments of Rostow, Mathias, Landes, and Milward and Saul, argued that trade and industrial production reinforced a divergence between Britain and France that was already there at the beginning of the eighteenth century.[8] Sidney Pollard posed a persuasive theory of regional complementarities to explain comparative industrial development across Europe. He adapted the Ricardian analysis of trade and economic growth, but shifted focus away from comparative advantage in manufactured over agricultural goods to primary and secondary products. Pollard argued that an advanced economy like Britain's relied on its technological advantage in key processes in primary goods production, for example, in iron and spinning, but that less advanced economies with the advantage of lower wages could produce many other secondary commodities more cheaply. With cheaper British inputs and deploying new British technologies along with their own low-wage labor, these less advanced economies took the opportunity to challenge Britain's hegemony.[9]

J. R. Hicks reduced the connection between trade and economic growth to its bare essentials, showing that one trading partner could benefit by the technological progress of its powerful neighbor. This was most likely to happen when the technological advance, resulting in a real cost reduction, took place in export industries. Such cost reductions brought not just cheaper goods to receiving countries but improved terms of trade. If receiving countries used these cheaper goods as inputs in their own import-substituting activities, they gained from their advanced trading partner's new technology. With their lower wages, they capitalized on cheaper inputs and transferring rival technologies. Britain effectively equipped its rivals. Pollard argued that Britain was thus driven from European markets: imports from newly industrializing countries of the rest of Europe formed the most rapidly growing item in Britain's trade balance. These countries also colonized the markets of their own backward neighbors.[10]

British historians have long claimed an industrial exceptionalism for the later eighteenth century based in a deep technological subculture of mechanics, metalworkers, millwrights, and engineers who adapted their skills to the challenges thrown up by new materials, energy sources, and power mechanisms. But Mokyr (chapter 4, this volume) challenges this "British"-biased Industrial Revolution: the "idea that the Industrial Revolution was 'British' and that Europe was just a 'follower' seems overstated." He draws attention to the high costs of the political upheaval on the Continent of the years between 1789 and 1815. Mokyr and Liliane Hilaire-Pérez also identify pan-European processes of technological advance: artisans and inventors traveled between Britain, France, and other parts of the Continent. British and French elites took an interest in science and technology; "useful knowledge" and economic improvement were part of the Enlightenment. The British had Smeaton and Watt, Harrison and Murdoch. But the French had Jacques de Vaucanson and Honoré Blanc. Smeaton went to France to study engineering techniques; Boulton, Watt, and Wedgwood all sent their sons to France to learn about French technology and products. German and Swedish travelers picked up quickly on the essential qualities of British technological advances and transmitted them to home manufacturers seeking to diffuse and adapt them to their own settings.[11]

This debate over technological ascendancy does not, however, acknowledge the association between technological development and the appearance of certain key manufactured products: these products were first and foremost textiles, but included other consumer goods, especially glassware, earthenware, metal goods, and machinery. The success of British technology was bound up with the success of its products—that is how contemporaries saw it at the time. Huntsman crucible steel produced the quality associated with English buckles; coining and minting machinery, along with presses and stamps, provided the variety and quality of British buttons, medals, and brass furniture ware. From the 1760s, Europe's consumers wanted these products above all others. During years of war, they smuggled them in where possible, and after the end of the Napoleonic Wars, they did indeed swamp Europe. They did so not because they were cheap and Europe was a good dumping ground, but because they were better, they were fashionable, and they were already recognizably branded. That branding covered all manner of commodities specifically named by region or place, but it meant that all of these products claimed recognized "British" attributes. This branding did not just happen as an offshoot of machines and technological processes; producers attended to the quality, variety, and novelty of their products as avidly as they did to how they were made.

Contemporaries saw themselves embarking on a national project to create quality consumer goods. They looked to the arts for the design and taste to make British goods that would substitute for luxury imports from Asia and the rest of Europe and would become exports in their own right. We need to discuss not only how the technology succeeded in Britain, but how a key group of products became desirable in international markets—indeed how they became globalized.

There is no doubt that mercantilist policies played their part in framing this approach to products. The state, projectors, and entrepreneurs made great efforts to promote and to start up these foreign forms of manufacture in England. It is difficult to ascribe intentions. Were they simply trying to make English copies, which, in the hothouse climate of tariff walls, were bound ultimately to fail in a freer international economy? Or was it the practice, if not the intention, to establish new products successful enough to generate not just domestic markets but their own international markets? Manufacturers put great effort into art and design and into sophisticated advertising; they fostered aristocratic patronage and by these means identified key goods with civility and modernity, thereby indicating a larger project. In the process, product development was linked to the development of national identities.

As we have seen, economists have recently reinserted product innovation into the analysis of manufacturing growth and productivity change.[12] It is now time to consider its central part in eighteenth-century invention. The Boulton and Watt steam engine has long ranked as the key indicator to historians of eighteenth-century invention and ingenuity. But pride of place during the eighteenth century itself was given to an explosion of new, intricate consumer goods, from silver-plated coffee pots to stamped brassware and japanned papier-maché tea trays. These consumer goods contributed as much to the newly advantageous place of Britain in the international economy as tools and machines; indeed both types of goods featured equally in patents and projects.[13]

The British identity acquired by these new products formed part of the wider development of Britishness in the eighteenth century.[14] Wider-world contacts through commerce and colonization also fostered British identities. Trade and empire meant a wider concept of Britishness. This was a time when the labor, commodities, and cultures of foreign and colonial peoples underwrote England's prosperity and "character."[15] Many perceived such global access as the key to a new extended Britishness.

Indeed Malachy Postlethwayt, in the third edition of his *Universal Dictionary of Trade and Commerce* in 1774, proposed a commercial union between Britain and the American colonies. Postlethwayt argued that Great Britain should encourage the colonies to produce whatever the

British imported from the East, as well as other goods the British took from Europe—"raw silk, cochineal and dyestuffs, silk, cotton and flax in particular and every other material we import from any other part of Europe or elsewhere." The American colonies could thus supply all those imports "we take for use, or convenience or even luxury from other states," and the British could reexport any others they did not use. Colonies encouraged to plant and raise all these products would enhance a commercial union.[16]

Such a union yielded more than commercial profit. There were common ties in customs, descent and "blood." David Hume maintained that "the same set of manners will follow a nation, and adhere to them over the whole globe, as well as the same laws and language." The American colonies meant the prospect of new opportunities for wider British identities. This was partly because the common manners and habits that reached across the Atlantic went with a common material culture. But as the American Revolution appeared on the horizon, fissures opened in the common connections across an "Atlantic border."[17]

"Britishness" conveyed both commerce and commodities produced in the British Isles and in its colonies. Goods from Britain came with particular associations, and considerable effort went into giving those goods a distinctive identity that set them apart from other European, and especially French, commodities, an identity that connected with trade and empire. The image of Britannia represented liberty and commerce. Britain had cast aside the old props of traditional society, modernized government and trade, nourished the national arts, and sent new, modern commodities out to the rest of the world. A national debate on design, argues Matthew Craske, appealed to patriotic impulses and invoked a competitive spirit.[18] Postlethwayt's very British *Universal Dictionary* was at one level an act of translation of Jacques Savary's *Dictionnaire du Commerce*. At another level, it was a reconfiguration, a commentary on French and other continental practices, and a plan for British national trading power. Britain's exploration of, commerce with, and in some cases dominion over other parts of the world opened routes to a new balance of power; they gave its manufacturers access to exotic materials and the platform to display virtuoso combinations of goods.

The national or "British" commodities, formed in the wake of this "design debate" as well as wider commercial practice, first joined, then dominated, international markets.[19] British identity spread a consumer culture connected to it, but success at home depended on success abroad. British goods were defined in this wider discourse by what they were not.

They were not French; they were not associated with aristocratic ostentation. They were new, and manufacturers made their markets on the desire to anticipate products and services not yet acquired. Their manufacturers and retailers played on anticourtly values and enlightened democratic principles, on simplicity over artifice; they harnessed antiluxury arguments to promote Britain's new consumer goods.

Yet ironically those new goods took their inspiration from a relatively recent global trade in luxury goods from Asia and the Americas. These goods opened tastes to a different material culture and also created a demonstration effect of high-volume production in a diverse range of goods; this was quality ware, but at accessible prices. Colonial groceries, especially tobacco, sugar, coffee, and tea, initially exotic luxuries, were also addictive, and elastic consumer markets stimulated plantation cultivation and the rapid expansion of trade. Groceries, moreover, packaged with cultures of eating and drinking, also transposed material cultures, especially of ceramics, but also of textiles, metals, and lacquerware. This was fine ware conveying civility, but it was produced in Asia on a large scale and in great variety, by labor-intensive methods to be sure, but, in the case of ceramics at least, in large-scale complexes deploying extensive division of labor. The challenge was not to produce an equivalent craft product but an equivalent factory good. Imitating the spirit of these goods from the East generated new products—products that were technical achievements, made out of new materials, fabricated with novel production methods and division of labor, and driven with hand, horse, water, and eventually steam power.[20] Their manufacturers gave some of these products Anglo-Indian or Anglo-Chinese names to remind customers of their Asiatic inspiration: there was Wedgwood's "Nankeen" chinaware, an earthenware substitute for Chinese export-ware, porcelain, and Lancashire-manufactured and branded "calicoes," "muslins," "dimities," and "cherryderries."

Look at the products that became international brands by the end of the eighteenth century: Manchester checks and Lancashire printed calicoes, English glass, Staffordshire china, Birmingham buckles, brassware and japanning, Sheffield plate and cutlery. British accessories—stamped buttons, plated buckles, commemorative medals, enameled pins and cameos—adorned men's and women's dress. British brass door handles, locks, and escutcheons were necessary accoutrements of furnishings and interiors in Europe and America. How was it that the new consumer goods were so quickly and effectively branded British, and then became world-class commodities by the last third of the eighteenth century?

These goods traveled; they were portable. Dress, long dominated by international fashions led by the French, gave way, for men's clothes at least, to English suits, where fabric and detail provided the branding, to English stockings, gloves, watch chains, buttons, and buckles. With international dress codes, there were internationally recognized architectural spaces, stage settings for social performances conducted with internationally accepted rules of etiquette. Furnishings, ornaments, dining and tea ware acquired international codes.

Successful goods achieved an international ascendancy; they were above all fashionable. The British knew that the route to successful consumer goods in markets abroad was not by adapting to local cultural frameworks, but by aggressively making British commodities fashionable and, in turn, identifying fashion with commercial modernity. Britain's trading middling classes defined their modernity by their possession of newly invented goods, made by mechanical techniques. Their possessions, according to Josiah Tucker, were a sign of their liberty: "England being a free country, where Riches got by trade are no disgrace, and where property is also safe against the prerogatives of either princes or Nobles, and where every person may make what display he pleases of his wealth."[21]

These new-fashion goods were technological achievements—"ingenious contrivances," attractive gadgets, sleek little masterpieces that worked.[22] This is what the classic British "toys" were all about: small consumer goods made of iron and steel, silver plate, and all manner of alloys—brass, tin, pinchbeck, and tutania. Birmingham polished-steel buckles and brass or enameled buttons were international fashion necessities by the 1740s, so much so that imitations made in Paris as late as 1810 bore false British identity marks.[23] Matthew Boulton knew the reasons for British domination of the international buckle trade: "That here are Secrets in the Buckle Trade, which Foreigners are Strangers to. . . . They cannot make Chapes in Spain and Portugal so good and cheap as in England, as they have no slitting or rolling Mills . . . their Iron is not so fit for Chapes."[24]

T. S. Ashton's "wave of gadgets" was indeed what this was. The gadgets demonstrated a technological prescience that reached out to scientific instruments, to tools, and to machines. The metals and the objects intrigued consumers. Sheffield plate was a mastery of art and industry. Buttons, tea urns, coffee pots, saucepans, tankards, and candlesticks, then dishes, jugs, spoons, and forks, could be made in copper fused between thin layers of silver—and look wonderful. The international tourist route took in showcase factories in Birmingham and Sheffield—the goods were about technology. Most knew the key to the success of these new decorative goods was the material they were made of and the machines that made their replication possible. Stamped brass and cast crucible steel were high-tech and fashion together. Mrs. Montagu, that paragon of the arts, saw a new patriotism in her patronage of the Soho ornament and challenged Matthew Boulton "to triumph over the French in taste & to embellish your country with useful inventions & elegant productions."[25]

Wedgwood invoked patriotic themes, drawing on history painting, and the narratives of virtue drawn from the Roman Republic. English glass, invented in an effort to imitate the glories of Venetian glass, turned to new forms and styles and was amenable to cutting, making for yet more products, such as dessert dishes and chandelier glass. Sheffield plate answered the fashion demand for light, highly ornamented, and engraved objects. British buckles, buttons, cameos, snuffboxes, watches, and watch chains answered the fashion demands of the London, Bath, and Paris seasons. These products also met another retail strategy of buying as collecting. Collections of medallions, cameos, and vases, produced in a series, envisioned markets far into the future. Beyond collection, there was context. Furniture makers, upholsterers, and architectural ornament makers developed strategies for displaying their goods in fashionable settings; in effect, they sold the setting as much as they did the furniture.

The goods conveyed taste and manners, and these, in Wedgwood's view, marked out British identity as much as did political culture or military and naval success.[26] Wedgwood branded his creamware and his Etruscan-inspired vaseware with British liberty and a British political discourse on classical Greek and Roman virtue and citizenship. His partner, Bentley, connected them to new enlightened values for simplicity in art and life espoused by Rousseau in France. Wedgwood debated the appeal of his utterly British designs to French consumers, girding himself for the product war to follow all the other eighteenth-century wars with France: "And do you really think that we may make a *complete conquest* of France? Conquer France in Burslem? My blood moves quicker, I feel my strength increase for the conquest. Assist me my friend & the victorie is our own. We will make them . . . our Porcelain after their own hearts, & captivate them with Elegance and simplicity of the Ancients."[27]

The vases were displayed on mantelpieces, beneath history paintings, and set in libraries, dining rooms, and public reception rooms. They symbolized masculine republican virtue, and in their series of imitation materials based on precious stones and ores, they took advantage of the fashion for geological collections. Later in the century, it was medallions, ceramic plaques, intaglios, and cameos that carried messages of British design and technological ascendancy. Medals, cameos, and intaglios, an affordable form of emblematic sculpture, were political statements, collectibles, and excellent gifts. Machinery for coining and for button and buckle making was easily adapted, as were the molding technologies developed in the earthenware manufactures.[28]

SELLING BRITISH TO AMERICA

The British ran their own Atlantic world shopping center. This was not, as Adam Smith thought, an artificial edifice of the Navigation Acts: "A great empire has been established for the sole purpose of raising up a nation of customers who should be obliged to buy from the shops of our different producers, all the goods with which they could supply them . . . the home-consumers have been burdened with the whole expense of maintaining and defending that empire."[29] Instead, this shopping center stocked the newest and best; Europe bought, and Britain itself shopped, because America bought. Staffordshire earthenware, English lead glass crystal, Sheffield plate, English light furnishings, especially tea tables and Axminster and Kidderminster carpets, scooped home and foreign markets because they were fashion leaders.

Merchant vessels plying the Atlantic carried British candlesticks, cutlery, tea ware, and a whole array of ingenious and mechanical toys. These newly invented products announced their presence in colonial newspapers, trade catalogs, shipping orders, and tradesmen's accounts. Family letters declared their desirability. This was Timothy Breen's "empire of goods."[30] National or British commodities first joined, then dominated, international markets. British as well as colonial consumers knew they were buying global products. Manufacturers saw to their international marketing as assiduously as they did to their home provisioning.

Between 1688 and the 1780s, the share of British gross domestic product sold overseas doubled from 8 to 16 percent; most of these goods were manufactured, and their most rapid movement into international markets took place in two periods: 1740–1760 and 1780–1801. By 1770, at least half of England's exports of ironware, copperware, earthenware, glassware, window glass, printed cotton and linen goods, silk goods, and flannels were sent to the colonies. In the case of hardware alone, North America provided 60 percent of the market for British goods, and between the later eighteenth century and 1820, the biggest export markets for

British ceramics shifted from Europe to America. By the 1790s, English earthenware made up most of the table, tea, and toilet wares used in the United States.[31]

Fashionable, high-quality wares were most in demand. Trade meant variety and choice, becoming a trope representing prosperity, civilization, and British goods. Listings of goods in shipping accounts, advertisements, and trade cards played a part in the making of a consumer consciousness. British consumer goods appropriated "choice," once the key attribute of luxury and trade goods brought from afar. Varieties of English broadcloth, calicoes, brass, ironware, and earthenware now displaced the wonders of the East. Variety was also "clustered" to convey British consumer culture. That consumer culture in turn represented personal and social values of respectability, civility, and gentility. American consumers perceived these British imports to offer alternatives by introducing categories of comfort and taste into the lives of the American middling sorts.[32]

Food and drink all over the North American colonies was served with the full paraphernalia of British glassware, china dinner services, and cutlery: Bristol glass and Staffordshire earthenware, with Wedgwood creamware, Birmingham and Sheffield plated cutlery and candlesticks, English japanned trays, English mahogany tea tables, chests and chairs—all of which conveyed English gentility and civility. Kendal cottons, Yorkshire kerseys, and Irish and Scottish linens brought wider British identities within their fabric. Brassware and cabinet-ware makers in Birmingham developed illustrated trade catalogs that not only sold their furniture handles and door knobs, but spread English design styles, fostering a demand for English furniture.[33] English floral designs on upholsteries and curtain material conveyed privacy, informality, and intimate relaxation.[34]

These goods were synonymous with fashion, quality, and variety shared across an Atlantic world that started, as Sir John Elliott argues, a "European lake," that is, as a European construct.[35] The British supplied their own diaspora: after the Seven Years War, they also enticed consumers in the rest of the Atlantic world, moving into France's earlier enclaves in the Caribbean and into the markets of Spanish America.[36]

Not Selling British to Asia

British manufacturers seriously sought out markets in Turkey, India, and China. Boulton avidly inquired about prospects for perfume containers in the private interiors of wealthy Turkish homes and discussed prospective markets in India. His inquiries yielded few prospects among the indigenous population: this "was a country where fashion and habits were deeply rooted." But markets might be built among the rising numbers of Europeans in India. And indeed this was the policy the British pursued in India. Policymakers promoted British goods among the Anglo-Indian population as a bulwark against "degeneracy" and "oriental" consumption.[37]

And yet the British government and the East India Company entertained hopes that the Chinese court would recognize the clear novelty and superiority of the assemblage of the British products of what was, in all but name, a trade mission. The Macartney expedition of 1792–1793 conveyed George III's aspirations to the Chinese emperor of "communicating the arts and comforts of life to those parts of the world where it appeared they had been wanting." The key goal of this diplomatic mission to China was to enhance conditions of trade in China for Britain, including "to excite at Peking a taste for many articles of English workmanship hitherto unknown there . . . [to] turn the balance of the China trade considerably in favour of Great Britain."[38] Macartney, as envoy of George III, also wished to convey to the emperor and Chinese authorities England's curiosity about the rest of the world and the desire to learn more about the morals and manners of other peoples. The specimens went with a "List" explaining the merit and use of the goods. Macartney saw such a list as just as important as the specimens.[39]

The goods the British took to China, in the eyes of Macartney's embassy, manifested modernity and scientific and economic ascendancy.[40] But Britain's products failed to impress another professed superior civilization. The Qianlong emperor sent Macartney back to his king with the message: "We have never valued ingenious articles, nor do we have the slightest need of your country's manufactures." From this time, British participants in empire, both at home and abroad, defined themselves against differences in creed, race, and "conditions of life."[41] Those conditions of life were about the goods. The goods taken out to other parts of the world represented the power of the nation; they also provided a defining material identity to those trading, traveling, and living far from their homes. More and better goods conveyed economic improvement.

The stonewalling of British missions to open trade for their goods in Asia was relatively unexpected after several decades of successful marketing in global markets, especially in North America, but also in the Caribbean and Latin America. But these were markets made in European-settled colonies and their Creole communities. American markets to an even greater degree than domestic markets provided the testing ground for the competitive success of products. Decades spent fostering these markets entailed endowing British products with all those desirable characteristics which then went into branding those goods at home as well as in wider world markets.

CONCLUSION

Even if the British failed in their efforts to breach Asian markets at this stage of development, their markets in the Atlantic world gave Britain a global advantage denied to Asian producers that had once prevailed in luxury goods production. Eighteenth-century Europeans responded to their commodity trade with the wider world. Inventing, producing, and consuming new European and especially British goods provoked changes in technologies, new uses for recently discovered materials and sources of energy, and the reorganization of labor that became the Industrial Revolution. Britain's Industrial Revolution was simultaneously a product revolution. Creating consumers and inventing technologies went together; they integrated product and process. The British products that seized global markets in the early nineteenth century were the success story of the Industrial Revolution; that industrial revolution, in turn, was made in the internationally recognized "identity" of British goods in the transatlantic world of the eighteenth century.

NOTES

1. Joel Mokyr, *The Gifts of Athena: Historical Origins of the Knowledge Economy* (Princeton, NJ: Princeton University Press, 2002).

2. *Lunar man* is a member of the learned Birmingham Lunar Society that brought together a number of key figures of the industrial Enlightenment.

3. The Birmingham Riots, Inventory of the House and Goods of Joseph Priestley, 399801/IIR30, City Archives, Birmingham Central Library.

4. Robert J. Gordon, *The Measurement of Durable Goods Prices* (Chicago: University of Chicago Press, 1990); Javier Cuenca Estaban, "British Textile Prices, 1770–1831: Are British Growth Rates Worth Revising Again?" *Economic History Review* 47 (1994), 66–105; William Nordhaus, "Do Real-Output and Real-Wage Measures Capture Reality? The History of Lighting Suggests Not," in *The Economics of New Goods*, ed. Timothy Bresnahan and Robert J. Gordon (Chicago: University of Chicago Press, 1997), 29–70; Joel Mokyr, "Accounting for the Industrial Revolution," in *The Cambridge Economic History of Modern Britain*, vol. 1, ed. Roderick Floud and Paul Johnson (Cambridge: Cambridge University Press, 2004), 1–28, 12–13.

5. See the debate over productivity change and the sources of economic growth during the Industrial Revolution in Nicholas Crafts and C. Knick Harley, "Output Growth and the British Industrial Revolution: A Restatement of the Crafts-Harley View," *Economic History Review* 45 (1992), 703–730; Esteban, "British Textile Prices," 66–105; Maxine Berg and Pat Hudson, "Rehabilitating the Industrial Revolution," *Economic History Review* 45 (1992), 24–50.

6. See, for example, Patrick K. O'Brien and Caglar Keyder, *Economic Growth in Britain and France, 1780–1914* (London: Routledge, 1976); François Crouzet, *Britain Ascendant: Comparative Studies in Franco-British Economic History*, trans. Martin Thom (Cambridge: Cambridge University Press, 1990); William Sewell, Jr., *Work and Revolution: The Language of Labor from the Old Regime to 1848* (Cambridge: Cambridge University Press, 1980).

7. More nuanced analyses of France's luxury industries were offered by Carlo Poni, "Fashion as Flexible Production: The Strategies of the Lyons Silk Merchants in the Eighteenth Century," in *Worlds of Possibilities: Flexibility and Mass Production in Western Industrialization*, ed. Charles Sabel and Jonathan Zeitlin (Cambridge: Cambridge University Press, 1997), 37–74, 41; Lesley Miller, "Paris-Lyons-Paris: Dialogues in the Design and Distribution of Patterned Silks in the Eighteenth Century," in *Luxury Trades and Consumerism*, ed. Robert Fox and Antony Turner (Aldershot: Ashgate Press, 1998), 139–167, 163–165. In a more recent direction, see the critique of Leora Auslander, *Taste and Power: Furnishing Modern France* (Berkeley: University of California Press, 1996) by Dena Goodman, "Furnishing Discourses: Readings of a Writing Desk in Eighteenth-Century France," in *Luxury in the Eighteenth Century*, ed. Maxine Berg and Elizabeth Eger (Basingstoke: Palgrave Macmillan, 2003), 71–88. See also Colin Jones, *The Great Nation: France from Louis XV to Napoleon* (London: Penguin Press, 2002), 349–363.

8. Crouzet, *Britain Ascendant*, 57–97, and "The Historiography of French Economic Growth in the Nineteenth Century," *Economic History Review* 56 (2003), 215–242.

9. Sidney Pollard, *Peaceful Conquest: The Industrialization of Europe, 1760–1970* (Oxford: Oxford University Press, 1981), 170–173.

10. John R. Hicks, *A Theory of Economic History* (Oxford: Oxford University Press, 1969); Pollard, *Peaceful Conquest*, 174, 184, 186.

11. Mokyr, "Accounting for the Industrial Revolution," 16–17, 25; Liliane Hilaire-Pérez, *L'Invention technique au siècle des Lumières* (Paris: Albin Michel, 2000); Reinhold R. Angerstein, *Illustrated Travel Diary, 1753–1755: Industry in England and Wales from a Swedish Perspective*, trans. Torsten and Peter Berg (London: Science Museum, 2001); Chris Evans and Gőran Rydén, "Kinship and the Transmission of Skills: Bar Iron Production in Britain and Sweden, 1500–1860," in *Technological Revolutions in Europe: Historical Perspectives*, ed. Maxine Berg and Kristine Bruland (Cheltenham: Edward Elgar, 1998), 188–206; Kristine Bruland, "Reconceptualizing Industrialization: the Scandinavian Case," this volume, chap. 7.

12. Gordon, *The Measurement of Durable Goods Prices*, 12. Maxine Berg, "From Imitation to Invention: Creating Commodities in Eighteenth-Century Britain," *Economic History Review* 55 (2002), 1–30.

13. Berg, "From Imitation to Invention," 1–30.

14. Linda Colley, *Britons: Forging the Nation, 1707–1837* (New Haven, CT: Yale University Press, 1992); Gerald Newman, *The Rise of English Nationalism: A Cultural History* (New York: St. Martin's Press, 1987), 68–84; Kathleen Wilson, *The Sense of the People: Politics, Culture, and Imperialism in England, 1715–1785* (Cambridge: Cambridge University Press, 1998), 187.

15. John G. A. Pocock, "British History, a Plea for a New Subject," *Journal of Modern History* 47 (1975), 601–621; Kathleen Wilson, *The Island Race: Englishness, Empire and Gender in the Eighteenth Century* (London: Routledge, 2003), 5.

16. Malachy Postlethwayt, "Discourse the First—The Ill State of Our Finances: With a Plan for a More Interesting Union between Great Britain and America," in *Universal Dictionary of Trade and Commerce* (London, 1774), xxv-xxx.

17. See Dror Wahrman, *The Making of the Modern Self: Identity and Culture in Eighteenth-Century England* (New Haven, CT: Yale University Press, 2004), 224, 248.

18. Matthew Craske, "Plan and Control: Design and the Competitive Spirit in Early and Mid-Eighteenth-Century England," *Journal of Design History* 13 (1999), 187–216, 189.

19. See Robert Dossie, *The Handmaid of the Arts*, vol. 1 (London: Routledge, 1764), vi, 26; David G. C. Allan, *William Shipley: Founder of the Royal Society of Arts: A Biography with Documents* (London: Scolar Press, 1979), 16, 46, 51.

20. I make this argument in "In Pursuit of Luxury: Global Origins of British Consumer Goods," *Past and Present* 182 (2004), 85–142.

21. Josiah Tucker, *Instructions for Travelers* (London: Royal Society of Arts, 1757), 26.

22. See Adam Smith on the attractions of "small conveniences" in *The Theory of Moral Sentiments* (1759), ed. David D. Raphael and Alan L. McFie (Oxford: Oxford University Press, 1976), book 4, chap. 1, 180, 183. On the aesthetics of the miniature, see Susan Stewart, *On Longing: Narratives of the Miniature, the Gigantic, the Souvenir, the Collection* (Durham, NC: Duke University Press, 1993), 237–269.

23. John R. Harris, *Industrial Espionage and Technology Transfer in the Eighteenth Century: England and France* (Aldershot: Ashgate Press, 1998), 202–203; Harry W. Dickinson, *Matthew Boulton* (Cambridge: Cambridge University Press, 1937), 32–33.

24. Smith, *The Theory of Moral Sentiments*, book 4, chap. 1, 421.

25. Elizabeth Montagu to Matthew Boulton, October 31, 1771, Matthew Boulton Papers, City Archives, Birmingham Central Library, 330/1.

26. Michael Vickers and David Gill, *Artful Crafts: Ancient Greek Silverware and Pottery* (Oxford: Oxford University Press, 1994), 27–28.

27. Wedgwood to Bentley, n.d., but written between February and September 1769, cited in ibid., 27.

28. Peter Jones, "'England Expects . . .' Trading in Liberty in the Age of Trafalgar," in *Enlightenment and Revolution: Essays in Honor of Norman Hampson*, ed. Malcolm Crook, William Doyle, and Alan Forrest (Aldershot: Ashgate Press, 2002), 187–202, 189–190.

29. Adam Smith, *An Enquiry into the Nature and Causes of the Wealth of Nations*, ed. Roy H. Campbell and Andrew S. Skinner, vol. 2 (Oxford: Oxford University Press, 1981), 66.

30. Timothy H. Breen, *The Marketplace of Revolution. How Consumer Politics Shaped American Independence* (New York: Oxford University Press, 2004).

31. Lorna Weatherill, "The Growth of the Pottery Industry in England, 1660–1815," *Post Medieval Archaeology* 17 (1983), 15–46, 28; George L. Miller, Ann Smart Martin, and Nancy S. Dickinson, "Changing Consumption Patterns: English Ceramics and the American Market from 1770–1840," in *Everyday Life in the Early Republic*, ed. Catherine E. Hutchins (Dover, DE: Henry F. du Pont Winterthur Museum, 1994), 219–248.

32. Breen, *The Marketplace*.

33. Maxine Berg, *Luxury and Pleasure in Eighteenth-Century Britain* (Oxford: Oxford University Press, 2005), chaps. 7, 8.

34. Beverly Lemire, "Domesticating the Exotic: Floral Culture and the East India Calico Trade with England, 1600–1800," *Textile* 1 (2003), 65–85, 80.

35. John H. Elliott, "Atlantic History: A Circumnavigation" in *The British Atlantic World, 1500–1800*, ed. David Armitage and Michael J. Braddick (Basingstoke: Palgrave, 2002), 233–250, 234.

36. Peter J. Marshall, "Britain without America: A Second Empire?" in *The Oxford History of the British Empire*, vol. 2, ed. Peter J. Marshall (Oxford: Oxford University Press, 1998), 576–595.

37. Matthew Boulton to George Millett to Matthew Boulton, Princess Amelia at Sea, July 31, 1796, Matthew Boulton Papers, City Archives, Birmingham Central Library; Elizabeth Collingham, *Imperial Bodies* (Cambridge: Cambridge University Press, 2001); Hosea B. Morse, *The Chronicles of the East India Company Trading to China, 1635–1834*, vol. 2 (Oxford: Oxford University Press, 1926), 213–214.

38. James L. Hevia, *Cherishing Men from Afar: Qing Guest Ritual and the Macartney Embassy of 1793* (Durham, NC: Duke University Press, 1995), 61; Sir George Staunton, *An Authentic Account of an Embassy from the King of Great Britain to the Emperor of China* (Dublin, 1798), 32.

39. Macartney to Dundas, November 9, 1793: "An Account of Sundry Articles Purchased by Francis Baring Esq. Chairman . . . Consigned to the Care of . . . Lord Viscount Macartney," Lord Macartney's Embassy to China, Miscellaneous Letters, 1792–1795, India Office Records, Factory Records China and Japan 1596–1840, G/12/92, 545–586, 578. For Boulton's list of goods, see "A General List of Goods

Manufactured at Birmingham and its Neighbourhood, c. July 22, 1792." A copy of the list was enclosed in Boulton's letter to Robert Wissett, Secretary to the East India Company, July 22, 1792, China Trade, Lord Macartney's Embassy, 1792. MS 3782/12/93, nos. 70 and 86, Matthew Boulton Papers, City Archives, Birmingham Central Library.

40. Hevia, *Cherishing Men from Afar*, 72, 179; China Trade, Lord Macartney's Embassy, 1792. MS 3782/12/93, 70 and 86. Matthew Boulton Papers, City Archives, Birmingham Central Library.

41. Cited in Linda Colley, "Britishness and Otherness: An Argument," *Journal of British Studies* 31 (October 1992), 309–329.

4

The European Enlightenment and the Origins of Modern Economic Growth

Joel Mokyr

The consensus is that modern economic growth was started by the British Industrial Revolution. As is well known, during the Industrial Revolution itself, growth was in fact fairly modest, but the sudden take-off of gross domestic product per capita after 1825 or thereabouts was made possible by a long period of laying the foundations.[1] The transformation was tantamount to a phase transition, a sea change in the mechanics of economic growth, with technological progress gradually coming to dominate the process, accounting for its novel features. But what were these foundations exactly? This chapter addresses this issue of foundations.

Before doing so, two central points must be made. The first is that events like a cluster of macroinventions such as happened in the first decades of the Industrial Revolution are not altogether unique in history, neither in Europe nor elsewhere. Moreover, growth, as Eric L. Jones and many others have noted, was not a new phenomenon in nineteenth-century Europe.[2] Many regions or groups had managed over the centuries to accumulate wealth, to produce surpluses beyond subsistence, as works of art, architecture, and science amply indicate. Yet none of these processes persisted; growth was always checked and eventually fizzled out. Often it was reversed, and societies declined and in a few cases, entirely lost their former wealth. The telling characteristic of modern growth is its sustainability, indeed its inextinguishability.

The second point is that the idea that the Industrial Revolution was "British" and that Europe was just a "follower" seems overstated and, in some sense, wrong. Some areas of Europe, such as Flanders, Alsace, and Switzerland, were able to follow Britain fairly quickly, and although a sense of inadequacy among contemporary continental Europeans in the first half of the nineteenth century when comparing their industrial achievements to Britain can be perceived, modern economic historians have been more cautious about this so-called continental backwardness. Such perceptions need to take into account the high toll that the political turmoil between 1789 and 1815 had on the economies of continental

Europe. Although these upheavals can be regarded as the price that the Continent had to pay to "catch up," the gap between Britain and the Continent was never on the order of magnitude of the gap between the West and China or Africa.

What, then, was behind this transformation? Historians have engaged the issue now for a century, and little consensus has emerged. Two significant recent contributions by David Landes and Kenneth Pomeranz have divided the causal factors between culture and geography.[3] Earlier, Jones provided a veritable smorgasbord of explanations, including the ingenious idea of the European "states system" which likened the fragmented political power in Europe to a competitive market, limiting the damage that rulers could inflict on their economies.[4] Others have focused on "Western science" as the crucial variable.[5] Still others blamed European imperialism, itself due to accident, and dismissed the entire event as epiphenomenal. These explanations have been vigorously criticized and vigorously defended.

It is odd that in this literature, the European Enlightenment plays such a minor role. In recent decades, the Enlightenment has not fared well in the view of historians, being held responsible for the horrors of the twentieth century by Theodor Adorno and Max Horkheimer and their contemporary postmodern epigones, such as John Gray. Among the oddest phenomena in modern historiography, indeed, are the vitriolic and nasty attacks on the Enlightenment, which, perversely, is blamed for modern-day barbarism but never credited for bringing about modern-day prosperity.[6] On the contrary, the European Enlightenment would seem to be a natural candidate in explaining the great divergence. After all, its timing took place approximately in the century before the beginning of modern growth in Europe, and it was clearly a Western phenomenon, its success more or less confined to the countries that by 1914 constituted the so-called convergence club of rich industrialized countries. Yet economic historians must have felt uncomfortable with the Enlightenment as an explanatory factor—perhaps because it is a relatively amorphous and hard-to-define intellectual movement, perhaps because the Enlightenment was believed to be primarily "French," whereas the Industrial Revolution was "British," and perhaps because the connections among beliefs, intellectual conventions, and economic events are poorly understood.

In this chapter, I argue that the Enlightenment played an important, perhaps crucial, role in the emergence of modern economic growth. This is not to denigrate other factors altogether. The cotton industry, one of the mainstays of the Industrial Revolution, could not emerge without access to sources of raw cotton, so international trade cannot be disregarded. Monetary and financial elements in the story are obscured, as are demographic and other factors. But the Enlightenment had two major effects that I emphasize: it transformed the motivation for and dynamics of technological progress, and it altered the institutional mechanism through which technological change affected the economy. These two formed a synergy that was the very foundation of the "European miracle."

The Enlightenment and the Growth of Useful Knowledge

The European Enlightenment was a multifaceted phenomenon concerning the natural rights of humans, concepts of religious and racial tolerance, political freedom, legal reform, and much else. At the deepest level, the common denominator was the belief in the possibility and desirability of human progress and perfectibility through reason and increasing knowledge. The material aspect of this belief followed in the footsteps of Francis Bacon's idea of understanding nature in order to control it. "Useful knowledge" became the buzzword of the eighteenth century. This term should not be associated simply with either "science" or "technology."[7] It meant the combination of different kinds of knowledge supporting one another. Not all of it was abstract science: the taxonomic work of Linnaeus and the descriptive writings of Arthur Young increased useful knowledge just as much as the abstract mathematics of Laplace or the experiments of Priestley and Lavoisier.

The eighteenth century marked both an acceleration of the pace of research and a growing bias toward subject matter that, at least in principle, had some practical value. Indeed, Peter Burke has argued that the eighteenth century saw the rise of "the idea of research" along with a sense that this knowledge could contribute to economic and social reform.[8] The change in the pace of the progress of knowledge after 1680 was indebted to the triumph of Newtonianism in the first half of the eighteenth century. The achievement of Newton did more than anything else to establish the prestige of formal science in the world of learning.[9] It was widely believed that the growth of useful knowledge would sooner or later open the doors of prosperity—to some extent with more hope than experience. But it was also clear that this growth could only be carried out collectively, through a division of labor in which specialization and expertization were carried out at levels far higher than before.[10] The way useful knowledge increased in the eighteenth century was a far cry from the processes of today's R&D (corporate and government). It might be better to say that much of it was by way of exploration and discovery. These were trial-and-error processes minimally informed by an understanding of the natural processes at work, inspired tinkering, and a great deal of serendipity and good fortune, albeit favored by prepared and eager minds. Over the course of the eighteenth century, these search processes became more systematic, careful, and rigorous. New technological methodologies were invented, such as the great engineer John Smeaton's development of the method of parameter variation through experimentation, a systematic way of making local improvements in a technique without necessarily understanding the underlying science.[11] To be sure, there were no truly fundamental scientific breakthroughs in the century between Newton and Lavoisier, but it was an age of consolidation, refinement, and organization of knowledge; the honing and sharpening of mathematical and experimental methods; and an age of observation, classification, and the jettisoning of doomed searches and projects.[12]

Hopes for a quick technological payoff to scientific research were, on the whole, disappointed in the eighteenth century. The "customary chasm" between science and the mundane details of production could not be closed in a few decades or even in a century.[13] One can, of course, find examples in which scientific insights did enrich the knowledge of key actors in the Industrial Revolution. Dexterity and mechanical intuition were, in many cases, complementary to certain critical pieces of scientific knowledge that guided and inspired the work. The scientific milieu of Glasgow in which James Watt lived contributed to his technical abilities. He maintained direct contact with the Scottish scientists Joseph Black and John Robison, and as H. W. Dickinson and Rhys Jenkins noted in their memorial volume, "one can only say that Black gave, Robison gave, and Watt received."[14] The introduction of chlorine bleaching and the solution of the longitude problem (i.e., how to determine longitude at sea) depended to some extent on advances in science, and formal hydraulics contributed to advances in waterpower.[15] Yet when all is said and done, much of the progress we associate with the First Industrial Revolution needed little more than the mechanics that Galileo knew, and innovation in manufacturing and agriculture before 1800 came without science providing indispensable inputs. William Cullen, the leading chemist of the mid-eighteenth century, was retained by Scottish manufacturers to help them solve a variety of problems. His self-serving prediction that chemical theory would yield the principles that would direct innovations in the practical arts remained, in the words of the leading expert on eighteenth-century chemistry, "more in the nature of a promissory note than a cashed-in achievement."[16] Manufacturers needed to know why colors faded, why certain fabrics took dyes more readily than others, and so on, but as late as 1790, best-practice chemistry was incapable of helping them much.[17] In medicine, in metallurgy, and in agriculture, to name just a few areas, the situation before 1800 was no different. The world may have been messier and more complex than the early and hopeful proponents of the Baconian program realized, as H. Floris Cohen has suggested.[18] Scientists did not know enough and lacked the tools to learn quickly. Tacit artisanal knowledge, such as mechanical dexterity, intuition, experience-driven insights, and similar abilities, drove many of the early inventions, although dismissing the contribution of science altogether is unwarranted.

And yet the belief that somehow useful knowledge was supposed to be the key to economic development not only did not fade as a consequence of such disappointments, it kept expanding on both sides of the Channel. The Baconian "program" was built on the belief that the expansion of useful knowledge would solve technological problems and that the dissemination of existing knowledge to more and more people would have substantial efficiency gains. These two notions formed the core of Denis Diderot's beliefs, and his admiration for Bacon, the first philosopher to lay out clearly a technological program for economic expansion, permeates his writing, as it does that of many other eighteenth-century *philosophes* and scientists. In Britain, of course, this belief was not only widespread but formed the explicit motive for the foundation of organizations and societies designed to advance it.[19]

Progress was limited simply by what people knew. The age of Enlightenment, for instance, never had a good concept of what "heat" really was. Its chemistry was, until the 1780s, anchored in phlogiston theory, and its understanding of biology and disease, despite some significant local advances, had progressed little beyond Galen. Newton's great insights, much as they supported the belief that rational argument and observation could help people understand the universe, were of limited practical value. Yet it was also readily recognized that very intelligent people, schooled in experimental science and mathematics, could make substantive contributions to technology even if they were not always quite sure why and how new techniques worked. Thus mathematicians were asked to solve mundane and practical problems, and sometimes they were successful.[20] Other examples are easy to find.[21] From the measurement of longitude (perhaps the best-defined single problem that the age of Enlightenment solved) to the improvement of waterpower by applying mathematics to the growing science of hydraulics, the knowledge of various "applied philosophers" was brought to bear on matters of technology.[22] The same is true for knowledge of plants and animals. Many scientists were concerned with the properties of steel: René Réaumur and Torbern Bergman wrote about them at length, recognizing their economic significance, and three of France's most learned men published a paper in 1786 establishing once and for all the differences among wrought iron, cast iron, and steel—even if the full effects of this insight were still decades in the future.[23]

Many men of science applied themselves to invention. Most of them applied notions of "open science" to their inventions and placed the knowledge in the public realm. Benjamin Franklin, Humphry Davy, Joseph Priestley, and Benjamin Thompson (Count Rumford), four of the leading scientists of the later decades of the age of Enlightenment, made numerous inventions, but refused to take out any patents, arguing that their efforts were made for the benefit of humanity, not for private profit. Such hybrid careers became common in the nineteenth century. Michael Faraday, besides his pathbreaking research on electricity, worked on various problems in materials, especially steel and glass.[24] Eda Kranakis emphasized the work of the French engineer and mathematician Claude-Louis Navier (1785–1836), who, among others, used the recently developed Fourier analysis to analyze the vibration in suspension bridges and did pioneering work in fluid dynamics for which he is still known. His work, and that of other *polytechniciens*, was highly abstract and mathematical, and it was of long-term rather than immediate applicability. Not so that of Lord Kelvin, a prolific inventor who owned seventy patents in electromagnetic telegraphy, marine navigation equipment, and electric instruments.[25]

The connection between the sphere of learning and the sphere of production has always been a sensitive spot in the history of economic growth. Narrowing this gap was perhaps the crowning achievement of the industrial enlightenment. Part of the contact between the two spheres took place through books and periodicals and part of it through direct contact and transfer of knowledge through teaching, imitation, and espionage. The publication most widely associated with the Enlightenment, Diderot and d'Alembert's *Encyclopédie*, contained numerous articles on technical matters that were lavishly illustrated by highly skilled artists who in most cases were experts in their fields.[26] Encyclopedias and indexes to "compendia" and "dictionaries" were the search engines of the eighteenth century. In order to be of practical use, knowledge had to be organized so that it could be selected from. Alphabetization was one way to do this, the organization of science into categories another.[27]

The eighteenth century also witnessed improved codification of formerly tacit knowledge. Part of this process was simply the improvement of the language of technology: mathematical symbols, standardized measures, and more universal scales and notation all added greatly to the ease of communication. Post-Lavoisier chemical nomenclature proposed by the Swedish chemist Berzelius in 1813 was agreed on after some hesitation. When new measures were needed, they were proposed and accepted. Thus, as is well known, James Watt proposed in 1784 the total amount of energy necessary to raise 33,000 pounds one foot in one minute as the fundamental unit of work, the horsepower.[28] Visual means of communication, above all diagrams and models, were vastly improved.[29] In addition, between 1768 and 1780, the French mathematician Gaspard Monge developed descriptive geometry, which made graphical presentations of buildings and machine design mathematically rigorous.[30] When human presence was required, travel became faster and more comfortable during the eighteenth century. The idea of the traveling expert or consultant was exploited by Boulton and Watt, whose patent-based monopoly on steam power extended to consulting on energy and mechanics. John Smeaton was perhaps the greatest consultant of all, founding the Society of Civil Engineers, but others followed his example.[31]

Knowledge was also transferred through personal contacts and lectures. The years after 1660 witnessed the founding of many state-sponsored, official academies such as the Royal Society, but these were always complemented by private initiative. Early in the eighteenth century, many of those lectures were informal and ad hoc, in pubs and coffeehouses.[32] After 1750, many of those informal meeting places crystallized into more formal organizations and societies, some of them with official imprimaturs. Of those, the Lunar Society of Birmingham is the best documented,[33] but the Chapter Coffeehouse in London was equally successful as a clearinghouse for useful knowledge.[34] Other organizations were more formal. The Royal Society of Arts, founded in 1754, encouraged invention by awarding prizes, publicizing new ideas, and facilitating communication between those who possessed useful knowledge and those who could use it. The Royal Institution, founded by Count Rumford and Joseph Banks in 1799, provided public lectures on scientific and technological topics. Its charter summarized what the industrial enlightenment was about: "diffusing the knowledge, and facilitating the general introduction, of useful mechanical inventions and improvements; and for teaching, by courses of philosophical lectures and experiments, the application of science to the common purposes of life." As James McClellan noted, the reason for all this institutional innovation was simple: it was perceived as useful.[35]

In the eighteenth century, alternatives to the universities emerged. The most dynamic elements in the English educational system were the dissenting academies, which taught experimental science, mathematics, and botany among other subjects.[36] On the Continent, new institutions training technical experts came into being, many of them under government sponsorship. Two of the famous French *grandes écoles* were founded in the eighteenth century: Ponts et Chaussées in 1747 and Mines in 1783. In Germany, the famous mining academy of Freiberg (Saxony) was founded in 1765, followed by others in the 1770s. All these institutions reached only a thin elite, though apparently that was enough. In general, the idea that the role of educational institutions was to create new knowledge rather than transmit existing knowledge to young generations took a long time to ripen. The belief that the Industrial Revolution in its early stages required mass education and literacy has long been abandoned. The British apprenticeship system with the educational institutions mentioned above was more than enough to supply British industry with the skills and craftsmanship it needed. The Industrial Revolution was an elite phenomenon: not, of course, just the handful of heroic inventors as worshipful or adulatory Victorian writers in the Smiles tradition of self-help would have it, but a few tens of thousands of clever and dexterous mechanics and skilled craftsmen who read blueprints, knew the properties of the materials they used, built parts according to specification within reasonable tolerance, had respect for precision, and had the experience to understand friction, torque, resistance, and similar concepts. For the rest of the labor force, education and literacy may not have mattered much, and Britain had no advantage in this domain.

The Enlightenment and Institutional Progress

Economists have lately realized what economic historians have known all along: that "good institutions" are essential to successful economic growth. In recent years, a genuine avalanche of empirical work has pointed to the centrality of property rights, incentives to innovation, the absence of arbitrary rule, and effective contract enforcement, to name but a few oft-mentioned institutional elements.[37] Yet these studies tend to exploit cross-sectional variation and do not bother much with how Europe acquired these good institutions.

The economic significance of the political and institutional reforms of the late eighteenth and early nineteenth centuries has not been fully realized in part because of the undue focus on the security of property rights without much attention to the exact content of these rights. This view overlooks the fact that ancien régime Europe was overgrown with secure and well-enforced local privileges, tax exemptions, monopolies, exclusionary rights, regulations, entry barriers, limited freedom of occupation, and similar arrangements that hampered markets, impeded technological progress, and threatened economic growth wherever it was attempted. In other words, what needed to be done was to eliminate bad rights and contracts.

Mercantilism, the organizing principle of the ancien régime economy, was based on the assumption that economic activity was zero sum.[38] Both at the aggregate level and at the levels of the firm and of the individual, the ruling economic paradigm was one of a fixed pie, and the more one player got, the less there was for others. The idea that production and commerce actually could expand as the result of free exchange ripened slowly in the age of Enlightenment, coming to a crashing crescendo with the Scottish Enlightenment of Hume and Smith and the French *économistes* of the physiocratic school. As observed persuasively by Robert Ekelund and Robert Tollison, the mercantilist economy was to a great extent a rent-seeking economy, in which the incentive structure was largely designed for redistributive purposes.[39]

It is possible to regard the age of Enlightenment partly as a reaction to the economic ancien régime. This is less far-fetched than it may sound. Enlightenment thought increasingly railed against the institutions that perpetuated rent seeking. It should be noted that many of those institutions had not originally been designed as rent-seeking institutions but eventually evolved into them. A paradigmatic example is the craft guild. Craft guilds in the eighteenth century, as Adam Smith argued forcefully, were costly to economic progress.[40] They erected artificial barriers to entry in order to reap exclusionary rents, and on the whole they were hostile to new technology.[41] The success of Britain, where guilds had been relatively weak since the mid-seventeenth century, seemed to confirm this belief. The literature on this matter has in recent years been subject to some serious revisionism, especially by S. R. Epstein.[42] Guilds were not invariably hostile to innovation, this literature argued, and in many ways, they helped in the formation and intergenerational transmission of human capital. Sheilagh Ogilvie has cast doubt on this revisionist literature and shown that, for Württemberg at least, the negative view of craft guilds is supported by a great deal of historical evidence.[43] The overall evidence is more mixed: some guilds were more powerful than others, and it seems that their actual functions changed over time. By 1750, in most places, they had become conservative and exclusionary, and it seems hard to imagine that radical innovation would have had much of a chance had they still been in control. Whenever guilds tried to maintain product-market monopolies, their incentives to innovate were lower than in a competitive market, and their incentives to protect their knowledge—through secrecy and limitations of the mobility of skilled labor—higher. This clearly had profound economic costs.

Abolishing or weakening craft guilds was a high priority for enlightened reformers, precisely because guilds were viewed as impeding efficiency and economic growth. Attempts to carry out such programs were in fact attempted before 1789 by reform-minded politicians such as Turgot in France, Sonnenfels in Austria, and Campomanes in Spain. Jeff Horn has pointed out that the reformist elements in the ancien régime in France needed to overcome the collective action of both masters and employees in French manufactures to create an environment more conducive to technological advance and productivity growth.[44] But all these attempts ran into stiff resistance, in part from the vested interests (of employees and industrialists alike) threatened by such reforms, but also in part because the rents that guilds collected were partially dissipated to the government and the fiscal consequences were often serious. Nothing but shock treatment could work, and on February 16, 1791, the French guilds were abolished by fiat of the National Assembly. When Revolutionary French armies advanced into the Low Countries, Italy, and Germany, this reform was invariably imposed. Although the suppression of the guilds did not lead to completely free labor markets and resistance to new technology in France could still be strong on occasion, by the time the dust settled on the Continent in 1815, this vestige of the economic ancien régime had been fatally weakened. By itself, the suppression of the guilds cannot be regarded as a necessary condition for economic growth: long before 1791, manufacturers were able in many cases to move out of towns controlled by guilds, employ women and children, and find other ways around guild restrictions. But as a symptom of a general change in the attitude toward rent seeking, the history of craft guilds is illustrative.

Commercial policy was at center stage of enlightened antimercantilist policy. Here too there was ambiguity. Not all Enlightenment writers were unambiguously pro–free trade.[45] Yet the theme of trade being a positive-sum game, so eloquently expounded by Adam Smith, had been advocated since the late seventeenth century and was becoming dominant in political economy by 1800. It is ironic, of course, that the wars of 1793–1815 caused by the French Revolution and its aftermath seriously disrupted international trade, leading David Ricardo, the greatest mind of early-nineteenth century political economy (an offshoot of the Enlightenment), to include an entire chapter devoted to this phenomenon.[46] With the exception of a brief interlude following the 1786 Eden Treaty, free trade was not to be seriously considered as a policy option until the 1820s. Smith himself was not optimistic about free trade being established in Britain any more than "that Oceana or Utopia be even established in it."[47] Yet the pax Britannica and the slow turn toward freer trade between 1820 and 1880 cannot be seen as the outcome of economic interests alone; persuasion on logical grounds was very much part of the story.[48] *The Wealth of Nations* may not have killed mercantilism with a single blow, but it clearly pushed it into a defensive corner.

What is not always realized, however, is that the main triumph of the free trade doctrine was the establishment of free internal trade. Enlightenment thinkers viewed internal tariffs as the rent-seeking abomination they were, and the elimination of the French internal tariff barriers followed the abolition of the guilds. The U.S. commerce clause had been inserted into the U.S. Constitution a few years earlier. Internal trade in Sweden was liberalized in the late 1770s.[49] In Germany, the matter was more complex, but the post-1815 movement toward a German *Zollverein* reflected the same sentiment. The system of tolls and duties on Germany's magnificent river system that hampered trade in the eighteenth century was dismantled. Arguably the lion's share of gains from trade was secured through internal rather than external trade.[50]

Did these ideological changes have an effect? It is hard, in the end, to be sure that Enlightenment thought was more than Saint-Exupéry's king who commanded the sun to rise every morning. John Stuart Mill's statement that a good cause seldom triumphs unless someone's interest is bound up with it does not imply that, at times, such good causes do not fail. Enlightenment-inspired reforms in the West came in four waves. First, there were the post-1750 reforms introduced by so-called enlightened despots, which were often inspired by the writings of the *philosophes*, but rarely had much staying power since they frequently ran up against deeply entrenched interests. Second, there were the "natural reforms" introduced in countries that had meta-institutions such as a Parliament with sufficient political adaptiveness to bring about induced institutional change. Britain was able to pass such "rational" legislation as the Turnpike Act, the East India Acts of 1784 and 1813, the abolition of the Statute of Apprentices and Artificers in 1809, and the Navigation Acts in 1849. In other countries, revolution, whether indigenous or imported, was necessary. Finally, there were reactive reforms in countries such as Prussia as a result of reforms in nations viewed as competitors. The Enlightenment's influence on the French and American revolutions needs no elaboration. Equally well documented is the enormous influence that the *Wealth of Nations* had on policymakers, especially after Dugald Stewart, Smith's successor at Edinburgh, turned the book into a fountainhead of wisdom.[51] Among Stewart's pupils were two future prime ministers, Henry Palmerston and John Russell, as well as other senior officials, such as William Huskisson, the prime mover in the British liberal reforms of the 1820s. His program was to remove all state support and protection for manufacturing and agriculture. Huskisson "zealously and consistently subscribed" to the theories of Adam Smith. "Smith's teaching is reflected in practically every reform in the twenties."[52] In Germany, the influence of "the Divine Smith" on Prussian reformers has been thoroughly documented.

In economic history, scholars often write of technological progress but rarely of institutional change, and for good reason. But it could be argued that in the century after 1750, there was something we might think of in those terms, because this was the age when rent seeking in Europe was losing ground to productive commerce and production, markets became a little freer of regulation, and taxation and economic policy became less distributive. That it did not produce laissez-faire economies, even in Victorian Britain, and that the movement was full of reversals and ambiguities requires no repetition. Britain's technological successes prompted a very unenlightened set of laws prohibiting the exportation of machinery and emigration of skilled artisans (which, however, did little to stop the flow of useful knowledge). The French Revolution, despite its overall commitment to Enlightenment values, triggered a serious reactionary backlash in Britain, and, in France itself, the Academy of Sciences was closed in 1793 by the Jacobins, who felt that "the Republic does not need *savants*." But "progress" there was all the same. The quarter-century between the Bastille and Waterloo was in some ways a *réculer pour mieux sauter* (take a step back in order to advance) kind of interlude. By the late 1820s, mercantilism had retreated, and serious growth could occur.

The long-run historical significance of this advance was that it eliminated the negative institutional feedback that had wiped out economic growth before 1700. It is easy to see a counterfactual scenario in which the economic gains of the mule, the Watt engine, and the puddling and rolling process were swallowed up by tax collectors, wars, protectionists, and distributive coalitions of various kinds. It is not hard to imagine the newly entrenched technological status quo becoming increasingly more conservative and resisting further technological advances through political action. That this did not happen is the result of the double action of the Enlightenment: while it increased useful knowledge and its effectiveness, at the same time it improved the incentives for its implementation and weakened the forces that would set it back. In that sense, Enlightenment-inspired technological progress and institutional change created a powerful synergy, which in the end was responsible for the sustainability of what started in Britain in the last third of the eighteenth century and its diffusion to the societies that shared the Enlightenment.

THE ROOTS OF THE EUROPEAN ENLIGHTENMENT

Attributing the emergence of modern growth in the West to the Enlightenment in Europe leaves the question of the roots of the Enlightenment itself unanswered. To put it bluntly, we need to ask why Europe had an Enlightenment and other cultures such as Islam or China did not. Answering this question satisfactorily would be a huge undertaking. Linking it to previous events such as the emergence of humanism in Renaissance Europe or the Reformation only pushes the question further back in time. An alternative approach is not to ask why Europe had an Enlightenment, but to postulate that "enlightened" ideas occurred in all societies and that only in the "West" was this movement successful in the fashion I have described. The victory of the Enlightenment was not just a case of a growing and cumulative store of knowledge, but the triumph of open and public knowledge over secret "arcane" knowledge, the victory of "mechanical" philosophy (e.g., verifiable knowledge about natural regularities) over "occult philosophy" dealing with mystical and unobservable entities. How, then, did the good guys win?

Europe's uniqueness was obviously not that it was monetized, commercialized, and enjoyed "good" governance. "Capitalism"—whatever may be exactly meant by that term in the context of early-eighteenth-century Europe—seems too vague a concept to be of much help. What seems unique to Europe in the period leading up to and including the Enlightenment is the growing opportunity for critics, skeptics, and innovators to try their ideas out in a marketplace for ideas and to survive the experience. The notion that Europe was deeply hostile to "heretics" based on the tragic experience of such figures as Giordano Bruno and Miguel Servetus is fundamentally mistaken. The picture of Europe in the period 1500–1750 is one in which innovative, often radical, intellectuals were able to play one political authority against another, different polities against each other, and, when necessary, to take advantage of central versus local power, the private against the public sphere, and spiritual against secular authority. By moving from one place to another when the environment became too hostile, the members of the intellectual class ("clerisy" as they are sometimes called) could remain active in the transnational community of scholars—the Republic of Letters. Iconoclastic scholars who brought the ire of the local establishment on themselves usually went elsewhere. Martin Luther and Paracelsus are the most famous rebels who successfully played this game. For the West as a whole, the salutary effects of this pluralism cannot be overestimated. David Hume, for one, felt that this was the main reason why the sciences in China "made so slow a progress." In China, he argued, "none had the courage to resist the torrent of popular opinion, and posterity was not bold enough to dispute what had been universally received by their ancestors."[53]

The fragmentation of power and the competitive "states system" (Jones's term) is slightly anachronistic for the principalities and bishoprics that enjoyed considerable political autonomy in the seventeenth and eighteenth centuries. Paul David has argued that many rulers competed to attract to their courts reputable scientists, in part because some skills could come in handy, but largely as a signaling device (that is, to show off). The competition for the "best" scientists between European rulers required open science as a solution to the asymmetric information problem that rulers faced: to identify the truly leading scientists of their generation. Only within communities ("invisible colleges") in which full disclosure was exercised, he argues, could credible reputations be established that would allow wealthy patrons to separate truly distinguished scientists from fraudulent ones. Open science then emerged as a better strategy for scientists competing for patronage. The competition of different institutions for the superstars of science meant that the very best could set their own research agendas and appropriate the benefits of research, such as they were, and that few governments had the power to suppress views they considered heretical or subversive.[54]

We may also point to specific institutional changes that encouraged both the growth of intellectual innovation and its growing bias toward "usefulness," though the latter term needs to be treated with caution. Perhaps the central development was a change in the relationship between the world of production—farmers, merchants, manufacturers, as well as government agencies engaged in military and infrastructural projects—and the world of intellectuals. The idea that *ars sine scientia nihil est* (practice is worthless without theory), first enunciated in Renaissance Italy, slowly won ground. Natural philosophers were increasingly retained and engaged in practical matters where, it was believed, they could use their knowledge of nature to solve problems and increase efficiency. The growing conviction that this knowledge had (at least in expectation) a positive social marginal product meant, of course, that the demand for useful knowledge increased. This created the standard problem of intellectual property rights for the spread of useful knowledge. The interesting way in which this was solved was by taking advantage of the fact that the creators of propositional knowledge sought credit rather than profit from their work. Such credit, in some cases, was necessary to assure them of some reservation price, mostly in terms of a sinecure: a pension, an appointment at a court or a university, or a sponsored job by an academy or scientific society.[55] The rules of the game in the Republic of Letters, as they were established in the second half of the seventeenth century, were credit by priority, subject to verification. This "credit" was a property right in that it attributed an innovation unequivocally to the person responsible.[56] Enhanced prestige was then often correlated with some appointment that provided the scientist with a reservation price, though the correlation was far from perfect. Others, such as Henry Cavendish, Joseph Banks, and Antoine Lavoisier, were financially independent and did not need or expect to be compensated for their scientific work.

The other factor that facilitated the success of the Enlightenment as an intellectual movement in Europe was the institutional fluidity of intellectual activity. No single set of institutions dominated thought in Enlightenment Europe the way the Roman Catholic Church had dominated in the medieval period and the way the Confucian mandarinate dominated Chinese thought. In Europe, such institutional domination was absent, and within the Republic of Letters there was free entry and furious competition for patronage and clients. Peter Burke has suggested that universities tended to suffer from "institutional inertia" and became conservative over time, so that only the founding of new ones kept them creative and lively.[57] Professor Martin Luther was teaching theology at an institution that was only fifteen years old, and the University of Leyden, founded in 1575 as a Calvinist University, became a major curricular innovator. But universities had to compete with the academies and courts of Europe to attract the best minds of Europe. The decentralized and multifocal distribution of wealth and power in Europe between Luther and Lavoisier led to a world of intellectual competition in which knowledge was both transmitted and augmented in ever more effective fashion.

There were other reasons for the success of the European Enlightenment. The *philosophes* of the eighteenth century were not a marginal group struggling for recognition. Despite their opposition to the existing arrangements and their dreams of reform and improvement, they were more often than not part of the establishment or, more accurately, part of some establishment. The triumph of the *philosophes* must be explained by their ability to act against the status quo from within the establishment. Many of the leading *philosophes* and political economists were well born and politically well connected. Even when they ran afoul of the regime, the relations rarely degenerated into hostility. This "cozy fraternizing with the enemy," as Peter Gay calls it, did not come without a price, but it allowed the *philosophes* to be politically effective without necessarily threatening the status quo.[58] In France, this relationship ultimately imploded (though it was soon restored), but elsewhere, it enabled their ideas to be adopted by the men who voted on policy decisions. All the same, throughout Europe, the Enlightenment was a decentralized and free-enterprise endeavor, sometimes tolerated but rarely managed or sponsored by governments. Yet it was not unorganized: enlightened ideas found expression in the myriad of friendly societies, academies, Masonic lodges, and similar organizations of people who shared beliefs and traded knowledge. To be sure, there were a few figures of political power who were associated with and influenced by the Enlightenment, the best known of whom were the so-called enlightened despots and some of their ministers. It stands to reason that an intellectual movement such as this can fail either because it is too close to the government or because it is so marginalized that it can be ignored. Much of the European Enlightenment fell in between.

The European Enlightenment pushed a dual platform that was radical and revolutionary: reform institutions to promote efficiency and innovation, and bring the full force of human knowledge to bear on technology.[59] Without that synergy, long-term economic growth in the West might not have happened either. The Enlightenment was an indispensable element in the emergence of modern economic growth. Its belief in social progress through reason and knowledge was shocked repeatedly as the superiority of reason was thrown in doubt.[60] But the idea of useful knowledge as an engine of social progress has not lost any of its power, even as it has been challenged, toned down, and refined in the two centuries since 1800. There was nothing preordained or inevitable about that course of history. Indeed, in hindsight, it seems rather unlikely, and any competent economic historian can point to a dozen junctures where the process could have been derailed. The fruits of these changes were, of course, very late in coming. Economic growth, in the sense that Robert Lucas had in mind, does not take off anywhere before 1830.[61] And yet from a long-term perspective, the striking thing is not that it happened so long after the necessary preceding intellectual changes but that it happened at all.

ACKNOWLEDGMENTS

The comments and suggestions of Kenneth Alder, Maristella Botticini, Wilfred Dolfsma, Margaret Jacob, Lynne Kiesling, Deirdre McCloskey, Edward Muir, Cormac Ó Gráda, Avner Greif, and Richard Unger are acknowledged. I am indebted to Fabio Braggion, Chip Dickerson, Hillary King, and Michael Silver for loyal research assistance.

NOTES

1. For the most recent estimates of growth during the Industrial Revolution, see Joel Mokyr, "Accounting for the Industrial Revolution," in *The Cambridge Economic History of Modern Britain*, vol. 1, ed. Roderick Floud and Paul Johnson (Cambridge: Cambridge University Press, 2004), 1–27; C. Knick Harley, "Re-Assessing the Industrial Revolution: A Macro View," in *The British Industrial Revolution: An Economic Perspective*, ed. Joel Mokyr, 2nd ed. (Boulder, CO: Westview Press, 1998), 160–205.

2. Eric L. Jones, *Growth Recurring* (Oxford: Oxford University Press, 1988).

3. David S. Landes, *The Wealth and Poverty of Nations: Why Some Are So Rich and Some So Poor* (New York: Norton, 1998); Kenneth Pomeranz, *The Great Divergence: China, Europe, and the Making of the Modern World Economy* (Princeton, NJ: Princeton University Press, 2000).

4. Eric L. Jones, *The European Miracle: Environments, Economies and Geopolitics in the History of Europe and Asia*, 2nd ed. (Cambridge: Cambridge University Press, 1987 [1981]).

5. Walt W. Rostow, *How It All Began: Origins of the Modern Economy* (New York: McGraw-Hill, 1975); H. Floris Cohen, "Inside Newcomen's Fire Engine: The Scientific Revolution and the Rise of the Modern World," *History of Technology* 25

(2004), 111–132; Richard G. Lipsey, Kenneth Carlaw, and Cliff Bekar, *Economic Transformations: General Purpose Technologies and Sustained Economic Growth* (New York: Oxford University Press, 2005).

6. Max Horkheimer and Theodor W. Adorno, *Dialectic of the Enlightenment* (New York: Continuum, 1971), and John Gray, *Enlightenment's Wake: Politics and Culture at the Close of the Modern Age* (London: Routledge, 1995). Eric Hobsbawm notes with some disdain that this literature describes the Enlightenment as "anything from superficial and intellectually naive to a conspiracy of dead white men in periwigs to provide the intellectual foundation for Western Imperialism." See his "Barbarism: A User's Guide," in *On History* (New York: New Press, 1997), 253–265.

7. This point has been well made by Ian Inkster, "Potentially Global: A Story of Useful and Reliable Knowledge and Material Progress in Europe, ca. 1474–1912," *International History Review* 18 (2006), 237–286, whose analysis parallels what follows in certain respects. Inkster proposes the term URK ("useful and reliable knowledge"), which is much like the term proposed by Simon Kuznets who preferred "testable." In my view reliability is an important characteristic of useful knowledge, but it seems less crucial than tightness, that is, the confidence and "consensualness" with which certain knowledge is held to be "true."

8. Peter Burke, *A Social History of Knowledge* (Cambridge: Polity Press, 2000), 44.

9. Margaret C. Jacob and Larry Stewart, *Practical Matter: Newton's Science in the Service of Industry and Empire, 1687–1851* (Cambridge, MA: Harvard University Press, 2004).

10. Joseph Priestley, *An Essay on the First Principles of Government and on the Nature of Political, Civil and Religious Liberty* (London: J. Doosley in Pall Mall, 1768), 7. Adam Smith, in the "Early Draft" to his *Wealth of Nations* in *Lectures on Jurisprudence*, ed. Ronald L. Meek, David D. Raphael, and Peter G. Stein (Oxford: Oxford University Press, 1978), 569–572, believed that the benefits of the "speculations of the philosopher . . . may evidently descend to the meanest of people" if they led to improvements in the mechanical arts.

11. Donald S. L. Cardwell, *The Fontana History of Technology* (London: Fontana Press, 1994), 195.

12. William H. Brock, *The Norton History of Chemistry* (New York: Norton, 1992), 37; Roy Porter, "Introduction," in *The Cambridge History of Science*, vol. 4: *Eighteenth-Century Science*, ed. Roy Porter (Cambridge: Cambridge University Press, 2003), 1–20.

13. The term is from Cohen, "Inside Newcomen's Fire Engine," 118, who adds that in the seventeenth century, useful applications of the new insights of science kept eluding its proponents.

14. Harry W. Dickinson and Rhys Jenkins, *James Watt and the Steam Engine* (London: Encore Editions, 1927), 16; Richard L. Hills, *Power from Steam: A History of the Stationary Steam Engine* (Cambridge: Cambridge University Press, 1989), 53.

15. Terry Reynolds, *Stronger Than a Hundred Men: A History of the Vertical Water Wheel* (Baltimore, MD: Johns Hopkins University Press, 1983), 233–248.

16. Jan Golinski, *Science as Public Culture: Chemistry and Enlightenment in Britain, 1760–1820* (Cambridge: Cambridge University Press, 1992), 29.

17. Barbara Whitney Keyser, "Between Science and Craft: The Case of Berthollet and Dyeing," *Annals of Science* 47:3 (1990), 222.

18. Cohen, "Inside Newcomen's Fire Engine," 123.

19. David G. C. Allan, *William Shipley: Founder of the Royal Society of Arts; A Biography with Documents* (London: Scolar Press, 1979), 192.

20. Judith V. Grabiner, "'Some Disputes of Consequence': MacLaurin among the Molasses Barrels," *Social Studies of Science* 28:1 (1998), 139–168.

21. Scottish chemists such as William Cullen and Joseph Black were much in demand as consultants to improving farmers and ambitious textile manufacturers. Joel Mokyr, *The Gifts of Athena: Historical Origins of the Knowledge Economy* (Princeton, NJ: Princeton University Press, 2002), 50–51.

22. Leonhard Euler, the most talented mathematician of the age, was concerned with ship design, lenses, and the buckling of beams; with his less famous son Johann, he contributed a great deal to hydraulics.

23. The three were Alexandre Vandermonde, Claude Berthollet, and Gaspard Monge, who jointly published their "Mémoire sur le fer," under the influence of the new chemistry of their master, Antoine-Laurent de Lavoisier.

24. Brian Bowers, *Michael Faraday and the Modern World* (Wendens Ambo (Essex): EPA Press, 1991).

25. Eda Kranakis, "Hybrid Careers and the Interaction of Science and Technology," in *Technological Development and Science in the Industrial Age*, ed. Peter Kroes and Martijn Bakker (Dordrecht: Kluwer, 1992), 177–204.

26. John R. Pannabecker, "Diderot, Rousseau, and the Mechanical Arts: Disciplines, Systems, and Social Context," *Journal of Industrial Teacher Education* 33: 4 (1996), 6–22, and "Representing Mechanical Arts in Diderot's Encyclopédie," *Technology and Culture* 39:1 (1998), 33–73.

27. Richard Yeo, "Classifying the Sciences," in *The Cambridge History of Science*, vol. 4: *Eighteenth-Century Science*, ed. Roy Porter (Cambridge: Cambridge University Press, 2003), 241–266.

28. Eugene S. Ferguson, "The Measurement of the 'Man-Day,'" *Scientific American* 225 (1971), 96–103; Svante Lindquist, "Labs in the Woods: The Quantification of Technology during the Late Enlightenment," in *The Quantifying Spirit in the 18th Century*, ed. Tore Frängsmyr, J. L. Heilbron, and Robin E. Rider (Berkeley: University of California Press, 1990), 291–314.

29. Maurice Daumas and André Garanger, "Industrial Mechanization," in *A History of Technology and Invention*, vol. 2, ed. Maurice Daumas (New York: Crown, 1969), 249; Kenneth Alder, "Making Things the Same: Representation, Tolerance and the End of the Ancien Régime in France," *Social Studies of Science* 28:4 (1998), 499–545.

30. Ken Alder, *Engineering the Revolution: Arms, Enlightenment, and the Making of Modern France* (Princeton, NJ: Princeton University Press, 1997), 136–146.

31. Paul Elliott, "The Birth of Public Science in the English Provinces: Natural Philosophy in Derby, c. 1690–1760," *Annals of Science* 57 (2000), 83; Robert Schofield, *The Lunar Society of Birmingham* (Oxford: Clarendon Press, 1963), 22, 201.

32. Larry Stewart, *The Rise of Public Science* (Cambridge: Cambridge University Press, 1992), and "The Laboratory and the Manufacture of the Enlightenment," unpublished manuscript (2004), as well as Jacob and Stewart, *Practical Matter*, chap. 5.

33. Schofield, *The Lunar Society*; Jenny Uglow, *The Lunar Men: Five Friends Whose Curiosity Changed the World* (New York: Farrar, Straus and Giroux, 2002).

34. Trevor H. Levere and Gerard L. E. Turner, with contributions from Jan Golinski and Larry Stewart, *Discussing Chemistry and Steam: The Minutes of a Coffee House Philosophical Society, 1780–1787* (Oxford: Oxford University Press, 2002).

35. James McClellan III, "Scientific Institutions and the Organization of Science," in *The Cambridge History of Science*, vol. 4: *Eighteenth-Century Science*, ed. Roy Porter (Cambridge: Cambridge University Press, 2003), 92.

36. Schofield, *The Lunar Society*, 195.

37. Examples of the literature are Dani Rodrik, "Getting Institutions Right: A User's Guide to the Recent Literature on Institutions and Growth," working paper, Harvard University (April 2004); Dani Rodrik, Arvind Subramanian, and Francesco Trebbi, "Institutions Rule: The Primacy of Institutions over Geography and Integration in Economic Development," NBER working paper 9305 (2002); and Daron Acemoglu, Simon Johnson, and James Robinson, "Reversal of Fortune: Geography and Institutions in the Making of the Modern World Income Distribution," *Quarterly Journal of Economics* 117 (2002), 1231–1294. A convenient summary is provided in Elhanan Helpman, *The Mystery of Economic Growth* (Cambridge, MA: Harvard University Press, 2004).

38. This is precisely captured by Adam Smith: "[N]ations have been taught that their interests consisted in beggaring all their neighbours. Each nation has been made to look with an invidious eye upon the prosperity of all the nations with which it trades, and to consider their gain as its own loss." *An Inquiry into the Nature and Causes of the Wealth of Nations*, ed. Edwin Cannan (Oxford: Oxford University Press, 1976 [1776]), 519.

39. Robert B. Ekelund, Jr., and Robert D. Tollison, *Mercantilism as a Rent-Seeking Society* (College Station: Texas A&M University Press, 1981), and *Politicized Economies: Monarchy, Monopoly, and Mercantilism* (College Station: Texas A&M University Press, 1997).

40. Smith, *Wealth of Nations*, 139–144.

41. The canonical statement is by the great Belgian historian Henri Pirenne: "The essential aim [of the craft guild] was to protect the artisan, not only from external competition, but also from the competition of his fellow-members." The consequence was "the destruction of all initiative. No one was permitted to harm others by methods which enabled him to produce more quickly and more cheaply than they. Technical progress took on the appearance of disloyalty." *Economic and Social History of Medieval Europe* (New York: Harcourt, 1936), 185–186. For similar statements, see Carlo Cipolla, "The Economic Decline of Italy," in *Crisis and Change in the Venetian Economy in the Sixteenth and Seventeenth Centuries*, ed. Brian Pullan (London: Methuen, 1968); Pierre Deyon and Philippe Guignet, "The Royal Manufactures and Economic and Technological Progress in France before the Industrial Revolution," *Journal of European Economic History* 9:3 (1980), 611–632; Jeff Horn, *The Path Not Taken: French Industrialization in the Age of Revolution, 1750–1830* (Cambridge, MA: MIT Press, 2006), chap. 2.

42. Stephan R. Epstein, "Craft Guilds, Apprenticeships, and Technological Change in Pre-Industrial Europe," *Journal of Economic History* 58:3 (1998), 684–713.

43. Sheilagh Ogilvie, "Guilds, Efficiency, and Social Capital: Evidence from German Proto-Industry," *Economic History Review* 57:2 (2004), 286–333.

44. Horn, *The Path Not Taken*, chap. 2.

45. David Hume, while certainly no mercantilist, was of two minds about it. "Of the Rise and Progress of the Arts and Sciences (1742)," in David Hume, *Essays: Moral, Political and Literary*, ed. Eugene F. Miller (Indianapolis, IN: Liberty Fund, 1985), 98. Alexandre Vandermonde, a noted mathematician and scientist who turned to economics late in life and taught it at the newly founded École Normale, and knew the *Wealth of Nations* inside out, never converted to free trade and preferred the protectionist doctrines of Smith's contemporary, James Steuart. Charles Coulston Gillispie, *Science and Polity in France: The Revolutionary and Napoleonic Years* (Princeton, NJ: Princeton University Press, 2004), 513.

46. David Ricardo, *Principles of Political Economy*, ed. R. M. Hartwell (Harmondsworth: Pelican Books, 1971 [1817]), chap 19.

47. Smith, *Wealth of Nations*, 493.

48. Charles P. Kindleberger, "The Rise of Free Trade in Western Europe, 1820–1875," *Journal of Economic History* 35:1 (1975), 20–55.

49. Gunnar Persson, *Grain Markets in Europe, 1500–1900: Integration and Deregulation* (Cambridge: Cambridge University Press, 2000), 139.

50. The Zollverein was preceded by the Prussian Maassen Tariff Law of 1818, which abolished all internal tariffs in Prussia and was influenced by a memorandum by G. J. C. Kunth, Beuth's mentor.

51. Arthur Herman, *How the Scots Invented the Modern World* (New York: Crown, 2001), 229–230. See also Emma Rothschild, *Economic Sentiments: Adam Smith, Condorcet, and the Enlightenment* (Cambridge, MA: Harvard University Press, 2001).

52. Alexander Brady, *William Huskisson and Liberal Reform: An Essay on the Changes in Economic Policy in the Twenties of the Nineteenth Century*, 2nd ed. (New York: Augustus M. Kelley, 1967), 133.

53. Hume, *Essays*, 122.

54. Paul David, "Patronage, Reputation, and Common Agency Contracting in the Scientific Revolution," unpublished manuscript (August 2004).

55. The economics of open science resemble in many ways the economics of open source software development, which has found that signaling to outsiders, peer recognition, and direct benefits all play a role. Josh Lerner and Jean Tirole, "The Economics of Technology Sharing: Open Source and Beyond," NBER working paper 10956 (December 2004).

56. In an earlier time, the absence of clear-cut rules discouraged open knowledge. Thus the architect Francesco di Giorgio Martini (1439–1501) complained that "the worst is that ignoramuses adorn themselves with the labors of others and usurp the glory of an invention that is not theirs. For this reason the efforts of one who has true knowledge is oft retarded." Cited by William Eamon, *Science and the Secrets of Nature: Books of Secrets in Medieval and Early Modern Culture* (Princeton, NJ: Princeton University Press, 1994), 88.

57. Burke, *A Social History of Knowledge*, 37, 48.

58. Peter Gay, *The Enlightenment: An Interpretation: The Rise of Modern Paganism* (New York: Norton, 1966), 24.

59. Cohen, "Inside Newcomen's Fire Engine," 131, raises a similar point.

60. Indeed, even during the Enlightenment, the supremacy of reason over sentiment and sensitivity has been shown to be a flawed concept. Jessica Riskin, *Science in the Age of Sensibility: The Sentimental Empiricists of the French Enlightenment* (Chicago: University of Chicago Press, 2002), 200.

61. Robert E. Lucas, "On the Mechanics of Economic Development," *Journal of Monetary Economics* 22 (1988), 3–42.

5

Avoiding Revolution: The French Path to Industrialization

Jeff Horn

France's path to industrialization was tortuous, prolonged, and unique. The French did not follow Britain's model of industrial development, but not for lack of trying. In the decades before 1789, French policymakers and entrepreneurs made serious and sustained efforts to emulate what they understood as the wellsprings of British industrial success: elite domination and the co-optation of the working classes while funneling working-class inventiveness into productive channels. French attempts to compete with Great Britain made great strides during this era, as witnessed by their willingness to sign a commercial treaty in 1786. Premature as this move proved to be, it signaled the belief among French policymakers that if the Bourbon state could not yet challenge the "first industrial nation," they expected to become genuinely competitive in the near term. French efforts to imitate their island rival's approach to technological development, labor relations, entrepreneurialism, and mode of state involvement in the economy were brought to a screeching halt by the French Revolution.

Although the new French regime struggled to maintain and even advance their emulation of the British model, the continuing threat of violence forced key French leaders to recognize that if they trod in Britain's footsteps, a thoroughgoing social revolution might result. In the decade after the end of the Terror, this recognition meant that French policymakers and entrepreneurs had to devise different means of fostering industrial development. After fits and starts, the institutions and modes of government support for technological innovation and industrial expansion created between 1794 and 1804 emerged as the regulatory framework for stimulating the French economy throughout the nineteenth century. A measure of the success of the French path to industrialization is that, by 1860, they could sign another commercial agreement bordering on free trade with Britain, then at the height of its industrial dominance, while continuing to avoid a social revolution.[1]

The French did not experience industrialization in the same way as any of their neighbors because of the greater threat of social revolution.

Forced by the "threat from below" to elaborate a unique path to industrial development, by 1914, French per capita income closely approximated that of its British rival As O'Brien and Keyder pointed out more than thirty years ago, from a longer-term perspective, France was not doomed to second-class industrial status.[2] As Sabel and Zeitlin have argued persuasively, there was more than one path to industrial society, as the intertwining of the parallel paths of these two nations in the twentieth century makes clear.[3] The term "Industrial Revolution" is a misnomer. Perhaps "Industrial Revolutions" is more apt; it is certainly far less teleological in that it does not assume the exclusive success of a singular model.[4] I am in full agreement with Kenneth Pomeranz's assertion that all national experiences must be considered as deviations; none should be elevated to the status of the normative, no matter the timing or power of the economy in question.[5] Anglocentrism and now America-centrism must not obscure technological, scientific, and industrial successes or failures outside these paradigmatic nations. The central goal of this book is to focus attention on the other factors, other choices, and other possibilities that must be considered and woven into the stories that the English-speaking world tells about the Industrial Revolution.

This chapter delineates briefly the contours of contemporary French understandings of the British model of industrialization and how attempts to emulate that model were derailed by the French Revolution. Then it focuses on the creation of new institutions and policies under Jean-Antoine Chaptal, minister of the interior, chemist, and entrepreneur at the dawn of the nineteenth century, who provided the framework for later French industrial success. This developmental trajectory provides essential support for O'Brien and Keyder's approach to comparative history and reveals both the limits of convergence and the origins of divergence in the unfolding of the Industrial Revolution in France.

The Path of French Industrialization

In the mid-eighteenth century, a new market-oriented approach to state-sponsored economic development emerged as part of the Enlightenment. This new approach concerned such vital subjects as the economic value of competition, the need for innovation in industrial production, the application of science to industrial matters, and a different but no less involved economic role for the state. Those most influenced by this new outlook were French state officials and scientists, most of whom were followers to one degree or another of Victor Riqueti, Marquis de Mirabeau, and François Quesnay. This group is usually referred to with the shorthand designation of the physiocrats. Their model for changing the French way of doing things was England, because of its more innovative means of applying scientific expertise to technical problems. A long line of influential French royal officials accepted the basic premise that England provided a useful model of how to employ the market to encourage economic growth and to jump-start industrial development. From 1750, they attempted to imitate that model, albeit selectively, in France.

This desire to follow the English economic path found practical application in major reforms such as the abolition of the guilds and the liberation of the grain trade by Anne-Robert-Jacques de Turgot in 1776. It was also essential to the thoroughgoing reform of the world of work between 1779 and 1781 under Jacques Necker and the dramatic tariff reduction embodied by the Anglo-French Commercial Treaty of 1786. Behind the scenes, these sweeping changes were supplemented by a systematic attempt by these ministers and their collaborators to use the impressive expertise of the French scientific establishment to solve practical technological problems.

Emulation of the English model enjoyed considerable success, and from the perspective of eighteenth-century economic actors, there were substantial reasons for believing that the French could beat the English at their own industrial game, even in the troubled years leading up to the outbreak of the French Revolution. It is worth remembering not only that French foreign trade was larger than England's in 1789, but also that France's manufacturing output was more than three times as great. Scientists, bureaucrats, mechanics, and entrepreneurs judged that the essential gaps between England and France—in developing and implementing industrial technologies, in innovative entrepreneurialism, and in enforcing worker discipline—were visibly narrowing, especially after 1785.

The confident steps taken down the English path of economic development were brought to a standstill following the outbreak of an unprecedented wave of machine-breaking in the summer of 1789. This outbreak was much more widespread and far more devastating than its better-known but far less significant English equivalent, the Luddite movement of 1811–1817. French machine-breaking coincided with two important nationwide events. First was the Great Fear in which rural populations, frightened by the specter of roving hordes of brigands and the possibility of an aristocratic reaction against the Revolution, decided to forestall this threat by sacking châteaux. This movement was directly responsible for the early passage of the *Declaration of the Rights of Man and*

Citizen and the abolition of feudal privileges on the night of August 4–5. The urban counterparts of the Great Fear were the municipal revolutions that broke out in twenty-nine of the thirty largest cities, which were sometimes violent but always accompanied by popular agitation. This multifaceted popular threat demolished French entrepreneurs' confidence that the working classes were sufficiently disciplined to permit English-style industrialization. The growth of the popular movement after 1791, and particularly the fifty thousand deaths that occurred during the Reign of Terror, suggested that the "threat from below" could be neither ignored nor contained.

The lingering "threat from below" derailed the French march down the trail blazed by England. From 1789 onward, the French state sought a different means of fostering economic development and nurturing industrial success. The impact of the French Revolution on industrial policy can be summed up as follows. After the summer of 1789, the English model, which had so dominated French thought and policy in the eighteenth century, became the path not taken. From that point forward, a different paradigm had to be developed not just in France but in all continental countries.[6] The consequences of the "threat from below" illustrate the distinctiveness of the British pattern of industrialization and why other countries' approaches converged around a divergent pathway.

This task was complicated by the collision course with the crowned heads of Europe embarked on by the Revolutionary state. In 1793, an embattled French government implemented sweeping wage and price controls known as Maximums, while systematically seeking to improve the application of advanced technical knowledge to productive practice as part of the formation of a vast military industrial complex almost from scratch. Only the Reign of Terror ensured the application of this groundbreaking wartime bid for economic transformation.

"Revolutionary government" ended in 1794. The Scylla and Charybdis of wartime economic dislocation and Revolutionary political unrest rendered French policymakers uncertain as to what economic path to follow. It was left to Napoleon Bonaparte and, more particularly, to his influential minister of the interior, Jean-Antoine Chaptal, noted chemist and wildly successful entrepreneur, to capitalize on the efforts made under previous regimes to forge a coherent conception of different means of fostering industrial competitiveness.

In the first decades of the nineteenth century, French policymakers recognized the many parallels with British experience, but unlike their eighteenth-century forebears, they did not believe that successes on one side of the Channel could be translated directly to the other. Although they may have paid lip-service to the ideas of Adam Smith or other conceptions of economic liberalism, French policymakers now understood that these ideas had to be adapted or modified in their application, just as the English themselves did. The abiding resonance of the "threat from below," coupled with the demands of war, meant that the market mechanism could not be entrusted with many essential economic functions. Instead, the government had to shape French industrial production much more systematically and far more directly than was the case in the English-speaking world until the American Civil War. The *dirigisme* inaugurated by Chaptal, which melded theory and practice to further an economic agenda powerfully influenced by liberalism, left the central state in a pivotal position in the economy. This approach characterized the French industrial landscape throughout the nineteenth century and beyond. The rest of this chapter traces the establishment of this unique approach to industrial development and its consequences.

The industrial policy fashioned by Chaptal and his collaborators and pursued by his successors was economically rational and market oriented. It concentrated on areas where France was internationally competitive: notably certain agricultural sectors and the production of a wide variety of luxury goods. State industrial policy was also prescriptive in that it sought to facilitate the application of French scientific predominance in chemistry, mathematics, and medicine to problems of production.[7] At the same time, French policies spread mechanical knowledge both more widely and more deeply among the laboring classes and furnished financial incentives for technical innovation to both tinkerers and entrepreneurs.

France's international military position forced it to spend vast amounts of time and money to develop adequate expertise in areas like iron smelting, cotton textiles, machine-building, and the use of coal as fuel, domains dominated by Britain. Such efforts must not be understood as the entire story. France developed considerable capacity, but as is well known, Britain's position in these areas was not challenged for generations. Britain's success in these areas overlooks the obvious: French industrial policy and success focused on discerning and supporting other products, other technologies, and other industries. Given France's natural resources and available sources of energy, the unique British model of industrialization based on iron and cotton textile production delineated by T. S. Ashton simply had less power to transform.[8] The long-term parity of French per capita economic growth and the rise of industrial powerhouses in the United States and Germany whose economies were configured differently from Britain's remind us that there was more than one avenue to an industrial revolution.

Jean-Antoine Chaptal and the Forging of a New Industrial Order

Jean-Antoine Chaptal (1756–1832) was Bonaparte's minister of the interior from 1800 to 1804.[9] Chaptal inaugurated a harmonious and creative approach to improving French industrial competitiveness that melded physiocratic notions concerning resources and liberal attitudes about the rights of the individual with an activist vision of the necessity of state action.[10] Thanks to the improved military and economic circumstances that prevailed under Bonaparte, Chaptal and his collaborators were able to found and develop a host of institutions designed to oversee the process of industrialization that had been inconceivable during the Revolutionary decade. Despite the impressive power wielded by the Bonapartist state, French policymakers, led by Chaptal, knew that France could not achieve the multifaceted and thorough domination of the working classes realized by Great Britain. Chaptal's institution building reflected his identification of Britain's ability to direct working-class inventiveness into productive channels as one of the major reasons for its industrial successes. In recent Anglo-American attention to Chaptal, an "Anglicized" version of his major achievements as the adaptation of English practice to continental realities predominates. Such a co-optation of Chaptal is demeaning to this accomplished public figure. As an entrepreneur, as a scientist, and, most important, as an administrator, Chaptal must be understood as quintessentially French in his outlook and endeavors. As a system builder, he mobilized and harnessed French industrial creativity in a time of war and revolution through distinctive means that went far beyond borrowings from across the Channel.

Born into a solid bourgeois family with close ties to the world of large-scale international commerce, Chaptal was educated within the vibrant medical community of the southern city of Montpellier. There he was influenced by vitalism, freemasonry, and physiocracy. He completed his education by studying medicine and chemistry in Paris. His reputation in chemistry was based on solving practical industrial problems. He founded a highly profitable manufacture of the salts and acids desperately needed by the local textile industry.

Ennobled on the eve of the Revolution, Chaptal revealed his hopes for political reform by becoming a member of Montpellier's initial Jacobin Club. Arrested as a moderate and a Federalist sympathizer during the

Terror, he was released and charged with overseeing gunpowder production by the Committee of Public Safety. With the easing of the war crisis, Chaptal returned to Montpellier to teach medicine until he became professor of chemistry at the École polytechnique in 1798. With the accession of Bonaparte, Chaptal became a councilor of state tasked with improving public education. He impressed Napoleon, who named Chaptal minister of the interior provisionally in November 1800, to replace Lucien Bonaparte.

When he left the ministry in July 1804, Chaptal entered the Senate and was named hereditary count of Chanteloup in 1808. Chaptal applied his ideas by expanding his manufacturing activities to include three large-scale chemical workshops around Paris. On his vast estate, he became vitally concerned—both scientifically and commercially—with improving the process of distilling sugar from grapes and the naturalization of both the sugar beet and merino sheep. In 1819, Louis XVIII named Chaptal a peer of the realm. The same year Chaptal published one of the first accounts of the nascent Industrial Revolution in France. Throughout the Restoration (1814–1830), Chaptal was active in promoting educational issues, particularly those related to science and technology, to improving the scientific basis of both agricultural and industrial production, and to employing his precepts to increase his personal fortune.[11]

Chaptal was strongly influenced by the physiocrats. He also read Adam Smith's *The Wealth of Nations* with approval not long after its publication in 1776. In hindsight, Chaptal saw himself as a life-long adherent of liberal economic theories, particularly "the most sacred and most inviolable of individual rights," that is, the right of individuals to make their own economic decisions, especially in disposing of their labor. He criticized earlier French administrations for their management of industrial matters, stating that their excessive regulation of production was the cause of French retardation vis-à-vis Great Britain, which left "our industry enslaved."[12]

According to Chaptal, the nineteenth-century government's role in managing industry had three basic components. First, state involvement was necessary to mend the damage done by the pre-1789 administration. By depriving those involved in the production and distribution of goods of their just "*considération*" in society, the Bourbon regime damaged French "public spirit" while alienating those involved in essential professions. Only active state sponsorship of the social value of commerce and industry could repair centuries of contempt. Second, in good Smithian fashion, Chaptal asserted that with regard to industry, "The actions of government ought to be limited to facilitating supplies, guaranteeing property, opening markets to manufactured goods, and to leaving industry to enjoy a most profound liberty. One can rely on the producer to pay attention to all the rest." Yet the reality of the situation was that no post-Revolutionary government could be quite so hands-off. To ensure that all French citizens could find gainful employment and to guarantee that the Revolutionary ideal of equality under the law existed in economic practice, the state must intervene. For Chaptal, the justification for this third precept was—again—French "public spirit." He contrasted the French emphasis on equality and its substantiation by the state with Great Britain where "private interest directs all actions" and "whose selfishness . . . offers us a terrible example of what I claim." According to Chaptal, even the most liberal of states had to mediate myriad private interests for the public good.[13]

It must also be remembered that market functionality was only a theoretical axiom—France was at war—on the battlefield, in the laboratory, and on the shop floor. Ultimately every high-ranking servant of the Bonapartist state had victory, not enhanced economic competition, in mind. Thus, the practical setting for liberal practices required a substantial revision of laissez-faire doctrines. Revision did not, however, mean abandonment. The institutions developed by Chaptal or with his support had a laissez-faire core. Sometimes this liberal marrow was sucked out, but the later perversion of an institution must not blind us to the original intent of the founders.

Educational reform was fundamental to Chaptal's program to improve France's competitive position technologically while enabling the market to function more efficiently. He perceived a twofold French educational deficiency susceptible to systematic state intervention. First, he wanted to create new means for French workers to experience advanced machinery while exposing them to the latest production techniques. Second, he hoped to increase the quantity and quality of interaction among scientists, tinkerers, artisans, and entrepreneurs to facilitate the development of practical solutions to shop floor problems. Both goals reflected a desire to imitate what Chaptal perceived as the wellsprings of English technological prowess.[14]

Spreading best practice techniques and providing access to advanced machinery were national priorities complicated by fragmented markets and disjointed transportation networks in a nation as large—and expanding as rapidly—as Napoleonic France. On May 24, 1802, the Musée des arts et métiers opened to provide hands-on instruction on a vast collection of machines. One workshop specialized in woodworking, another in iron making and steelmaking, and a third in making precision instruments. In June 1804, Chaptal revived the free spinning schools (Écoles gratuit de filature) common under the ancien régime where foremen and experienced spinners (usually women) taught groups of twenty-five students (a mix of men and women) the use of new machines copied or acquired from Great Britain and how to maintain them. When some of his proposals to focus the curricula of primary schools on mathematics, design, and mechanics were rejected by the legislature in February 1801, Chaptal did what he could by ordering instructors to teach girls spinning, knitting, and sewing. All students at elite specialized technical schools like the École des mines and the École des ponts et chaussées also began to spend part of the year in the field learning to apply their mathematical skills, an aspect of their curriculum that is underappreciated.[15]

Chaptal was never the sole architect of the new institutions uniting educational and industrial goals. An experimental school founded in 1780 by the Duke de La Rochefoucauld-Liancourt to provide a mix of primary education and hands-on mechanical experience was turned into a technical training school for skilled workers, foremen, engineers, and scientists. At Chaptal's urging, Bonaparte visited the school in 1799 and decreed that the school would be reborn in 1803 as the École des arts et métiers to "train petty officers for industry." With a curriculum designed by the eminent French scientists Gaspard Monge, Pierre-Simon Laplace, and Claude Berthollet that was half theoretical and half practical, the school was so successful under La Rochefoucauld-Liancourt's oversight that a second school was established in 1811 and a third was planned. Such a trajectory illustrates the collaborative nature of the institutional edifice created during the Consulate.[16]

If education was essential to Chaptal's strategy to improve France's technological prowess and industrial know-how, it was also the aspect of his program that focused most on the long term. It would take at least a generation for this knowledge to diffuse through the productive population. In the meantime, Chaptal understood the need for more immediate means of spreading technical information and providing access to advanced machinery. He wanted to provide expert advice to the technologically ignorant and heavy-handed administrators responsible for implementing the state's economic *dirigisme*. In a sense, Chaptal hoped to make France a unified technological and productive environment that would parallel the national political unity fashioned by the Revolution.

Among Chaptal's institutional goals, spreading technical knowledge and facilitating the interaction of scientists, innovators, entrepreneurs, and bureaucrats to solve problems of production took pride of place. The Conseils d'agriculture, des arts et commerce founded in each department in June 1801 were so active that a Conseil supérieur du commerce with significant provincial representation was set up in December 1802 to deal with the flood of suggested actions and improvements. Moreover, twenty-three municipal Chambers of Commerce (December 1802) were revived in the largest cities (Paris was exempted for political reasons). In addition, more than 150 Chambres consultatives de manufactures, fabriques, arts et métiers (April 1803) were founded in small and medium-sized urban areas as institutional means of ensuring that technical knowledge and market opportunities were available to those who might put them to use and "to express the needs and methods of improving manufacturing."[17] Chaptal also supported the formation of the Society for the Encouragement of National Industry (November 1801). Ostensibly a private organization, most of the initial three hundred members were important state officials recruited by Chaptal.[18]

In addition to bringing together "officials, scientists, merchants, manufacturers, artisans and inventors," the Society sought to "excite emulation, spread knowledge and support talent." At the first meeting in January 1802, Chaptal was elected president, and he was reelected unanimously each year until his death. The Society contributed to a number of French technological advances and improved techniques with early successes in the perfection of the Jacquard loom for silks in 1808 and the naturalization of the sugar beet. Later recipients of funds from the Society included Louis Pasteur, Alphonse Beau de Rochas, and the Lumière brothers.[19]

From a political perspective, these institutions were a response to the fact that despite all of the benefits stemming from his ideas and programs, there was widespread resistance among entrepreneurs and workers to government intervention that favored either the unfettered rule of market forces or technical innovation. According to Chaptal, Bonaparte was infuriated by such defiance.[20] But thanks to Chaptal's influence, the lack of immediate success in overcoming resistance by government fiat led not to more dictatorial methods but, at least initially, to a renewed effort to co-opt local people and local methods to accomplish the goals of the state.[21] In Chaptal's institutional framework for the French economy, the institutions discussed above emulated the English model of technological development anchored by the Royal Society for the Encouragement of Arts, Manufactures and Commerce (1754) and the network of provincial scientific societies. The need to tackle these issues from outside the state's direct control is also a clear indication of the difficulties Chaptal experienced in creating an institutional environment that would improve technological competence, animate entrepreneurs to innovate, and stimulate the French desire to compete directly with Britain.

Nowhere is this difficulty more apparent than in the legislation regarding the world of work enacted under Napoleon. Chaptal adopted from Smith a dedication to the essential right of individuals to make their own economic decisions, particularly with regard to the disposal of their labor. Unsurprisingly therefore, his conception of labor relations was fundamentally individualistic–on both sides of the bond. The infamous law of 22 Germinal, Year XI (April 12, 1803) outlawed all forms of coalition. This regulation included an explicit prohibition of *patrons* uniting against workers, an interdiction that did not fit the model of laissez-faire Britain in the mid-nineteenth century.

As a result of popular restiveness, the French state was concerned about the excessive mobility of labor that had materialized late in the Revolutionary decade; workers were apt to leave employers who enforced industrial discipline, understood by all to be necessary to maximizing the efficiency of production. A second difficulty was that entrepreneurs, pressed to find competent and disciplined laborers, enticed them away from other entrepreneurs with promises of higher wages and better conditions, and more. Vociferous complaints from all over France prompted the government to intervene as a means of improving French international competitiveness.[22]

On 9 Frimaire, Year XII (December 1, 1803), the *billets de congé* (discharge papers) used under the ancien régime were dusted off and revived as the *livret* (report book) to ensure the "honesty" of workers and to minimize their nomadism, while making it more difficult to poach the workers of another entrepreneur. The livret, issued by the municipality, served as identification papers. All laborers had to present their livrets before beginning employment. The date they began a job was inscribed in the livret by the employer, who then took charge of the document. If the worker wished to leave, the employer had to certify that all financial and work commitments had been fulfilled before filling in the date and returning the livret. The worker then had to inform the mayor or the deputy mayor where he was going next.[23]

The disciplinary apparatus thus deployed by the Napoleonic state was apparent, but repression was not the sole intent. The law instituting the livret delegated the power to adjudicate many kinds of conflicts between employers and laborers to the mayors, making justice more accessible to workers as well as less expensive. To foster conciliation and to minimize the employers' advantage, lawyers were explicitly forbidden to appear before this body. A further step in this direction was the creation of a *conseil de prud'hommes* (arbitration board) on March 18, 1806, composed of both employers and skilled workers to arbitrate conflicts in the Lyon silk industry. Imitated informally throughout France, other *conseils* were established in other industrial centers on June 11, 1809, that had the authority to adjudicate disputes up to 100 francs.

Chaptal was no longer in office when these last measures were enacted, but his activities clearly formed the basis for them. For example, although he saw some clear technical benefits from certain types of worker organization, namely *compagnonnage* (journeymen's organizations) and the *tour de France* (when journeymen traveled the country learning skills and techniques), he believed that workers, particularly skilled workers, should not be permitted to dictate terms to their employers. To that end, he sought to replace the job-placement functions of organized labor by sponsoring the creation of worker employment bureaus in major industrial centers (November 20, 1802). Neither the market nor workers themselves could be trusted to allocate France's precious stock of skilled workers with maximum efficiency and a minimum of social unrest.[24]

Chaptal understood and appreciated the vital role that markets had to play in improving France's industrial competitiveness. He believed that reserving the domestic market to French producers could form the economic basis for international competitiveness just as it did for Britain. In 1802–1803, he intervened repeatedly to raise tariffs. But the domestic market alone would not suffice; Chaptal had an imperial vision of French potential. To realize that potential, France would have to recover its position as a great trading nation. Chaptal established seven trade entrepôts in major transit centers in July 1803. In addition, he revived the great month-long commercial fairs of Saint-Denis and Beaucaire.[25]

Chaptal improved on many of his predecessors' ideas. He revived an earlier French innovation, the industrial exposition, on November 13, 1800, with the announcement that "the government's devotion to this noble cause stems from liberal principles, which have been ignored or neglected [during the Revolution]. Now, this task has become the most important responsibility of the [new] government."[26] With Bonaparte's support, Chaptal intended for industrial expositions to become a regular event to cap the nationwide effort to heighten French competitiveness. Industrial expositions in 1800, 1801, and 1806 were a popular and public means of displaying French successes, informing people of new technologies, and publicizing the work of the new industrial and scientific institutions. Finally, the expositions were also perceived to be an important means of generating orders for entrepreneurs who developed or adopted advanced technology, thereby providing an important profit incentive for innovation.[27]

This résumé of Chaptal's industrial role is far from exhaustive. Scanning the information-filled pages of the *Bulletin de la Société d'Encouragement pour l'industrie nationale* or *Annales des Arts et manufactures ou Mémoires technologiques sur les Découvertes modernes concernant les Arts, les Manufactures, l'Agriculture et le Commerce,* both of which received considerable subsidies from the ministry and were distributed by it, can provide further insight. These journals depict a wide variety of Chaptal's activities ranging from creating prize contests for specific inventions that would improve industrial competitiveness, to announcements of grants to inventors or entrepreneurs who adopted advanced machinery. He targeted certain industries for development, such as woolens, and supported the construction of English-model machines and subsidized their employ with impressive results for French competitiveness.[28]

From an institutional standpoint, Chaptal was unquestionably the father of the nineteenth-century French economy. For a quarter-century, his successors continued his policies and regularly asked his advice.[29] The distinctive deployment of state *dirigisme* melding theory and practice used to further a liberal agenda, yet leaving the government in a stronger position than before it intervened, remained characteristic of the French industrial landscape for many decades. It is this enduring mix of short-term practicality and a longer-term understanding of how industrial "catch-up" must be undertaken on the Continent that is most often overlooked by Anglocentric commentators from Landes to Mokyr to Jacob.

THE SUCCESS OF THE FRENCH PATH TO INDUSTRIAL SOCIETY

What were the effects of the divergent industrial path embarked on by Chaptal and his collaborators? As always, the figures vary widely, but the consensus is that French industry grew relatively rapidly in the period 1815 to 1850 when the era of war ended and the generation raised with the institutions created earlier came of age. Estimates of annual industrial expansion range from 2.5 percent on the low end to 3.4 percent on the high end. Agricultural growth was also strong at 1.2 percent annually in the half-century from 1820 to 1870. Over the long term, France averaged a 1.4 percent annual increase in per capita economic growth between

1815 and World War I.[30] In large part because of a spurt during the *belle époque*, by 1914, on a per capita basis, French economic performance was broadly comparable to that of Great Britain.[31] British per capita income remained higher by about 20 percent, but the gap between Britain and France did not widen even at the height of British industrial dominance in the first half of the nineteenth century.[32] France enjoyed impressive long-term growth, both overall and per capita, particularly in light of its slower population expansion and less lucrative colonial opportunities. French society and its structures changed dramatically as a result of industrialization during this period. France had an industrial revolution, albeit more gradual and less abrupt than Great Britain's. Under Chaptal's stewardship, French short-term industrial performance did not quite equal Great Britain's. This is understandable given the limitations in French factor endowments, domestic markets, imperial advantages, capital stock, and the transport system. When the basis of industrial success changed at the end of the century, France outperformed Britain handily. As a number of historians have observed, the economic structures that made England successful in the first half of the century later held it back.[33] According to Martin Daunton, "Arguably [after 1850], a gap was opening up between the production *institutions* which were developed in Britain—small family firms, a reliance on subcontracting between and within firms, a highly formalized system of collective bargaining—and the needs of production *technology*."[34] On the other flank of the Channel, the flexible web of institutions created by Chaptal and his collaborators and maintained by his successors enabled France to survive English industrial dominance and to take advantage of the opportunities of the Second Industrial Revolution. As contemporary signs in the London Underground remind us, it is important to "mind the gap."

O'Brien and Keyder's analysis of comparative economic performance has, in the thirty years since they published their controversial analysis, garnered widespread support from French economists and historians. Why hasn't this consensus influenced Anglo-American versions of how technological change affects economic performance or comparative studies of the process of industrialization?[35] Viewed from the Continent or indeed from almost anywhere else, the dominant Anglo-American version of comparative industrialization with its emphasis on the cultural aspects of the British model appears terribly parochial. As Peter Mathias pointed out so clearly, British industrialization shifted the context for those who came behind. Following precisely in Britain's footsteps was impossible. Why would anyone expect France or any other country to industrialize on the same pattern as a pioneer, especially when they lack the same mix of resources or expertise and have to contend in international markets with competition from the cradle?[36] It is this commonsense question that gets us to the heart of the matter. As Martin Wiener, W. D. Rubinstein, and Nicholas Crafts, among others, have suggested in one way or another, the underlying question or implicit challenge in this literature is to understand whether and how late-Victorian or post-1945 Britain lost its industrial edge, an issue made more potent in the 1970s and 1980s when a surging Japanese economy made many Americans fearful of eclipse.[37] French performance during the Industrial Revolution must be considered on its own impressive merits, not through the lens of later eras. The French path to industrial society diverged from Britain's, but, over the long term, its seemingly tortuous route produced broadly comparable success.

What made this longer passageway necessary was the impact of the French Revolution and the lingering threat of social revolution. The consequences of this historical distinctiveness help us to understand the wellsprings of a great deal of French particularity in the nineteenth century. They complement our understanding of France's characteristic interaction of economic and scientific institutions, political policies and developments, and the evolution of the long cross-Channel rivalry. An emphasis on alternate pathways and their effects should also frame comparisons to nineteenth-century industrialization in other continental countries, especially the coalescing nation-states of Germany and Italy.

The practical effects of the "threat from below" were long-lasting and profound. They received powerful stimulus in the Revolutions of 1830, 1848–1851, and 1870–1871, not to mention lesser-known but significant events such as the wave of labor unrest, destruction of property, and violence that shook Normandy, France's industrial heartland, in 1825.[38] In such a political environment, the French "underinvestment" stressed by Paul Bairoch and the type of entrepreneurialism denigrated by David Landes made good sense. The slower and more dispersed pattern of French industrialization was a major factor in the low rate of urbanization in France and encouraged the persistence of regional economies and identities. Despite a relatively powerful state structure, as Eugen Weber pointed out more than thirty years ago, the French did not seek systematically to overcome regionalism until the late nineteenth century. How can such "failures" be understood other than as an unwillingness to facilitate social and political mobility and the attendant threat of revolution, especially given how efficient the French labor market was during this era?[39] That these efforts followed the creation of a national transportation network by at least a generation and paralleled the efforts of other continental states are issues that merit further research.

French population growth lagged dramatically behind its peers. Unlike its neighbors, France did not export its people. As an industrial society, France relied heavily on easily dominated immigrants to make up labor shortfalls. Wages remained low, and immigration generally facilitated the success of a divide-and-conquer strategy in dealing with the working classes. Demand in the national market never performed the same function in France that it did in Great Britain, Germany, and the United States. The "rational" economic assumptions underlying the standard-of-living debate so important to analysts of the British Industrial Revolution cannot explain French behavior. The political possibilities inherent in a revolutionary tradition combined with a powerful nationalism and diminished income inequality kept the French at home despite the discipline exercised by a seemingly omnipresent central state. The contradictions and limitations of French economic and political development during the process of industrialization were the result of far more than purely economic or technological factors.[40]

As many commentators have noted, Great Britain avoided a political revolution through industrialization. France's political restiveness greatly affected its process of industrial development. If Peter Mathias's question "First, and, therefore, unique?"[41] is applied to the French Revolution rather than British industrialization, then the divergent paths to industrial society followed on either shoulder of the Channel seem commonsensical rather than irrational or culturally determined. If we take the notion of political economy seriously, then perhaps we need to refocus attention on the subject raised with such despair by E. P. Thompson, namely, how Great Britain came to dominate its working classes during early industrialization.[42] Given the impressive performance of the French economy during the nineteenth century and the depth of the attendant structural changes, we must recognize that it was Britain, not France, that truly avoided a revolution.

NOTES

1. The themes discussed in this chapter are discussed at length in Jeff Horn, *The Path Not Taken: French Industrialization in the Age of Revolution, 1750–1830* (Cambridge, MA: MIT Press, 2006), where the full range of archival and secondary sources that support this interpretation can be found.

2. Patrick K. O'Brien and Caglar Keyder, *Economic Growth in Britain and France, 1780–1914: Two Paths to the 20th Century* (London: Allen & Unwin, 1978).

3. Charles Sabel and Jonathan Zeitlin, "Historical Alternatives to Mass Production: Politics, Markets and Technology in Nineteenth-Century Industrialization," *Past and Present* 108 (1985), 133–176.

4. Margaret C. Jacob, *Scientific Culture and the Making of the Industrial West* (New York: Oxford University Press, 1997); David S. Landes, *The Unbound Prometheus: Technological Change and Industrial Development in Western Europe from 1750 to the Present* (Cambridge: Cambridge University Press, 1969); Joel Mokyr, "Editor's Introduction: The New Economic History and the Industrial Revolution," in *The British Industrial Revolution: An Economic Perspective*, 2nd ed., ed. Joel Mokyr (Boulder, CO: Westview, 1999), 1–127.

5. Kenneth Pomeranz, *The Great Divergence: China, Europe, and the Making of the Modern World Economy* (Princeton, NJ: Princeton University Press, 2000).

6. Jeff Horn, "Machine-Breaking in England and France during the Age of Revolution," *Labour/Le Travail* 55 (Spring 2005), 143–166.

7. See Charles Coulston Gillispie, *Science and Polity in France: The Revolutionary and Napoleonic Years* (Princeton, NJ: Princeton University Press, 2004).

8. T. S. Ashton, *The Industrial Revolution, 1760–1830* (Oxford: Oxford University Press, 1948).

9. On Chaptal's importance and the cultural components of his industrial vision, see Jeff Horn and Margaret C. Jacob, "Jean-Antoine Chaptal and the Cultural Roots of French Industrialization," *Technology and Culture* 39:4 (1998), 671–698. Additional material that diverges from Jacob's interpretation is from *The Path Not Taken*, chap. 6.

10. Joel Mokyr sees Chaptal as a paradigmatic figure of the "industrial Enlightenment" that bridged the Scientific and Industrial Revolutions: *The Gifts of Athena: Historical Origins of the Knowledge Economy* (Princeton, NJ: Princeton University Press, 2002), 28–77, esp. 31, 36, 52, 64, and 74. The reference to Chaptal as one of the paradigmatic figures of the industrial Enlightenment is from a personal communication in February 2003 that Mokyr has generously allowed me to cite.

11. On Chaptal's life and career, see Michel Péronnet, ed., *Chaptal* (Paris: Privat, 1988), and the sources cited in Horn and Jacob, "Jean-Antoine Chaptal."

12. Jean-Antoine Chaptal, "Quelques réflexions sur l'industrie en général, à l'occasion de l'exposition des produits de l'industrie française en 1819" (Paris: Corréard, 1819), cited in Péronnet, ed., *Chaptal*, 243–246.

13. Jean-Antoine Chaptal, *De l'industrie française*, ed. Louis Bergeron (Paris: Imprimerie nationale, 1993 [1832]), 364, 370–372.

14. Jean Pigeire, *La vie et l'oeuvre de Chaptal (1756–1832)* (Paris: Ses, 1932), 275.

15. Jean-Antoine Chaptal, *Circulaire aux Préfets des départemens*, (29 Thermidor, Year XIII [August 17, 1805]), Archives Départementales de la [henceforth AD] Somme, M 80028, and the *Arrêté portant organization d'une École d'arts et métiers à Compiègne*, in the *Bulletin des Lois* 7: 220–262, 3rd series (Year XI [1802]), 484–494; Le Brun, *Notice sur les Écoles impériales d'Arts et Métiers*, 1863, Archives Nationales de France [henceforth AN], F17 14317; Pigeire, *La vie et l'oeuvre de Chaptal*, 239, 270–275, 350–354. For a different interpretation of French curricula, see Margaret C. Jacob and Larry Stewart, *Practical Matter: Newton's Science in the Service of Industry and Empire, 1687–1851* (Cambridge, MA: Harvard University Press, 2004).

16. Charles R. Day, *Education for the Industrial World: The Ecoles d'Arts et Métiers and the Rise of French Industrial Engineering* (Cambridge, MA: MIT Press, 1987).

17. Charles Ballot, *L'Introduction du machinisme dans l'industrie française* (Geneva: Slatkine, 1978 [1923]), 30–32, and Péronnet, ed., *Chaptal*, 196–204.

18. Chaptal claimed credit for founding the institution. Jean-Antoine Chaptal, *Mes souvenirs sur Napoléon* (Paris: Plon, 1893), 100.

19. For the early activities and statutes, see E. J. Guillard-Senaineville, *Notice sur la Société d'Encouragement pour l'industrie nationale*, 1818 [1814], AN F12 2333. See also, Péronnet, ed., *Chaptal*, 191–195.

20. Chaptal, *Mes souvenirs*, 274–277.

21. See Péronnet, ed., *Chaptal*, 244. On the initial willingness of the Bonapartist regime to conciliate its opponents, particularly workers, see Jeff Horn, "Building the New Regime: Founding the Bonapartist State in the Department of the Aube," *French Historical Studies* 25:2 (2002), 225–263.

22. Leonard N. Rosenband's important article "Comparing Combination Acts: French and English Papermaking in the Age of Revolution," *Social History* 29:2 (2004), 165–185, situates such movements in their proper international and sectoral context.

23. Denis Woronoff, *Histoire de l'industrie en France du XVIe siècle à nos jours* (Paris: Seuil, 1994), 199; Chaptal, *De l'industrie*, 438–449. The quote comes from Claude-Anthelme Costaz, *Mémoire sur les moyens qui ont amené le grand développement que l'industrie française à pris depuis vingt ans* (Paris: Firmin Didot, 1816), 16–17.

24. The two preceding paragraphs are based on Chaptal, *De l'industrie*, 438–444; Costaz, *Mémoire sur les moyens*, 17–19; and Pigeire, *La vie et l'oeuvre de Chaptal*, 170.

25. Pigeire, *La vie et l'oeuvre de Chaptal*, 404–407; Chaptal, *De l'industrie*, 463–500.

26. Jean-Antoine Chaptal, *Circulaire*, 1 Floréal, Year IX (April 21, 1801), AD Loire 75M 1.

27. Chaptal, *Circulaire*, 1 Floréal, Year IX.

28. Chaptal, *De l'industrie*, 256–57; Horn and Jacob, "Jean-Antoine Chaptal."

29. Pigeire, *La vie et l'oeuvre de Chaptal*, 424–425.

30. Jean-Charles Asselain, *Histoire économique de la France du XVIIIE siècle à nos jours*, vol 1., *De l'Ancien Régime à la Première Guerre mondiale* (Paris: Seuil, 1984), 130; Patrick Verley, *La Révolution industrielle* (Paris: Gallimard, 1997), 317; Maurice Lévy-Leboyer, "Capital Investment and Economic Growth in France, 1820–1930," in *The Cambridge Economic History of Europe* vol. 7, part 1, *The Industrial Economies: Capital, Labour, and Enterprise: Britain, France, Germany, and Scandinavia*, ed. Peter Mathias and Michael M. Postan (Cambridge: Cambridge University Press, 1978), 267; Jean-Pierre Daviet, *La société industrielle en France, 1814–1914: Productions, échanges, représentations* (Paris: Seuil, 1997), 17.

31. For a recent economic analysis, see André Louat and Jean-Marc Servat, *Histoire de l'industrie française jusqu'en 1945: Une industrialisation sans révolution* (Poitiers: Bréal, 1995), 280–281.

32. Philip T. Hoffman and Jean-Laurent Rosenthal, "New Work in French Economic History," *French Historical Studies* 23:3 (2000), 451; Christian Morrisson and Wayne Snyder, "The Income Inequality of France in Historical Perspective," *European Review of Economic History* 4 (2000), 72; François Crouzet, *Britain Ascendant: Comparative Studies in Franco-British Economic History*, trans. Martin Thom (Cambridge and Paris: Cambridge University Press and Éditions de la Maison des Sciences de l'Homme, 1990), 342.

33. See, for example, William Lazonick, "What Happened to the Theory of Economic Development?" in *Favorites of Fortune: Technology, Growth, and Economic Development since the Industrial Revolution*, ed. Patrice Higonnet, David S. Landes, and Henry Rosovsky (Cambridge, MA: Harvard University Press, 1991), 267–296, esp. 281–286.

34. Martin J. Daunton, *Progress and Poverty: An Economic and Social History of Britain, 1700–1850* (New York: Oxford University Press, 1995), 564–565. Even a British triumphalist like François Crouzet agrees, although he shifts the date for the institutional change to after 1870. See his "France," in *The Industrial Revolution in National Context: Europe and the USA*, ed. Mikulás Teich and Roy Porter (Cambridge: Cambridge University Press, 1996), 59–60.

35. See, for example, the lack of reference to France in Alfred D. Chandler, Jr., "Creating Competitive Capability: Innovation and Investment in the United States, Great Britain, and Germany from the 1870s to World War I," in Higonnet, Landes, and Rosovsky, eds., *Favorites of Fortune*, 432–458, and the more negative evaluation of French performance in Colin Heywood, *The Development of the French Economy, 1750–1914* (Cambridge: Cambridge University Press, 1992).

36. Peter Mathias, *The Transformation of England: Essays in the Economic and Social History of England in the Eighteenth Century* (London: Methuen, 1979), 3–20.

37. Nicholas Crafts, "Forging Ahead and Falling Behind: The Rise and Relative Decline of the First Industrial Nation," *Journal of Economic Perspectives* 12 (Spring 1998), 204–207; W. D. Rubinstein, *Capitalism, Culture & Decline in Britain, 1750–1990*

(London: Routledge, 1993); and Martin Wiener, *English Culture and the Decline of the Industrial Spirit, 1850–1890* (Cambridge: Cambridge University Press, 1981).

38. On events in Normandy and the negative effect on local industry see Horn, *The Path Not Taken*, chap. 8; William M. Reddy, *The Rise of Market Culture: The Textile Trade and French Society, 1750–1900* (Cambridge and Paris: Cambridge University Press and Éditions de la Maison des Sciences de l'Homme, 1984), 113–125.

39. Paul Bairoch, *Révolution industrielle et sous-développement* (Paris: Éditions de l'EHESS, 1963); David S. Landes, "French Entrepreneurship and Industrial Growth in the Nineteenth Century," *Journal of Economic History* 9 (1949), 45–61; Eugen Weber, *Peasants into Frenchmen: The Modernization of Rural France, 1870–1914* (Stanford, CA: Stanford University Press, 1976); Yves Lequin, "Labour in the French Economy since the Revolution," in Mathias and Postan, eds., *Industrial Economies,* 307; Hoffman and Rosenthal, "New Work in French Economic History," 452.

40. The political significance of the "threat from below" in France provides the context for Sabel and Zeitlin's argument about the emergence of flexible means of mass production.

41. Mathias, *The Transformation of England*, 14.

42. Edward P. Thompson, *The Making of the English Working Class* (New York: Vintage, 1963), 605.

6

The Political Economy of Early Industrialization in German Europe, 1800–1840

Eric Dorn Brose

The four decades after 1800 represent a critical period in the economic and technological history of Prussia—the era of early industrialization. For many contemporaries, however, this era was confusing and, in the end, often disappointing. Indeed, whether bureaucrats or businessmen, few individuals accurately predicted the country's industrial future. Some, like the technocrat Peter Beuth, wanted an "aesthetic industrialization" replete with country factories, neoclassical in design, producing fabrics, prints, and other items with ancient motifs. The captains of the government mining corps (*Oberberghauptmannschaft*) adhered to centuries-old techniques, confident that these traditional ways promised the most prudent path of industrial development, while the leaders of the state-run Overseas Trading Corporation (*Seehandlung*) hoped to imbue industrialization with Freemasonic principles of harmony and human perfectibility. All of these visions would be lost and buried as Prussia entered the frenetic era of railroads, heavy industry, and rapid growth (in these sectors) during the 1840s and 1850s. There were those, however, who had seen this coming, like industrialist Friedrich Harkort, who as early as 1825 envisioned the "smoking colossuses" of the yet-unborn railway age.[1]

Given the usual insecurities felt throughout societies experiencing rapid change, it should come as no surprise that disappointed opponents of Prussia's modern industrialization tried to stop it in its tracks. All too often, unfortunately, histories of this period tend to gloss over the debates and struggles that contemporaries waged, or simply skip over them entirely, employing a type of *post hoc ergo propter hoc* logic: Prussia industrialized; therefore it must have been inevitable. As I have demonstrated elsewhere, however, close scrutiny of the history of this country's political economy reveals how real the possibility of snuffing out or seriously inhibiting industrialization remained until the 1830s.[2]

This chapter builds on my earlier discussion by comparing the Prussian experience with industrialization debates that raged in other parts of "German Europe," that large portion of the Continent situated roughly within the area Aachen-Berlin-Freiburg-Vienna.[3] When taken together with the Prussian story, what do the histories of Saxony, Baden, Württemberg, Bavaria, Austria, and other German states tell us about the political economy of early industrialization? I argue that three critical turning points were reached and passed as industrialization-related crises were affirmatively resolved: a crisis over licensing of factories in the 1810s and 1820s, a crisis over establishment of a German customs union in the early 1830s, and a crisis over the concessioning of railroads in the mid- to late 1830s. Industrialization moved forward after the pro-industrial resolution of these crises, but the outcome was never a foregone conclusion. Finally, I widen the focus by briefly comparing the politics of industrialization in German Europe with circumstances in China, Russia, and France and make one comparative comment on Britain, the United States, and Japan.

The Licensing of Factories

British competition represented the greatest force pushing German Europe in the direction of industrialization in the early nineteenth century. By the 1820s, a new technological system featuring fossil fuels, reciprocating steam engines, coke blast and puddling furnaces, mechanized spinning and weaving of textiles, and lead chambers for mass-producing sulfuric acid had revolutionized the means of production in the "first industrial nation."[4] These developments brought cheap British iron, textiles, and chemical products spilling into Central European markets. In 1826, for example, British coke pig iron undercut Rhenish charcoal pigs by 25 percent in the Rhineland, while British exports of cotton yarn enjoyed a 43 percent advantage in the same market—and price differentials were widening. In 1829, English worsteds sold in Elberfeld at a price equivalent to that paid by domestic manufacturers to their spinners.[5] "It is impossible to compete with this country,"[6] said one prominent industrialist. Unless superior quality insulated them from this stiff competition, German businessmen found it hard to survive.

Although pleas for tariff protection abounded, the most enterprising and insightful German businessmen realized that they would have to adopt British techniques. Smuggling—one stark fact of business life that receives little attention in the literature for the obvious methodological reason that its secretiveness leaves historians with little data—certainly contributed to this realization. Britain negotiated low transit duties with German states like Hanover, Brunswick, Oldenburg, Hesse-Kassel, Nassau, and Saxony in order to move goods illegally into neighboring states. Thus, the British representative at the Diet of the German Confederation in Frankfurt/Main referred openly to these transit deals as "affording immense facilities for carrying on the contraband trade in the dominions of Prussia, Bavaria, Württemberg, and Hesse-Darmstadt."[7] From the South German states, contraband moved illegally into Austria, making that country's prohibitive tariff wall much more porous than pro-industrial officials desired. Self-help measures to adopt British technology proved difficult, however, because British efforts to block the export of machines and the emigration of skilled workers complicated schemes to hire British mechanics and engineers to install and maintain little-understood machines. The eccentricity of British mechanics, exacerbated by the language barrier, created additional headaches.[8] So German manufacturers turned to their governments for help, asking them to obtain and spread information about the revolutionary new technologies evolving in Britain, teach subjects like mathematics and science in schools, provide tax breaks and subsidies, improve transportation, rescind restrictive banking laws, modify regulations against deep mining, adopt liberal licensing policies for new businesses, and stifle the antitechnological efforts of the old guilds.

Pro-industrial advocates ran headlong, however, into a determined opposition. Guild spokesmen argued that it was wrong to abandon corporations whose rights and privileges were centuries old and provided stability in turbulent times. Social observers of both liberal and conservative persuasion agreed. Visitors to England witnessed pollution, lower-class squalor, and disorder in the cities. Would it not make more sense to shore up the old social order rather than sacrifice it for such unfortunate and precarious circumstances? Soldiers also cast a wary eye at a British-style industrial future. Long hours in unsafe factory conditions weakened the mind and body, thereby limiting the potential of these wretched souls as fighting men. Opponents of industrialization therefore pressured governments to erect tariff barriers and catch the smugglers as means to protect struggling craftsmen, to enforce guild monopolies on the training of labor and the production and pricing of goods, and to refuse licenses to factories—that is, to ban them.

Bavaria provides an excellent example of the kind of policy vacillations that could result from such contradictory pressures.[9] Between 1804 and 1811, the state removed the right to license new businesses from guilds, but then reversed itself from 1811 to 1825, with the result that new ventures, mostly guild shops, expanded a mere 3 percent in fourteen years. Liberal policies were readopted in 1825, facilitating a 22 percent increase in factories in five years. The violence of the European Revolutions of 1830 affected another *volte face*, however, as Bavaria now turned its back on the new industrial world emanating from England. "There you will see living machines driven into battle with their industrial masters," said one shocked minister. "You will see the opposing camps of owners and workers—and in the background blood and cannon fire."[10] Between 1815 and 1840, the state licensed only sixty-two factories—most during the quinquennium after 1825—but in 1840 there were actually fewer factories than in 1830.

Württemberg and Saxony offer a subtle but significant distinction from Bavaria.[11] Both of these middle-sized states created a very inhospitable environment for modern industry, which they saw as a trendy and unreliable source of livelihood that did not compare favorably with the traditional ideal of handicraft. Accordingly, factories were taxed more heavily than guild operations, and entrepreneurs' pleas for protective tariffs, tax breaks, and subsidies for new machinery, elimination of internal tariffs and road tolls, and abolition of still valid guild privileges were ignored. "Out of concern that certain [traditional] things might be destroyed," observed one critic of Saxon policy, "everything, even the completely antiquated, was retained."[12] However, out of the need for revenue, ministers in Stuttgart and Dresden collected fees during the 1810s and early 1820s for the licensing of scores of factories, whose numbers nearly doubled. Guild privileges, in other words, were also ignored. But officials clearly had their qualms about the allegedly ugly, unaesthetic aspects of the newcomer ventures. In Württemberg, for instance, the state held special exhibits where machines, "not actually art in a higher sense," were displayed with objects of "high art."[13] Only in 1828 did Württemberg and Saxony change these attitudes and adopt somewhat more sympathetic pro-industrial policies that accelerated the pace of licensing—Württemberg even abolished guild laws. Counting both large and small-scale operations, both states counted around three hundred factories in 1830.

As in other states, factory licensing policy in Baden provides insights to the development, or lack thereof, of industry.[14] Officials in Karlsruhe remained basically skeptical about abandoning the guilds and their preindustrial order for the brave but uncertain world of factories and machines. Therefore local officials tied to the guilds retained the right to rule on petitions for new factories—with the result that few found approval. Making matters harder for businessmen, Baden imposed discriminatory taxes on factories, which were raised even higher in 1825, and then offered no tariff or patent protection. All too often, moreover, the aid that industrialists received from bureaucrats was unsolicited and unwanted. Beginning in 1821, for instance, the state sponsored high art and industrial artistry exhibits designed to raise the cultural level of the unpleasing new world of business. Small wonder that the number of factories rose at a snail's pace from 146 in 1809 to 153 in 1829. Factories that went bankrupt were barely outbalanced, in other words, by new ventures. Baden represented something very close to the guildsman's ideal.

Most of the smaller German states, in fact, could make the same claim. Officials in Nassau and Hesse-Kassel remained so sure about the evils of industry that factories almost never received licenses. Hesse-Darmstadt waxed somewhat more liberal, but even there, conservative rulers gave approval only rarely and grudgingly, interpreting periodic recessions as proof that industrialization was unstable, unreliable, and unworthy of significant state aid. Elsewhere, from the Mecklenburgs, Hanover, Brunswick, and the Thuringian states to city-states like Frankfurt/Main or the former Hanseatic towns, reactionary, anti-industrial patterns persisted as the norm. This backlash of reactionary conservatism was particularly evident in the old merchant citadels of Bremen, Hamburg, and Lübeck, for France had annexed them in 1810, transplanting the Empire's antiguild, free enterprise legislation. The collapse of French power in Germany after 1813 was followed immediately, however, by a return to the old ways. As Hans-Ulrich Wehler observed in general about these smaller German states, "every in-depth socio-historical investigation of the upper and old urban bourgeoisie confirms what widespread tradition-blessed obstinacy there was in the effort to keep economic activity within customary bounds."[15] If German Europe were to avoid becoming an economic backwater, falling further and further behind Britain and France, the decision lay seemingly with Prussia and Austria.

Austria struggled with the question of mechanization for many decades.[16] From 1794 to 1809, decrees barred the importation or use of flax spinning machines, forbade the opening of new factories in Vienna and other cities, and encouraged the dispersion of existing urban plants to the countryside where workers would represent less of a revolutionary threat to one of the most autocratic regimes in Europe. Finally, in 1809, a pro-industrial faction in the bureaucracy convinced Kaiser Francis II of the financial merits of lifting the antifactory ordinances. Liberals strove to maintain prohibitive tariff walls to create a huge internal free trade zone incorporating the Austrian, Bohemian, Moravian, and Polish provinces, as well as Hungary, the Tyrol, and northern Italy. This protected market would stimulate investment in newly licensed state-of-the-art factories. Supporters of industrialization also facilitated technological innovation with direct subsidies to innovative entrepreneurs. A technological institute disseminated information on new devices, and a permanent machine exhibit in Vienna, which eventually numbered twenty thousand pieces, encouraged adoption of new techniques.

These industrializing policies triggered a powerful backlash from guildsmen throughout the empire who seethed with anger over the creation of factory competition that violated their legal privileges. During the late 1810s and early 1820s, they succeeded in creating a bureaucratic, obstructionist nightmare for aspiring factory owners as each licensing petition was challenged and investigated. In 1822, the handicraft lobbies convinced the kaiser that industrial technology would impoverish small shop owners, undermine a still important part of the tax base, and perhaps provoke riots. After a near halt to concessions for two years, Francis agreed in 1824 to ban all new factories. The decree was enforced for three years. Finally, in 1827, pro-industrial ministers, incredulous at the guilds' apparent death blow to modern industry, managed to lift the ban.

In the end, the guilds lost their bid to halt Austrian industry in its tracks. Indeed, by 1830, the empire's output of pig iron exceeded Prussia's and amounted to about three-quarters as much as all Germany to the north. Austria's 115 large cotton-spinning factories dwarfed those of any other German kingdom and had 25 percent more spindles than the rest of Germany. Despite rampant smuggling, moreover, protection and the government's promotional efforts facilitated qualitative advance, for the empire's spinning mules, steam engines, and blast furnaces more than held their own with everything north of the border. Such progress certainly sheds a questionable light on what David Good describes disapprovingly as "the reigning interpretation of the Habsburg Monarchy's economic failure."[17] To repeat the point, however, this progress was not inevitable; political battles had to be won first.

Prussia too fought over the costs and benefits of modern industry.[18] Free enterprise ordinances of 1810–1811 swept away the old mercantilist policy of supporting economic growth with direct subsidies, monopolistic charters, and prohibitive tariffs. The legal privileges of the guilds were also revoked. The new programs aimed to push existing plants out of the towns into the countryside where cheap waterpower, raw materials, and labor would allow Prussian entrepreneurs to compete without significant state aid. With the exception of the left-bank Rhineland, where French free enterprise laws had been on the books since the 1790s, it took until about 1818 to install the new system.[19] Bureaucrats like Peter Beuth offered free advice and information on the latest foreign technologies and the most aesthetic surroundings for the new machinery. The government also eliminated internal tolls and invested in road construction. External tariffs were moderate: significantly higher than most of the smaller and middle-sized states, but much lower than Austria, Russia, France, and England. Patents were hard to receive, furthermore, and subsidies next to impossible. Guild privileges seemed to be a relic of the past.

In the years after 1815, however, a determined guild initiative backed by powerful conservatives around King Frederick William III seriously threatened free enterprise. By 1818, the king had been moved by these entreaties to order his chancellor to gut the earlier legislation. Intrigues and counterintrigues kept the issue in doubt for the next decade. In 1824, for instance, Frederick William ordered the acceleration of "guild reforms," but seemed to be leaning away from his anti-industrial decision due to the state's growing dependence on the increasing tax income generated by new industrial ventures. The state would modify existing laws in accordance with the "reasoned wishes" of conservatives, but budget concerns did not allow it "to take back [free enterprise] institutions on which it relies for revenue without replacing them with others which guarantee the same financial results."[20] But as I have argued elsewhere, it was not until 1828 that the conservative campaign had lost all chance of changing the king's mind.[21]

Thus, German Europe nudged toward industrialization as market forces pressed from below and political actors resolved critical political controversies from above. Although most German states opposed modern industry, several crucial states in the end did not: Austria by 1827 and Prussia, Württemberg, and Saxony by 1828. The 1820s, a decade usually passed over as insignificant, are in fact highly significant in the history of Germany's embryonic Industrial Revolution.

THE ZOLLVEREIN

The story of the creation of the German Customs Union, the Zollverein, need not be repeated here in great detail.[22] Suffice it to say that liberal ministers in Prussia began the process with a tariff reform in 1818 that erected a moderate external tariff and eliminated internal tolls, thereby facilitating trade between the Brandenburg heartland and the isolated provinces of Rhineland-Westphalia. There followed six free-trade treaties with adjacent small states as well as foreign states landlocked by Prussia.

A more significant treaty with Hesse-Darmstadt in 1828 completed the free trade bridge from eastern Prussia, across enclaves and neighboring states, to Rhineland-Westphalia. Bavaria and Württemberg joined the Prussian trade bloc in 1829 and Hesse-Cassel and Saxony-Weimar in 1831. After agreements were reached with the centrally positioned Kingdom of Saxony and the remaining Saxon duchies, a Zollverein of eighteen contiguous states that commercially united most of Germany north of Austria went into effect on New Year's Day 1834. From its inception—and even before—the Zollverein had a significant impact on industrialization.

These commercial developments were by no means inevitable. In fact, they were the result of a complex political struggle within and between Prussia and Austria, and between these states and other trading interests in lands of the so-called Third Germany competing with (and wary of) Vienna and Berlin. The core lands of the Third Germany stretched from Baden through the Hessian states to Württemberg and beyond to Bavaria. At the center, King William of Württemberg, a monarch who agreed to parliamentary reforms in 1818 just as Baden had done that year and Bavaria in the next, had wanted these states to break free from the control of the powerful absolute monarchies of Prussia and Austria, only to see these plans crushed by the reactionary Carlsbad Decrees of 1819 that brought the thirty-six-member German Confederation established in 1815 back into the conservative wake of Vienna and Berlin. Formation of a Third German free trade bloc, however, represented one means to revive this agenda. It came closer to realization in 1828 when Stuttgart and Munich agreed to unite their tariff systems with free-flowing trade across their common borders. The extension of this agreement to neighboring states came unraveled only weeks later, however, when Hesse-Darmstadt, desirous of export markets in Prussia for Hessian wine and farm products, but also secretly (and correctly) assuming that Prussia would become the dominant political and military force in the Confederation, joined the Prussian bloc. This unpleasant surprise essentially forced Württemberg and Bavaria to abandon the commercial portion of the Third Germany agenda. The lucrative allure of free access to the large Prussian market convinced both states to follow Hesse-Darmstadt's lead in 1829.

Contrary to the longstanding interpretation, the Prussian trade policy agenda involved more than fiscal improvements and diplomatic leverage against Austria.[23] It originated with economic and political liberals who were motivated initially by the hope that increased commerce would contribute to aesthetic and rural industrialization, and later, with the appointment of Finance Minister Friedrich von Motz in 1826, the knowledge that an expanding trading bloc would promote modern "bigger industry" and "superior fabrication"[24] as well as transportation improvements like railroads. Much better known is the fact that Motz and his allies also saw the critical treaties with Hesse-Darmstadt, Württemberg, and Bavaria (1828–1829) as an excellent means to drive a wedge between Austria and a North Germany controlled by Prussia. The finance minister thought war with Austria a distinct possibility in the late 1820s; if it came, he wanted non-Austrian Germany on Berlin's side, and commercial treaties could facilitate such cooperation. There seems to be considerable truth to Millward and Saul's statement, therefore, that "for all its economic importance the creation of the Zollverein depended more on political pressures than on economic ones and the long history of its successful completion is more meaningfully told by historians of diplomacy than by economic historians."[25]

Clemens von Metternich, the arch-conservative first minister to Kaiser Francis II of Austria, saw right through the Prussian intrigue, which worried him all the more because his spies reported that the economic liberalism of the free traders in Berlin was coupled with a liberal political agenda: they wanted to promote parliamentarization throughout a Germany largely dominated by Prussia.[26] In 1829, therefore, Metternich urged his emperor to discard Austria's prohibitive tariff system in order to be able to compete with Prussia for the economic, political, and perhaps military favor of the other German states. The veteran schemer clearly wanted to integrate the Austrian economy more closely with the rest of German Europe, but each state would stand alone with its own moderate tariffs rather than join Prussia's free-trade bloc. With each tub on its own bottom, Austria could better manipulate German affairs. The plot failed, however, because the kaiser yielded to the entreaties of Austrian industrialists who had no desire to risk reduced domestic sales for uncertain export gains against other German states.

But Metternich still had an ace in the hole: his allies at the court of Frederick William III of Prussia. Indeed, to these conservative politicians, all liberal schemes to link Germany politically, economically, and militarily to Berlin were anathema, for it would all happen at the expense of Austria, the enemy of parliaments and free trade. After the Revolutions of 1830 struck fear in the minds of German monarchs, especially in Prussia and Austria, the conservatives mounted a concerted effort to convince Frederick William to sack the liberals and roll back commercial treaties whose transparent political and military agenda so clearly threatened the conservative monarchical principle. A few high-ranking officers who had promoted parliamentary ideas in Prussia and southern Germany were indeed demoted or transferred. But a king who liked to play the conservatives and liberals off against each other had no incentive to fulfill the entire conservative program. The result was that Motz's successor in the Finance Ministry, Georg Maassen, managed to convince his sovereign that completion of the Zollverein would be beneficial to the economy, and hence to the overall welfare of the kingdom. So Frederick William signed the last set of treaties. It was a typical balancing act for him, but one that had far-reaching economic repercussions.

Economic historians of recent decades are returning to the traditional view that the Zollverein contributed mightily to German industrialization. Contemporaries like the famous proponent of industrialization, Friedrich List, had trumpeted this thesis, one repeated by late-nineteenth-century writers like Heinrich von Treitschke and Gustav Schmoller. It had grown into the orthodox view by the 1960s.[27] Long before this, however, skeptics had begun to swing the historiographical pendulum in the opposite direction. John Clapham, for instance, rejected "the fallacious argument" of Zollverein-induced industrialization: "Post hoc, ergo propter hoc—Germany began to prosper about 1835; therefore the events of 1834 caused her prosperity."[28] More recently, Rolf H. Dumke, on the basis of in-depth statistical analysis, denied that trade accelerated or industrial investment took off in the mid- to late 1830s.[29] But within a short time of the appearance of his antithesis in 1984, a new synthesis emerged as historians pointed to two significant longer-term effects. First, the creation of a common external tariff and the distribution of tariff receipts to member states soon pressured participating governments to standardize their currencies.[30] The Munich Coinage Treaty of 1837 and the Dresden Coinage Convention of 1838 fixed the silver content of South German gulden coins and made these coins legal tender throughout the Zollverein. The latter treaty linked guldens to the Prussian thaler at a fixed silver exchange rate expressed in so-called Cologne fine marks. Furthermore, the Dresden agreement also introduced a new common silver coin of the Zollverein worth two thaler (or three and a half gulden). The simplifying, transaction-cost-reducing effects were something akin to recent developments in the European Union. While monetary unification advanced, a second significant result of the Zollverein became evident: it promoted railroad-building schemes. Whether trade actually grew rapidly after

1834—and even Dumke admits that the elimination of tariff walls meant the disappearance of government data on the volume of inner-German trade, and hence a more difficult task for historians to prove anything one way or the other—contemporaries nevertheless predicted and eagerly anticipated positive benefits from the creation of a huge free trade zone. Germans assumed that trade would rapidly intensify and investments in new expanded plants would multiply as economies of scale were achieved. Moreover, the remunerative advantage of having improved transportation to capture a bigger share of business along yet-to-be-determined inner-German trade routes would greatly increase. Thus the day of railroads dawned before New Year's Day 1834, for railroad building schemes, wrote Hans-Ulrich Wehler, "fed on the expectation of an expanding volume of trade and a concentration of business contacts."[31] Or, as Friedrich List put it at the time, the railroads and the Zollverein were Germany's "Siamese Twins."[32]

As Richard Tilly and Rainer Fremdling have demonstrated, railroad construction, the "leading sector" of German industrialization, rapidly transformed the country in modern ways.[33] In the 1840s alone, for example, agriculture's share of net investment declined from 57.8 percent to 28.6 percent as construction, transport, and industry increased proportionately. Within the industrial sector, whose share of net investment rose from 2 to 3 percent to 13 percent, the share of heavy industrial producer goods (e.g., metals, fuels, and industrial leather) grew from 8 to 16 percent.

The dramatic nature of the transition and changeover to modern times can be all the more appreciated if one focuses on technology. It is significant, furthermore, that these technical changes first became discernible in the immediate aftermath of the Zollverein and the first railroad line concession petitions as novel productive forces tooled up to better supply millions of tons of coal and iron. Indeed, as would seem to be reflected in the heavy industry investment figures cited above, deeper, more ambitious mine shafts powered by bigger steam engines, ingenious new coking ovens, heavy-duty lathes and milling machines, and taller blast furnaces using coke iron had begun to appear already in the mid-1830s.

THE RAILROADS

List's "Siamese Twins" would not have counted for much, however, if Germans had continued to place political barriers in the path of the exciting new transportation technology. As with factory licenses in the 1820s and the unfolding of the Zollverein in the late 1820s and early 1830s, in fact, there was a critical political dimension to the coming of the railroads—they were not inevitable, but rather a matter of choice—and for many years, the answer was no.

This is not the appropriate forum for a full discussion of the political blockages to Prussian railroad development.[34] Suffice it to say here that as early as the late 1820s, officials considered proposed railroad lines in Westphalia, the Ruhr, and the lower Rhineland. The lines were approved, but the lack of state help nixed most of these ventures. By 1835, even more ambitious projects were under bureaucratic scrutiny: railroads linking Cologne and Aachen, Cologne and Minden, Hamburg and Berlin, and Potsdam and Berlin. Official councils were bitterly divided over the pros and cons of railroads, however, with the majority leaning against. Civilians like Peter Beuth feared that railroads would divert investment monies from textiles and aesthetic country factories, for example, and many in the mining corps expressed alarm that railroad-induced demand for coal would deplete the nation's mineral reserves. Many army officers also worried that the French Army would seize the lines and strike deep into the heartland of a surprise-attacked Prussia. For three years, in fact, these worries halted Prussian railroad development as factions and parties in and out of government fought and schemed. Finally, in late 1838, with military leaders increasingly of the mind that railroads, far from being a liability, would in fact be an indispensable tool of war, construction of the stalled lines went forward. By 1841, Prussia boasted 375 kilometers of completed track.

The significance of the uncertainty of railroad expansion in Prussia is underscored and magnified when we widen the focus to all of German Europe. Turning first to southern Germany, pride of place in German railroad history traditionally goes to Bavaria, which opened the first European railway powered by steam locomotives in December 1835. But this short stretch connecting Nürnberg and Fürth was destined to be the only Bavarian line that decade, for private initiative could not muster the capital required for construction in hilly terrain and an anti-industrial government that considered waterways a more sensible investment for farmers and small craftsmen refused to help. Thus, Munich poured 17 million florins into the Ludwig Canal (between the Main and Danube Rivers) while the desperate promoters of the Augsburg-Munich Railroad, who needed far less, pleaded in vain for state support. Similar stories unfolded in Baden and Württemberg. Small towns away from the projected railroad routes objected to government expenditure that would hurt them and benefit others. State officials in Karlsruhe and Stuttgart, like those in Munich and, for many years, Berlin too, refused to support a new, costly, unproven, politically divisive transportation technology. When the 1840s dawned, the 6-kilometer run outside Nürnberg was the only operating line in all of southern Germany.

Political decisions also delayed the debut of railroads in Austria. In this case, early proposals foundered on the deeply ingrained conservatism of Kaiser Francis II. "Railways," he said, "will only bring revolution into the country."[35] Reinforcing the monarch's prejudice was the skepticism of leading technocrats "who were convinced of the fleeting nature of the whole railroad mania and were [therefore] staunchly against approving any extensive experiments in this direction."[36] When Francis died in 1835, Austria had only the short horse-drawn line between Budweis and Linz. The death of the old emperor, however, created a more propitious environment for railroad schemes as bankers got the ear of Clemens von Metternich, who was now more powerful than ever, for the new kaiser was mentally incapacitated and unable to rule. By the late 1830s, concessions had been granted for lines connecting Vienna with Galicia, Budapest, and Trieste. The empire boasted 500 kilometers of track in 1841—more than any other continental state at that time.

The exception to the rule throughout German Europe for most of the 1830s was Saxony. The Revolutions of 1830 induced a bureaucratic shake-up that brought a new set of innovative thinkers to the fore. Comprehensive reforms followed, including a much more systematic promotion of agriculture and industry. These programs, together with the economic stimulus of membership in the Zollverein, triggered an industrial boom: 242 new factories were established between 1834 and 1836. Saxon railroad policy was consistent with this newly found bureaucratic enthusiasm for the acceleration of industrial development. Officials in Dresden encouraged the promoters of the Leipzig-Dresden Railroad as early as 1833, and when the company formed in early 1835, it received generous government aid. The state guaranteed a 3.5 percent return to investors and pledged to buy shares itself should capitalization problems arise. The bureaucracy also facilitated the expropriation of land and placed state engineers at the disposal of the railroad company. The staggered opening of the 116-kilometer line during 1837 and 1838 brought German Europe—indeed all of Europe—its first railroad line of significant length.

Thus, the railroad age, as we have come to know it, remained a highly uncertain and iffy proposition for many years. Baden, Württemberg, and Bavaria rejected railroads in the 1830s. We should not overlook the fact that their "tradition-blessed obstinacy" to this technology reflected a similarly deeply entrenched opposition to modern industrialization in most of the states of the German Confederation. Until 1835, only Saxony avidly backed railroads. The death of Francis II facilitated a more positive approach in Austria, with the result that concessions were granted two to three years later. Prussia turned the corner in 1838, passing Austria as the leading railroad builder on the Continent by the late 1840s. With this resolution of the controversy over railroads, a decades-long struggle over industrialization stretching back to the pre-1815 period came to an end. We know the outcome, but this does not diminish the historical significance of the struggle—or the need to examine it closely.

Positive policies mixed with negative in Austria, Prussia, and Saxony—to name the three most important states—but the influence of the government programs over the years studied here was ultimately positive and benign, and thus contributed to the acceleration of industrialization in German Europe after the mid-nineteenth century. Although statistics on rates of growth remain, after decades of research, patchy estimates based mainly on North German states like Prussia, the available data point to impressive economic performance. Real national product per capita rose gently, almost imperceptibly in the early 1800s, but then quickened to a 2.0 to 2.5 percent annual range in the 1840s and 1850s and averaged 2.5 percent from 1850 to 1873. Not surprisingly, investment drove much of this growth, increasing from 5 to 6 percent of net national product early in the century to 10 to 11 percent in the 1860s. If we leave these aggregate figures aside and isolate industry and transport (i.e., railroads), however, the impressiveness of German economic performance becomes even more apparent. Industry's share of net investment, for example, rose from 2 to 3 percent in the early 1800s to around 20 percent in the 1850s. Accordingly, industrial production averaged a torrid 4.8 percent annual increase from 1850 to 1873, while railroads shot up a meteoric 14 percent—faster than early industrialization rates in Britain and France and roughly equal to that of the United States.[37] By the late nineteenth century, moreover, Germany was leading the way into an era of new science- and education-based steel, chemical, and electrical technologies—breakthroughs that seem to justify the expression "Second Industrial Revolution."[38] Indeed, no other European nation could match German economic and technological prowess, a stark fact that would have dire diplomatic and military effects as the Continent drifted ominously toward war after 1900.

Notes

1. Eric Dorn Brose, *The Politics of Technological Change in Prussia: Out of the Shadow of Antiquity, 1809–1848* (Princeton, NJ: Princeton University Press, 1993).

2. Ibid.

3. Ibid.

4. See Eric Dorn Brose, *Technology and Science in the Industrializing Nations, 1500–1914*, 2nd ed. (Amherst, NY: Prometheus Books, 2006).

5. Brose, *Politics of Technological Change*, 118–123, and "Competitiveness and Obsolescence in the German Charcoal Iron Industry," *Technology and Culture* 26:3 (July 1985), 545.

6. Cited in Brose, *Politics of Technological Change*, 120.

7. Cited in William O. Henderson, *The Zollverein* (Chicago: Quadrangle Books, 1959), 66.

8. William O. Henderson, *Britain and Industrial Europe* (Leicester: Leicester University Press, 1965).

9. See Dirk Götschmann, *Das Bayerische Innenministerium, 1825–1864* (Göttingen: Vandenhoeck & Ruprecht, 1993).

10. Cited in Edward Lazare Shorter, "Social Change and Social Policy in Bavaria, 1800–1860" (Ph.D. diss., Harvard University, 1967), 196.

11. See Hubert Kiesewetter, *Industrialisierung und Landwirtschaft: Sachsens Stellung im regionalen Industrialisierungsprozess im 19. Jahrhundert* (Cologne: Böhlau, 1988); Paul Gehring, "Das Wirtschaftsleben in Württemberg unter König Wilhelm I (1816–1864)," *Zeitschrift für Württembergische Landesgeschichte* 9 (1949–1950), 196–257; Paul Gehring, "Von List bis Steinweiss: Aus der Frühzeit der Württembergischen Industrialisierung," Zeitschrift für *Württembergische Landesgeschichte* 7 (1943), 405–444.

12. Cited in Kiesewetter, *Industrialisierung und Landwirtschaft*, 94.

13. Cited in Gehring, "Das Wirtschaftsleben," 215.

14. See Wolfram Fischer, *Der Staat und die Anfange der Industrialisierung in Baden, 1800–1850* (Berlin: Duncker & Humblot, 1962).

15. Hans-Ulrich Wehler, *Deutsche Gesellschaftsgeschichte*, vol. 2 (Munich: Verlag C. H. Beck, 1987), 183.

16. See David F. Good, *The Economic Rise of the Habsburg Empire, 1750–1914* (Berkeley: University of California Press, 1984); Johann Slokar, *Geschichte der österreichischen Industrie und ihrer Förderung unter Kaiser Franz I* (Vienna: F. Tempsky, 1914).

17. Good, *Economic Rise*, 12.

18. See Barbara Vogel, *Allgemeine Gewerbefreiheit: Die Reformpolitik des preussischen Staatskanzlers Hardenberg (1810–1820)* (Göttingen: Vandenhoeck & Ruprecht, 1983); Brose, *Politics of Technological Change in Prussia.*

19. Prussian policy here stands in sharp contrast to city-states like Bremen, Hamburg, and Lübeck. Berlin's liberalism was particularly welcome in left-bank mining districts, where the strict regulations of the Prussian Mining Corps had no validity, unlike right-bank areas, where harshness, especially against deep mining shafts, prevailed. For a discussion of the contrast, see Brose, *Politics of Technological Change*, 145–149.

20. Cited in Brose, *Politics of Technological Change in Prussia*, 59.

21. See note 1.

22. See Arnold H. Price, *The Evolution of the Zollverein* (Ann Arbor: University of Michigan Press, 1949); Henderson, *Zollverein*; Wolfram Fischer, "German Zollverein: A Case Study in Customs Union," *Kyklos* 13 (1960), 65–89; Hans-Werner Hahn, "Hegemonie und Integration: Vouraussetzungen und Folgen der preussischen Führungsrolle im Deutschen Zollverein"; Rolf H. Dumke, "Der Deutsche Zollverein als Modell ökonomischer Integration," in *Wirtschaftliche und politische Integration in Europa im 19. und 20. Jahrhundert*, ed. Helmut Berding (Göttingen: Vandenhoeck & Ruprecht, 1984), 45–71, 72–102, respectively.

23. See Brose, *Politics of Technological Change*, 67–68, 87–97. For the earlier view, see Tom Kemp, *Industrialization in Nineteenth-Century Europe* (London: Longman, 1969), 84; Takeo Ohnishi, *Zolltarifpolitik Preussens bis zur Gründung des deutschen Zollvereins* (Göttingen: Vandenhoeck & Ruprecht, 1973), 227; Martin Kitchen, *The Political Economy of Germany, 1815–1914* (London: Croom Helm, 1978), 40–43.

24. Cited in Brose, *Politics of Technological Change*, 68.

25. Alan Milward and S. B. Saul, *The Economic Development of Continental Europe, 1780–1870* (London: Allen & Unwin, 1973), 374.

26. See Adolf Beer, *Die österreichische Handelspolitik im neunzehten Jahrhundert* (Vienna: Manz, 1891); Robert D. Billinger, *Metternich and the German Question: States Rights and Federal Duties, 1820–1834* (Newark: University of Delaware Press, 1991), 43–44.

27. For a fine discussion of the literature, see Dumke, "Der Deutsche Zollverein," 78–79.

28. John H. Clapham, *The Economic Development of France and Germany, 1815–1914* (Cambridge: Cambridge University Press, 1966), 97.

29. Dumke, "Der Deutsche Zollverein." Also see his "The Political Economy of German Economic Unification" (Ph.D. diss., University of Wisconsin, 1976).

30. See Carl-Ludwig Holtfrerich, "The Monetary Unification Process in Nineteenth-Century Germany: Relevance and Lessons for Europe Today," in *A European Central Bank? Perspectives on Monetary Unification after Ten Years of the EMS*, ed. Marcello de Cecco and Alberto Giovannini (Cambridge: Cambridge University Press, 1989), 216–241.

31. Wehler, *Deutsche Gesellschaftsgeschichte*, vol. 2, 134.

32. Cited in Gehring, "Das Wirtschaftsleben," 243.

33. Richard Tilly, "Capital Formation in Germany in the Nineteenth Century," in *The Cambridge Economic History of Europe*, vol. 7, part 1, *The Industrial Economies: Capital, Labour, and Enterprise: Britain, France, Germany, and Scandinavia*, ed. Peter Mathias and Michael M. Postan (Cambridge: Cambridge University Press, 1978), 382–441; Rainer Fremdling, *Eisenbahnen und deutsches Wirtschaftswachstum 1840–1879* (Dortmund: Ardey-Verlag, 1975).

34. Brose, *Politics of Technological Change*, 209–240. Also see Dietrich Eichholz, *Junker und Bourgeoisie vor 1848 in der preussischen Eisenbahngeschichte* (Berlin: Akademie-Verlag, 1962); James M. Brophy, *Capitalism, Politics, and Railroads in Prussia, 1830–1870* (Columbus: Ohio State University Press, 1998).

35. Cited in Carlile A. Macartney, *The Habsburg Empire, 1790–1918* (New York: Macmillan, 1969), 259.

36. Hermann Strach, ed., *Geschichte der Eisenbahnen der Oesterreichisch-Ungarischen Monarchie* (Vienna: K. Prochaska, 1898), 1:132.

37. Reinhard Spree, *Die Wachstumszyklen der deutschen Wirtschaft von 1840 bis 1880* (Berlin: Duncker & Humblot, 1977); Richard Tilly, *Vom Zollverein zum Industriestaat: Die wirtschaftlich-soziale Entwicklung Deutschlands 1834 bis 1914* (Munich: Deutscher Taschenbuch Verlag, 1990), and "Capital Formation in Germany in the Nineteenth Century," 382–441.

38. See the discussion in Brose, *Technology and Science*, chap. 3.

7

Reconceptualizing Industrialization in Scandinavia

Kristine Bruland

> To get the historical record straight . . . means giving credit to just those dull, everyday, pragmatic, honest betterments in simple technology, routine services, law, and administration that were taking place in Europe before the Industrial Revolution.
>
> Eric L. Jones[1]

The Scandinavian countries (Denmark, Finland, Norway, and Sweden) have undergone spectacular development processes and are now among the most successful economies in the world in terms of both economic indicators and social and health outcomes. However, for many centuries, Scandinavia was a poor region. It was certainly not obvious to eighteenth- or nineteenth-century observers that it would become rich. As Norwegian professor Anton Schweigaard remarked in 1848, "Industry is most backward in this country."[2] Although these economies had resource advantages—in timber, fish stocks, waterpower, and ferrous and nonferrous ores, for example—the resource base was arguably no greater than in other economies that have achieved nothing like the Scandinavian development record. Scandinavia also faced serious obstacles: small populations, shortage of arable land, formidable communication barriers, and severely unfavorable climatic conditions. Poverty and rural demographic pressures led to out-migration of a large proportion of the working-age population in the late nineteenth and early twentieth centuries, and this too was an impediment to growth.

Yet Scandinavia succeeded in industrializing beginning in the mid-nineteenth century and also in sustaining its growth. This was not simply a matter of creating domestic industrial bases that could deploy modern technologies. These economies have also managed to build significant businesses that have occupied world-leading market positions in such sectors as telecommunications, shipping, chemicals, paper, transport equipment, power engineering, mining equipment, and household goods. These capabilities have continued, with entry into such advanced technologies as mobile telephony, a sector in which Swedish and Finnish companies currently have dominant global positions in base stations, infrastructure equipment, and handsets. An important historiographical question about Scandinavian industrialization, then, is: What characteristics of these economies made it possible to create such high-income industries in the growing world economy?

Although industrialization has multiple dimensions and cannot be reduced to technological change, new technologies are an irreducible part of the process and can be used to map the extent of industrialization. An adequate conceptual approach to industrialization should therefore rest on an understanding of the scope of technological change across Scandinavia and then on the factors that impelled and shaped it. In particular, was it a narrowly focused or a broadly spread phenomenon? The answer to this question is likely to have important implications for industrial history. A narrow industrialization process, focused perhaps on the application of hydrocarbon-based power or the introduction of machinery, or even on the development of specific industries, is likely to need an equally narrow explanatory approach. For example, such technologies might be introduced by political or economic elites independent of any wider change in the social or economic environments. On the other hand, broadly distributed processes of technological change cannot be accounted for by elite decision making. Broad-scope development seems to require more general explanations, focusing on economy-wide incentives and propensities to change and on social processes of capacity building. A further crucial aspect of the Scandinavian situation was the extent to which learning and industrial growth rested on responses to foreign technological developments in terms of both learning from abroad and direct technology import. Here too there are questions of scope. Was technology importation a broad or a narrow process? If it was wide in scope, what institutions supported it? The institutional aspect is important because both technology creation and absorption rest on the acquisition of knowledge, and this, in turn, requires institutional support and the commitment of resources.

The underlying argument of this chapter is that Scandinavian industrialization should be understood as a broad process based not on the rapid development of a few key sectors, but rather involving multisectoral learning and growth. The absorption of foreign technologies across a wide range of sectors was a key element in this process. The structural and technological changes through which Scandinavia grew were equally wide: they were enabled and supported by institutional and cultural changes and by political initiatives that fostered the acquisition and use of new technologies from abroad. The conceptualization of industrialization in Scandinavia should rest on an understanding of the scope and dimensions of these institutional changes, many of which were highly specific to regional circumstances. It is important to note that the modern Scandinavia nation-states either did not exist or changed significantly during the nineteenth and early twentieth centuries.[3] Although government initiatives played important roles in industrialization, much broader processes of social change contributed to the multifaceted developments associated with industrialization. This is not a story of a heroic "big push" toward industrialization led by dramatic technological breakthroughs. The industrialization of Scandinavia was instead a complex process of sustained transition across many aspects of the economy and society. Our challenge is to grasp the dimensions of this complexity and to delineate the social commitments that made it possible.

The Scope of Scandinavian Industrialization

Only limited sources provide insight into the technological breadth of economic activity in Scandinavia during the eighteenth century, but those we have suggest a very broad process, often affecting simple activities, shifting formerly household production into a manufacturing or even an industrial framework.

An illustration of the breadth of development can be found in the existing evidence on prizes for invention. From the mid-eighteenth to the mid-nineteenth century, both Denmark and Norway had "knowledge academies" that awarded prizes for developments in handicraft and manufacturing. These prizes were actively sought after and discussed. From 1760 to 1860, approximately 120 prizes were applied for across such activities as household manufacture, small industry, agriculture, and hunting. Most of the 3,300 applications were for incremental improvements to existing practices rather than for full-scale innovations, but they were nevertheless broad in character.[4]

In terms of production, the scope of change also appears to be small scale but broadly based. Manufactories producing textiles and clothing emerged in Sweden in the seventeenth and early eighteenth centuries. Klas Nyberg has argued that these shops were, from the beginning, technologically internationalized (in dyeing for example) and acted as channels through which international production standards and tastes entered Sweden.[5] Starting in the mid-eighteenth century, there were also significant expansions of larger-scale activity, particularly in textiles, as governments in all of the Nordic countries sought to establish state textile manufactories, mainly using modified traditional technologies. In Iceland, for example, textile manufactures were developed throughout the eighteenth century, with larger-scale production in the end rejected in favor of domestic fabrication.[6] Small-scale textile production and clothing manufacture also appeared in southern Finland. Extensive trading networks developed in Norway.[7] These trading networks involved both land travel and coastal shipping, with peddlers carrying on significant trade between rural and urban areas, and acting as channels for the distribution of manufactures.[8] Brewing industries grew as this product (beloved by Scandinavians) shifted from domestic manufacture to large-scale integrated production and distribution.[9] In Sweden, agriculture, mining, and iron production all expanded significantly from the mid-eighteenth century and were, moreover, "spatially diffused over the country to a remarkably large extent."[10] Studies within a "proto-industrial" framework for Denmark, Finland, Sweden, and Norway all suggest widespread development in textiles, pottery, metals (and smithing), sawmills, wood processing, agricultural equipment, paper, shoemaking, glassmaking, and fishing.[11] This process of development was so extensive that it gave rise to serious ecological problems, most notably deforestation in Sweden. This became a key technological issue: the quarterly publications of the Swedish Academy published 105 articles between 1739 and 1815 dealing with technological issues related to better use of energy resources, especially timber.[12] From a strictly domestic point of view, it seems reasonable to see the evolving Scandinavian economy as resting on a broad range of developing activities and technologies starting in the early eighteenth century.

International Technology Acquisition by Scandinavian Economies

Industrialization in Scandinavia has at least one distinctive technological characteristic: high levels of technology importation. Even today, relatively few of the technologies in use in Scandinavian industry have been developed within its boundaries. Many individual studies suggest an important role for foreign technology in particular sectors or with respect to particular technologies. This section maps the acquisition of foreign technology in various sectors in Norway, Denmark, and Sweden. A general picture of key foreign technology inputs to growing sectors in the Scandinavian economies, particularly from 1750 to 1850, has developed on the basis of historical studies of Scandinavian growth, which consists mainly of case studies of particular industries or cases of technology transfer. The history of technology has, for some time, been an important part of Scandinavian historiography. Although we lack a complete empirical picture of the sectoral structure of economic growth in Scandinavia, there are enough detailed studies to permit some generalizations. It is possible to investigate the direct and indirect importation of technology and to construct a summary of the recipient sectors, the technologies transferred, and the enterprises that benefited and their locations. The overview, even in summary form, is far too extensive to include here. To give a flavor of the technology flows, see table 7.1.[13]

The key point emerging from table 7.1 is that there was substantial foreign technology acquisition across all sectors of the Scandinavian economies. It should be emphasized that this overview rests on industries that have been the object of specific studies by historians, so the numbers of cases, and the scope of acquisition, are significantly understated. Tables 7.2 and 7.3 provide a more detailed picture of two industries: mining and food processing. Mining of both ferrous and nonferrous metallic ores was (and remains) an important sector in both Sweden and Norway, and technology transfer into this sector began very early and continued over many years. It appears that key elements of the core technology of mining were imported: pumps, explosives, and general mining equipment. In many cases, it was the ability to use these technologies that made mining possible in otherwise unfeasible conditions. In the food-processing industry, the transition away from domestic production to manufactured food products was also accompanied by a wide set of cases of imported inputs and production technologies.

In principle, this type of overview is possible for all of the sectors listed in table 7.1. It supports the argument that Scandinavia undertook an industrialization process characterized by both breadth and complexity. For Denmark, Norway, and Sweden, the technology import process began early and continued over a long period. Two large points can be drawn from this kind of survey. The first is that technology import appears to have been integral to industrialization. Wherever detailed sectoral studies have been carried out, we find the presence of technology transfer, often on a very significant scale. Based on the current evidence, it is extremely difficult to link technology transfer and specific growth effects: certainly not all transfer processes were economically successful. But the general prevalence of transfer, especially early in the industrialization process, suggests that it played a central role. The second point concerns the nature of the growth process. If the technological changes mapped by

Table 7.1
Scandinavian Inward Technology Acquisition by Sector

Sector and number of cases	Time period	Technologies and recipient country studied
Agriculture (15 cases)	1740–1800	Sieving (N), potato (S), stone sheds (N), elevators (N), English sheep (N)
	1800–1850	Draining (S,D), horse walk (N), clog wheels (N)
	1850–1890s	Reaping (N,D), cutting machine (D), dairy equipment (S to N,D)
Construction and building (17 cases)	18th century	Brickworks (N), horse walk (N)
	1830–1850	Brick technology (D)
	1850–1890s	Drainpipe machinery (D), firing grates (D), blunging machinery (D), steam engines (D), clay cleaning machinery (D), brick machinery (D), cement (D), kilns (D), clay packing machinery (N)
Engineering (95 cases)	1769	Firm founded (S)
	1800–1850	Firm founded (S), steam engines (S), all types of milling equipment (S), lathes (N), steam hammers (N), blowing machine (N), drawings (D), nail machine (N)
	1850–1890s	All types of machinery (N), sewing machine (S), steam hammer (N), tools (D)
Food and drink (27 cases)	16th century	Spirits (N), sugar and fruits (S,D)
	17th century	Sugar and fruits (N), baking ovens (N)
	1736–1800	Saltwater distillery (N), horse walk (N), salt refinery (N), windmill (N)
	1820	Spirit-making construction (N)
	1846–1890s	Beers (D,N), steaming techniques (N), turbines (D), yeast (D to N), flour milling technology (N,D), spirits apparatus (D), tobacco spinning machine (D), centrifugal flour sieves (D), margarine (N), beer cooling machinery (D), sugar technology (D)
Forestry (6 cases)	16th century	Saws (N,S), tar resin (S)
	1730–1800	Saws (S,D), saw blades and frames (D), machinery (N,S,D)
	1800–1850s	Circular saw (N), machinery (N,S,D)
Glass (13 cases)	16th century	Glass huts (S,D)
	17th century	Potash production (S), glass huts (N)
	1755–1800	Window glass (N), kelp making (N) soda ash production (N), fensterglass method (N), furnace technology (N)
	1840–1845	Stretched glass method (N), firing technology (N)
Iron and steel (15 cases)	1760–1800	Foundries (S), reverbatory furnace (S), molding (S)
	1800–1850	Crucible steel (S), puddling (S), forging methods (S), castings (S), general machinery (S)
	1850–1890	Bessemer (S), Siemens Martin (S), open hearth (S)

Table 7.1
(continued)

Sector and number of cases	Time period	Technologies and recipient country studied
Mining (30 cases)	16th century	Extraction methods (N), blast furnace (N), double blast furnace (S)
	17th century	Blast furnace (N), power transmission (N), equipment (N), steel (N,S), horse walk (N)
	18th century	Gunpowder (S,N), refining (N), irrigation (N), steam engines (S), cannon boring works (D), wind furnace (N), molds (N)
	1800–1850	Steam engines (S,N), rails (N), water machine (N), blasting and elevation techniques (N), refining (N), wire (N), water turbine (N)
	1850–1880s	Water lifting machinery (N), blasting techniques (S,N), steam engines (N)
Printing (2 cases)	1825–1829	High-speed presses (D).
Pulp and paper (23 cases)	17th century	Paper production (N)
	1800–1850	Bleaching (D), steam (D), presses (D), paper machines (D,S), waterwheel (D), turbines (D)
	1850–1899	Pulp machinery (N), wood grinding machinery (S), complete plant (N), papermaking machines (N), turbines (N), generators (D)
Pottery and clay (2 cases)	1847–1851	Full plant and equipment (N), drainpipe machine (D)
Textiles (more than 50 cases)	18th century	Knitting frames (S,D), flying shuttle (D), knitting machine (N).
	1800–1850	All preparatory, spinning, weaving, and finishing equipment (N), wool machinery (D), Jacquard (D), firm establishment (S), linen technology (D), cotton weaving (D)
	1850–1890s	All preparatory, spinning, weaving, and finishing machinery (N,D,S), gloves machinery (D), shoes machinery (N,S)
Transport and infrastructure (9 cases)	1700–1800	Axles (N), axle grease (N), dock equipment (D)
	1800–1850	Canal building equipment (S), steamships (N)
	1850–1870s	Dry dock (D), railways (N), locomotives (N,S), naval vessels (S)
Utilities (4 cases)	1847–1890s	Gasworks (N,D)
Weapons (1 case)	1867	Remington rifles (S)

Note: D = Denmark, N = Norway, S = Sweden.

Table 7.2
Technology Import to Scandinavia: Mining Sector

Technologies	Period	Source country	Products	Recipient country
Extraction	1538–1622	Germany	Direct extraction of iron	Norway
Extraction	1545	Germany	Direct extraction of iron	Sweden
Furnace	1570–1580	UK	Blast furnace	Norway
Furnace	1598	Wallonia	Double blast furnace	Sweden
Furnace	1622	Germany	Blast furnace	Norway
Power transmission	1643	Germany	Flat rod power transmission	Norway
Equipment	1640s	Germany	Water-driven mining equipment	Norway
Steel method	1653	Germany	Steel cementing and burning	Sweden
Steel method	1661	Germany	Steel cementing and burning	Norway
Steel method	Late 18th century	UK	Steel cementing and burning	Norway
Blasting/explosives	c 1630	Germany	Gunpowder	Sweden
Power	c 1680	Germany	Horse walk	Norway
Refining	18th century	Germany	Bar iron	Norway
Irrigation	1720	Germany	Irrigation system	Norway
Power	1727–1736	UK	Newcomen steam engine	Sweden
Blasting/explosives	1732–1816	Germany	Gunpowder	Norway
Techniques	1732–1816	Germany	Various mining techniques	Norway
Blasting/explosives	1732–1816	Germany	Gunpowder	Norway
Enterprise	1751	France	Bronze cannon boring works	Denmark
Furnace	1776	UK	Wind furnace	Norway
Techniques	1799	UK	"Flasks"	Norway
Power	1790s	UK	Newcomen steam engines	Sweden
Power	1804	UK	1 Watt steam engine	Sweden

Table 7.2
(continued)

Technologies	Period	Source country	Products	Recipient country
Power	1804	UK	4 Watt steam engines	Sweden
Rails	1808	?	Iron rails for transport of logs	Norway
Machine	1824	Germany	Pressure water machine	Norway
Blasting/explosives	1830s	UK	Safety wick	Norway
Elevation/lifting	1830s	UK	Elevation techniques	Norway
Power	1840s	Sweden	Steam engines	Norway
Refining	1842	UK	Lancashire process	Norway
Equipment	1844	UK	Wire, pipes	Norway
Power	1847	Germany	Schwamkrug's water turbine	Norway
Elevation/lifting	1853	Germany	Water lifting machine	Norway
Blasting/explosives	1863	Italy	Nitroglycerin	Sweden
Blasting/explosives	1872	Sweden	Dynamite	Norway
Power	1880s	Germany	Steam engine	Norway

the transfer process were associated with industrialization and economic expansion, then growth was multisectoral, occurring across a wide range of activities and technologies. The sectoral studies are comprehensive enough to suggest that inbound technology transfer was common across the entire industrial economy of the Scandinavian area. Understanding growth in the Scandinavian region requires an understanding of impulses or incentives shared by many activities. Since the most significant common features of industrial development are the underlying institutions on which economic activity rests, this, in turn, suggests that institutional processes were associated with multisectoral technology transfer.

The Scandinavian economies were open to both the importation of technology and external trade. From the early seventeenth century, Denmark established colonies and had extensive trading relations in Asia, Africa, and the Americas and was also actively involved in the slave trade. Beginning in the late eighteenth century, Sweden was engaged in

Table 7.3
Technology Import to Scandinavia: Food and Drink

Technologies	Period	Products	Source country	Recipient country
Preservation	16th century	Sugar, preservatives	India, Spain, Italy	Sweden, Denmark
Preservation	17th century	Sugar, preservatives	India, Spain, Italy	Norway
Baking	17th century	Bread oven	Netherlands	Norway
Salt making	1736	Saltwater distillery	Germany	Norway
Power	1745	Horse walk	?	Norway
Power	18th century	Steam engine	UK	Norway
Salt making	1774	Salt refinery (enterprise) design Stone salt	UK	Norway
Power	1788	Windmill	?	Norway
Beer	1846	Bayer beer	Germany	Denmark
Beer	1846	Lager beer	Germany	Denmark
Beer	1840s	Beer	Denmark	Norway
Power	1850–1880s	Fourneyron and Jonval turbines	Germany	Denmark
Yeast	1850s	Yeast	Denmark	Norway
Flour milling	1866		Germany	Norway
Cigar machinery	1870s	"Vikkel" cigar machines	Germany	Denmark
Spirits	1820	Spirit-making construction	Germany	Norway
	1870s	Distillation apparatus	Germany	Denmark
Tobacco machinery	1870s	J. E. H. Andrew's tobacco spinning machinery	UK	Denmark
Flour sieving	1870s	Centrifugal sieves	Germany	Denmark
Margarine	1876	Margarine production	France	Norway
Beer cooling	1879	Cooling machine	Germany	Denmark
Sugar making	1870–1880s	Basic technology	France Germany	Denmark
Flour milling	1880s	Grinding machine factory	Germany	Norway
Flour milling	1880s	Grinding machine	Hungary	Norway
Beer cooling	1882	Cooling machines	Germany	Denmark
Flour milling	1895	Automatic rain mill	Hungary	Norway

Note: ? = country not specified.

transatlantic trade in iron. Both the Danish Asiatic Company and the Swedish East India Company conducted extensive trade in South Asia and China and reexported such products as tea throughout Western Europe and the Baltic countries.[14] These trading patterns were important in the development of financial systems because they contributed to the emergence of discount houses and quasi-banking operations around the major trading cities, Gothenberg in particular.[15]

Institutional and Policy Foundations of Industrialization

The impressive scope of technological change and industrial growth suggests a broad social process of change in Scandinavia that supported the building of capabilities and the creation of an industrial culture. There was general technological advance rather than rapid development confined to a few sectors. So the question that arises is: What kinds of change, in institutions or organization or policies, underpinned this process? In the historical literature on Western economic development, considerable stress is placed on changes in the institutional framework in the sense that Douglass North used the term—that is, a stable set of rules that regulate economic behavior such that innovation is promoted or at least not hindered. Most of this literature strongly emphasizes the emergence and role of one particular set of institutions, namely a framework of law relating to property rights, leading to the enforceability of contracts, to financial systems, to the regulation of entry and exit conditions, and to a framework for employment.

This type of change was certainly present and important in Scandinavia. As Fritz Hodne pointed out, Scandinavian "governments provided protection for private property and contracts, and jealously guarded this reputation."[16] However, it is important to look beyond this general framework of institutional change into the specifics of what was involved in capability building. Conceptualizing the broad underlying changes that promoted industrialization is not simply a matter of identifying large-scale institutional shifts, such as the emergence of a property rights regime, but of identifying the particular institutional changes that underlay new technologies. This means looking into the organizations and institutions of knowledge acquisition and learning to identify those aspects of the social and organizational environment that promoted and supported industrialization.

What were the main ways in which technologies were accessed and adapted? This seems to require a general differentiation between domestic and foreign technology acquisition. Domestic innovative activity certainly occurred broadly, but it was primarily focused on the deployment, adaptation, modification, and use of ideas, information, technological concepts, and artifacts from abroad. The institutional mechanisms that mattered were those that promoted the acquisition and adaptation of foreign technologies. This process encompassed knowledge about both new technologies (e.g., knowledge about their existence and capabilities) and of those technologies themselves (meaning knowledge of how to build and/or operate them).

Joel Mokyr has emphasized that technology has a clear knowledge dimension, not just in terms of skills or tacit understanding but also in terms of formal knowledge.[17] What is often called the Enlightenment, associated with new modes of inquiry, the growth of bodies of formal knowledge, and the diffusion of knowledge (through encyclopedias, for example), was much more than a phase in the history of ideas. From Mokyr's perspective, it was critical to industrialization, since it diffused various forms of prescriptive knowledge that made technology creation possible. During the Enlightenment, Scandinavia had a wide range of formal and informal means for monitoring foreign developments, including societies, journals, and newspapers that played an extensive role in acquiring basic factual information concerning the technological frontier in the larger economies. These institutions also played an important role in the learning processes through which they were accessed and operated.

By the mid-nineteenth century, all of the Scandinavian countries possessed technical societies that focused strongly on foreign developments. During the eighteenth century, Denmark saw the creation of the Royal Danish Academy of Sciences and Letters in 1742, the Royal Danish Society for Natural History in 1745, the Society for the Promotion of Fine Arts and Practical Sciences in 1759, and the Royal Danish Agricultural Society in 1769. These societies, which published, held meetings, and even established schools, were closely connected.[18] At least five important societies were established in Norway in the late eighteenth century, one focused on running a textile school employing British technology.[19] During the second half of the nineteenth century, the Norwegian Polytechnic Society averaged twenty-five meetings per year, addressing issues such as "Portraits of Brunel and Stephenson," "Notes from a Journey to England," and "Cotton and Cotton Spinning Mills."[20] Societies were initiated by both governments and industrialists. Most industrialists and many engineers were members of these societies, which also sponsored visits to foreign countries for the purpose of examining industry and acquiring information.

The societies conducted regular meetings, published journals, and facilitated discussion of technological issues. A central aspect of this activity was an outward-looking orientation: they channeled information on foreign developments in general and sometimes played direct roles in transferring technology. For example, the Royal Danish Agricultural Society "took on the transfer of agrarian technology, models of machinery and awarded prizes to pave the way for land reform."[21] The Danish technical journal, the *Ursins Magazine,* first published in 1826, typically contained reports on foreign developments drawing on British sources such as *The Mechanic's Magazine* and *The Repertory of Arts, Manufactures & Agriculture.*[22] A Danish travel report on a visit to London, published in 1816, allegedly prompted the mechanic Winstrup to invent a paper machine, which he subsequently sought to have patented. This was nine years after the patent specifications of a similar machine had been published in *The Repertory of Arts, Manufactures & Agriculture*, which had been unavailable in Denmark at the time because of the war.

In Sweden, the Royal Academy of Sciences and various technical societies played a central role, according to Jan Hult, Svante Lindqvist, and others, in the overall development of technological capability:

> The technical changes during the 18th century can to a large extent be characterised as qualitative, and these changes were to be important for the industrialisation process which started during the 19th century. Generally, we can say that the largest technical change during the 18th century was that "technique became technology." This was because at this time a comprehensive and systematised knowledge about technique developed, and towards the end of the period we also see, for the first time, the word "technology" in the literature. This knowledge was sustained, developed and communicated through new institutions such as the Iron Office, the Academy of Sciences (Kungeliga Vetenskapsakademiet) and various agricultural societies (*hushållningssälskap*). These institutions raised technical problems which earlier had just been local questions to a general level and made them national concerns. Through journals and books which were published by these institutions the knowledge was spread to new groups. This institutional development was perhaps the most important technical change in Sweden during the 18th century.[23]

The processes that these scholars identified were not peculiar to Sweden; such developments can be traced across all of these economies.

Human mobility was a key channel of the transfer of technological knowledge. This included travel abroad by Scandinavian mechanics and engineers and immigration. Like many European governors, Scandinavian legislators were anxious for close contact with industrial developments abroad and actively promoted foreign travel. The eighteenth-century development of the Danish naval dockyard at Holmen into a leading European establishment was, in the main, built on "new knowledge and technique" from abroad. Intense foreign contact, repeated visits, and spying were integral to this growth and to the development of the dockyard into a key center of learning. The "works master" F. M. Krabbe traveled to France, Holland, and Belgium. He made his way into the naval dock in Portsmouth, but although attired in newly purchased English clothing, he was discovered and thrown out. Yet he succeeded in entering several famous British and French yards, and he met with the renowned French theoreticians Pierre Bouguer and Henri-Louis Duhamel du Monceau.[24] When the Danish government decided to construct a new dock in the mid-nineteenth century, foreign travel again facilitated the acquisition of essential knowledge and technology developed abroad.[25]

Foreign travel was subsidized by all the Scandinavian governments. In 1765, the Danish state established a special fund to support foreign travel by artists and artisans: it was supplemented by the Reiersenske fund in 1795 and the Industry fund in 1797. From 1850, traveling scholarships for a total of 10,000 riksdaler were distributed every year.[26] During the second half of the nineteenth century, the Norwegian government granted 1,006 travel stipends, of which 187 went to "mechanics."[27] In Sweden, where the gathering and dissemination of knowledge "tended to be centralized under the auspices of the Bergskollegium and the Ironmasters Association (Jernkontoret)," these institutions "sponsored technical experts to travel to other important iron-producing regions of Europe and report upon new developments." This accounted for the "succession of Swedish observers [touring] Britain between the late seventeenth and mid-nineteenth centuries." Such support was an important initiative because of the restrictions put on Swedish miners' mobility at the time.[28] Jörberg points to the significance of "the many journeys that Swedes made to British enterprises" to the introduction of innovations to Sweden.[29] Ljungberg, who was Swedish-born, was hired as a spy by the Danish government and spent fifteen years in England; the Dane Bidstrup resided six to seven years there (1787–1794), including a period as an apprentice, and then he ran his own engineering workshop in Leicester Square; and Nordberg, who was born in Sweden but grew up in England, worked in the British textile industry.[30] Most industrialists spent considerable time abroad, with many gaining extended work experience.[31]

Traveling and working abroad were important mechanisms for gathering specific information about developments and for building up individual technological capacity. To keep abreast required frequent visits, as did the learning involved in mastering the new techniques. For example, Nordberg's travels to England were undertaken to stay up to date and to solve persistent problems linked with reconstructing and running textile equipment he had smuggled to Denmark.[32] Bidstrup entered a course of what can be thought of as a British education in instrument making, which perhaps involved learning some basic scientific and technological principles.[33] His visits were necessary because "he realized that he had to master all phases of the division of labour himself, since the various skills were absent in Copenhagen."[34] Presumably he succeeded in this, since he went on to establish modern Danish instrument making.

In some cases, work experience and training abroad made bringing foreign workers to Scandinavia redundant. Examples include Danish instrument making at the end of the eighteenth century[35] and the successful introduction of the Huntsman method to the Swedish iron industry in the 1790s by two Swedes who had labored in Sheffield: ("after their study tours . . . they needed no further help from Britain"). Two decades earlier, the introduction of new production methods in steel had involved Britons "in important functions . . . as owners, technical supervisors and skilled workers."[36]

Foreign travel and work experience seem to have been pervasive in the case of the Danish woolen industry, sometimes preceding the establishment of firms. The dyer M. K. Brandt spent several years in Germany as an apprentice in the 1820s. His son traveled in Germany and worked in a textile firm in 1865. Upon his return, he managed the installation of machinery at the family's new textile factory, which he later directed with his brother. Before the establishment of the Bruunshåb factory in 1821, the owner-manager sent his son to Hamburg, Berlin, and Bohemia to study modern wool weaving. His son later worked two years in a factory in Belgium and became a manager of the family firm. J. L. Binder, who established a woolen factory in Denmark in 1883, had also visited Germany, and his son traveled in Germany for three years in the early 1890s, gaining work experience at several German factories.[37] Visits were made to England in the 1870s and 1880s. In the 1840s and 1850s, a number of visits were supported by Danish Reiersenske money.

The development of Scandinavian industry in this period relied heavily on foreign workers. Workers were recruited to the glass, leather, salt, steam engine, cobalt, porcelain, textile, iron and steel, and mining industries. An extensive program for developing domestic industry involved bringing approximately sixty manufacturers from Britain, Holland, and Germany to Denmark during the period 1730 to 1746, for which a large sum was set aside annually.[38] In Norway, the Bærum ironworks successfully operated steelmaking equipment smuggled from Sheffield's Carron Iron Works with the aid of two British workers.[39] Such instances seem to be rather general. Table 7.4 demonstrates a range of known cases of foreign skilled labor input. Like table 7.1, it is drawn from a survey (by no means complete) of the literature on the history of technology in Scandinavia.

Dan Christensen concluded in his study of technology transfer to Denmark and Norway that "expertise from abroad made up the foundation on which the development of modern Danish and Norwegian industries could be built."[40] He noted, however, that "in view of the significant British emigration to France and the USA it is remarkable how few came to Denmark-Norway."[41] Nevertheless, immigrant workers are a persistent feature of the picture in sectoral studies of Scandinavian industrialization.

As the industrialization process accelerated in the middle of the nineteenth century, the communication and demonstration of industrial innovations took new forms. Formal exhibitions became particularly important. These presented, in effect, a review of the state of the art in mechanical and industrial technique. Such exhibitions appear to have been very important for the diffusion of knowledge in Scandinavia. During the nineteenth century, exhibitions were held in a number of large European cities (including, as well as the major capitals, Prague, Lyon, Mulhouse, and Birmingham), but they were not confined to major cities. This desire for industrial exhibitions spread to Scandinavia. Between 1823 and 1849, for example, no fewer than sixteen exhibitions were held in Sweden.[42]

The combined Kingdom of Norway and Sweden made serious efforts to participate in international exhibitions. Participation in the 1862 exhibition in London was organized by a prestigious committee, chaired by Prince Oscar (the Swedish king's brother), and members were drawn from the academic elite. The committee held approximately fifty meetings, and local and regional subcommittees were established all over Sweden and in Norway. Sweden had 511 exhibitors and Norway 216.[43] These exhibitions became clearly linked to transfer activities. For example, the Norwegian Parliament funded travel by two mechanics to visit the

Table 7.4
Some Immigrant Workers in Scandinavia

From	To	Sector	Numbers/name	Date	Function
Holland	Norway	Paper	"Experts"	Late 17th century	To introduce methods
Holland	Sweden	Paper	"Experts"	18th century	
UK, Holland, Germany	Denmark	Various	60	1730–1746	"Manufacturers"
France	Denmark	Engineering	Etienne Jandin de Peyrembert	1751–1756	To introduce making iron cannons
France	Denmark	Engineering	Lorenz Juncker	1756	Boring master; makes cannons
UK	Sweden	Iron	2 workers. Samuel Houlder Sr.	1829	Smiths; teach British forging methods
UK	Sweden	Iron	"Smiths from Monmouthshire"	1830	Continue 1829 activity; refining; steel
UK	Sweden	Iron	2 workers	1830s	Teach blister steel production
UK	Sweden	Engineering	Several workers	1830s	Teach; work; manage
UK	Sweden	Textiles	Several workers	1830s	Foremen; various
Germany	Denmark	Building and construction	Many	1830s	Brick and tile workers and managers

Table 7.4
(continued)

From	To	Sector	Numbers/name	Date	Function
Germany	Denmark	Glass	21 workers	1833	
UK	Norway	Engineering	John Wilson	1840s	Taught drawing and mill construction
UK	Denmark	Textiles	Hugh F. Rennie	1840s	Engineer and manager
Germany	Denmark	Textiles	Several workers	1840s	
UK	Norway	Engineering	At least 33 workers	1843–1890	
UK	Norway	Textiles	86 workers	1845–1871	Skilled workers and foremen
UK	Norway	Utilities	James Malam	1847–1848	Constructed first gasworks in Oslo
UK	Denmark	Utilities	At least 14 workers	1850–1890s	Workers and managers
UK	Norway	Utilities	James Small	1851–1853	Built first gasworks in Trondheim
Germany	Norway	Food and drink	Engineer	1866	Set up machinery for flour milling
UK	Denmark	Building and construction	Master	1860	Organized cement production
Germany	Norway	Food and drink	Engineer	1866	Set up machinery for flour milling
Sweden	Denmark	Building and construction	Workers	1872	Brick and tile workers

1851 exhibition in London: this led to the importation of reaping equipment diffused in Norway through demonstrations at local farm fairs.[44] The organized effort to participate in exhibitions and make use of their results was thus a component of the wider transfer endeavor.

John R. Harris has argued that espionage was central to knowledge diffusion during the industrialization of Western Europe.[45] It was certainly a key element of knowledge acquisition in Scandinavia: individuals, firms, and governments sponsored industrial espionage on a large scale. Espionage occurred in textiles, where "the Danish and Swedish governments sent hand-picked agents to Britain;"[46] in Norwegian glass production, where a large part of the development of this industry rested on the work of "a young citizen . . . on an assignment of industrial espionage to Britain;"[47] in metalworking, where the Dane Jørgen Dalhoff bribed his way into plants in Britain and Germany (paying skilled artisans for specific information);[48] and in iron production, where a Swedish production manager was advised to invite British mill foremen out and ply them with drink, since "that usually opens the Englishman's heart."[49] Other fields included weapons, pottery, instruments, general engineering machines and tools, canal building, mint production, and steam technology. Christensen maintained that the international competitive strength of Danish and Norwegian salt, cobalt, glass, and porcelain production all depended on a continuous flow of information on foreign developments gained through repeated spying in Britain and on the Continent. He has argued that it is "difficult to see how Danish textile production could have been established without the espionage of Ljungberg and Nordberg."[50] The information so acquired related not only to process innovations but also to new products.

There seems to have been an important element of "know who" in all this—that is, social knowledge about whom to contact and seek information from.[51] At various times between 1750 and 1850, fourteen Norwegian and Danish spies operated in a wide range of fields, primarily in Britain.[52] They brought back knowledge about salt, steam technology, tools, iron and steel, cobalt, textile technology, mining, weapons, porcelain, glass, and various other products and sectors. Eight spies were members of Masonic lodges with extensive international links. For example, it appears that members of London lodges significantly aided the access to technological information for three important Danish-Norwegian spies.[53] Scandinavian spies were frequently members of the key societies and associations in the target countries, and their espionage efforts were aided by interaction with co-members, as well as with members of similar organizations abroad. Christensen argued that transfer to Denmark and Norway "was undertaken by a small cosmopolitan elite." Most industrial spies were postgraduates in or professors of natural philosophy. They were members of the Danish or Norwegian Societies of Sciences and Letters (and a few of the Royal Society in London) or of the Society of Agriculture (and some of the London Royal Society of Arts): "they established networks with corresponding organisations abroad."[54] Finally, important input came from contact with Scandinavian officials stationed abroad. During his visit to England in 1807, a Danish state-supported spy named Marstrand contacted diplomats and other Scandinavians in London, who told him that he would get nowhere without British connections and their letters of introduction. In the event, his efforts were thwarted, and he left after the British bombardment of the Danish fleet.[55] Industrial espionage was widely regarded as a threat by the British, as David Jeremy has shown.[56] This concern was justified—and as a channel of general information, it was unquestionably important.

CONCLUSION

This chapter has focused on knowledge creation and knowledge flows and on the specific institutions that supported such flows. For reasons of space, some important dimensions of the creation of knowledge capabilities have been omitted. Perhaps the most central of these is education, particularly having to do with literacy and the social value placed on it. Certainly human capital, often reflected in literacy and education, has become an important part of recent growth theory. How did Scandinavia perform with respect to literacy? Scandinavia generally had high levels of literacy from an early era. Much of this was driven by religious impulses, with churches and pastors placing a high value on religious literacy featuring a wide circulation of religious texts. States in all parts of the region began literacy campaigns in the late seventeenth century. This initiative developed into comprehensive education systems that no doubt have had powerful long-term economic impacts.[57]

A further issue that requires greater attention is tariff policies and their impact on knowledge creation and industrialization. It is sometimes argued that protection is central to industrialization, but Scandinavia had no clear or consistent form of tariff protection across the various countries or regions during this period. The variety of the Scandinavian experience suggests that neither tariff policies nor direct state interventions were particularly central to the knowledge-acquisition processes on which industrialization really rested. Tariffs were political measures aimed at affecting income and wealth distribution rather than industrialization, and although tariffs protected some sectors in some parts of Scandinavia during some periods, they did not facilitate the broad process of industrial technology acquisition across Scandinavia as a whole.[58]

This chapter has suggested that although industrialization is ultimately a technological process, technological change requires explanations that are appropriate to its character and scope. Technology in itself explains little, except in the most proximate way. Rather, it is necessary to look behind the process of technological change to explain its preconditions and underlying mechanisms. This chapter has understood technology as knowledge and industrialization as a process of knowledge acquisition and accumulation, which means that it is necessary to explore what forms knowledge acquisition took, as well as how it was shaped and organized.

Explanations of this development need to focus on how a wide range of more or less contemporaneous institutional and organizational changes created the knowledge that enabled Scandinavia to innovate and grow in a multisectoral way. Learning and knowledge accumulation, and hence the general propensity to innovate, cannot be reduced to any single institution of the market economy, such as property rights regimes. Rather, there were quite specific areas of institutional and organizational change that interacted to create the conditions for industrialization. One possible conclusion is that the human capital dimensions of industrialization operated as an interrelated complex. No specific sector, mode of learning, or technological change underlay Scandinavian industrialization. What mattered was an overall set of activities that, partly by accident, but also to a serious extent by design, worked together to create a differentiated economic system capable of sustained growth. Technical societies, scholarly institutions, publishing behavior, foreign travel by engineers, espionage, industrial exhibitions, and the growth of schooling were far from accidental: they required serious commitments of time, energy, and money. A large part of this commitment was by the state, but much also depended on the wider social capital, such as technical societies and even churches, which promoted education. These related developments made it possible for competence to be acquired and expanded at many levels and in many ways. These phenomena interacted, suggesting that it was the overall impact rather than specific activities that mattered. Although the concept of "the knowledge economy" is very recent, it may be rather old in practice, as much of the material presented in this chapter suggests.

In effect, the Scandinavian region created the institutions of a learning economy, building capacities for technological change that ultimately reached into every corner of these economies.

ACKNOWLEDGMENTS

Thanks to Ragnhild Hutchison for invaluable research assistance in preparing this chapter and to the University of Oslo for financial support.

NOTES

1. Eric L. Jones, *The European Miracle: Environments, Economies and Geopolitics in the History of Europe and Asia* (Cambridge: Cambridge University Press, 1981), 69

2. Anton M. Schweigaard, cited in Trond Bergh et al., *Norge fra U-Land til I-Land. Vekst og Utviklingslinjer, 1830–1980* (Oslo: Gyldendal Norsk Forlag, 1983), 140.

3. The national borders of the Nordic countries have changed frequently and radically at least since the late medieval period. In the nineteenth century, these shifts included the Treaty of Kiel (1814) that marked the end of a long-lasting subordination of Norway to Denmark and the start of a Norwegian-Swedish union under the Swedish Crown; the Swedish surrender of Finland to Russia (1808–1809); and Denmark's loss of Schleswig-Holstein to Germany (1864). In the early twentieth century, Norway became independent from Sweden (1905), Iceland from Denmark (1918), and Finland from Russia (1917–1919).

4. See Monica Aase, "Patrioter og bønder. Det Kongelige Norske Videnskabers Selskabs arbeid med landbrukspremier, 1772–1806," *Vitenskap & Kultur* (Trondheim: NTNU, 1998), 1–227, and "Premiering av bøndenes flid. Premiesøknader til Det Kongelige Norske Videnskabers Selskab," *Heimen* 33 (1996), 103–110. In Sweden, the Royal Patriotic Society (established in 1766) similarly rewarded innovation from 1772.

5. Klas Nyberg, "Staten, manufakturen och hemmemarknadens framväxt" (paper presented at the Att vara på modet. Den förste tvärvetenskapliga konferencen om kläder ock mode från medeltid til moderne tid," October 24–25, 2003, Södertörns högskola).

6. Hrefna Robertsdottir, "Manufaktur og reformpolitikk. Nye arbeidsmetoder og opplæringstiltak innenfor ullproduksjonen i 1700-tallets Island," *Scandia* 66:2 (2000), 358–375.

7. Kirsi Vainio-Korhonen, "Handicrafts as Professions and Sources of Income in Late Eighteenth and Early Nineteenth Century Turku (Åbo): A Gender Viewpoint to Economic History," *Scandinavian Economic History Review* (*SEHR*) 48:1 (2000), 40–63; Stein Tveite, "The Norwegian Textile Market in the 18th Century," *SEHR* 17: 2 (1969), 161–178.

8. Gunvor Trætteberg, "Omfarshandel. Skreppekarer, Driftekarer og jekteskippere i Hordaland," *Norveg* 2 (1952), 104–131 ; Sven Dahl, "Traveling Pedlars in Nineteenth Century Sweden," *SEHR* 8: 1 (1960), 166–178.

9. Kristof Glamann, "Beer and Brewing in Pre-Industrial Denmark," *SEHR* 10 : 1 (1962), 128–140; Lars Magnusson, "Drinking and the Verlag System 1820–1850: The Significance of Taverns and Drinks in Eskiltuna before Industrialisation," *SEHR* 34: 1 (1986), 1–19.

10. Johan Söderberg, "A Long-Term Perspective on Regional Economic Development in Sweden, ca. 1500–1914," *SEHR* 32: 1 (1984), 11.

11. Ove Hornby and Erik Oxenbøll, "Proto-industrialisation before Industrialisation? The Danish Case," *SEHR* 30: 1 (1982), 3–33; Edgar Hovland, Helge Nordvik, and Stein Tveite, "Proto-industrialisation in Norway 1750–1850: Fact or Fiction?" *SEHR* 30: 1 (1982), 45–55; Kai Hoffman, "Sawmills–Finland's Proto-Industry," *SEHR* 30: 1 (1982), 35–43; Lennart Schön, "Proto-Industrialisation and Factories: Textiles in Sweden in the Mid-Nineteenth Century," *SEHR* 30: 1 (1982), 57–71; Lars Magnusson and Maths Isacson, "Proto-Industrialisation in Sweden: Smithcraft in Eskiltuna and Southern Dalecarlia," *SEHR* 30: 1 (1982), 73–99; Ragnhild Hutchison, "Enigheten–tekstilfabrikken i Østerdalen. Fabrikkdrift og teknologioverføring i det norske bondesamfunn på slutten av 1700 tallet" (Hovedfag thesis, University of Oslo, 2003); Arne Espelund and Ole Evenstad, *Iron Production in Norway during Two Millennia* (Trondheim: Arketyp, 1995); Ingrid Lowzow, *Det gamle skomakeri* (Oslo: Instituttt for etnologi, University of Oslo, the Forbundsmuseet in Akershus, 1989); Birgitta Conradson, *Pappermästrana. Om holländarna på 1700-talets Tumbla bruk* (Stockholm: Nordiska Museet, 1994).

12. Svante Lindqvist,"Natural Resources and Technology. The Debate about Energy Technology in Eighteenth-century Sweden," *Scandinavian Journal of History [SJH]* 8 (1983), 83–110.

13. The detailed sources and references for this table are available from the author.

14. Ole Feldbæk, "The Danish Asia trade 1620–1807. Value and Volume," *SEHR* 34:1 (1991), 3–27; Erik Gøbel, "The Danish Asiatic Company's Voyages to China, 1732–1833," *SEHR* 27: 1 (1979), 22–46; Erik W. Fleisher, "The Beginning of the Transatlantic Market for Swedish Iron," *SEHR* 1:2 (1953), 178–192.

15. Bertil Andersson, "Early History of Banking in Gothenburg Discount House Operations 1783–1818," *SEHR* 31: 1 (1983), 50–67; Kurt Samuelsson, "International Payments and Credit Movements by the Swedish Merchant-Houses, 1730–1815," *SEHR* 4: 1 (1956), 161–202; Elsa Britta Grage, "Capital Supply in Gothenburg's Foreign Trade," *SEHR* 29: 2 (1981), 97–128.

16. Fritz Hodne, "Transfer Patterns of Technology: Theory and Evidence," in *Technology Transfer and Scandinavian Industrialisation*, ed. Kristine Bruland (New York: Berg, 1991), 177.

17. Joel Mokyr, *The Gifts of Athena: Historical Origins of the Knowledge Economy* (Princeton, NJ: Princeton University Press, 2002). See also chapter 1, this volume.

18. Niels Clemmensen, "The Development and Structure of Associations in Denmark, c. 1750–1880," *SJH* 13 (1988), 355–370.

19. Rolv Petter Amdam, "Den organiserte jordbrukspatriotismen 1769–1790–ei jordbrukspolitisk reformrørsle?" (Hovedfag thesis, University of Oslo, 1985); Hutchison, "Enigheten—tekstilfabrikken I Østerdalen."

20. Kristine Bruland, *British Technology and European Industrialization: The Norwegian Textile Industry in the Mid-Nineteenth Century*, 2nd ed. (Cambridge: Cambridge University Press, 2003 [1989]), 58–59.

21. Dan Christian Christensen, *Det Moderne Projekt. Teknik & Kultur i Danmark-Norge 1750–(1814)–1850* (Copenhagen: Gyldendal, 1996), 801.

22. Ibid., 500, 460.

23. Jan Hult, Svante Lindqvist, W. Odelberg, and Sven Rydberg, *Svensk Teknikhistoria* (Värnamo: Gidlunds Bokförlag, 1989), 121.

24. Frank Allan Rasmussen, "Holmen og Orlogsværftet. Et teknisk innovativt center?" *Fabrik og Bolig. Det Industrielle miljø i Danmark* 2 (1991), 3, 3–16.

25. Ibid.

26. Poul Strømstad, "Artisan Travel and Technology Transfer to Denmark," in Bruland, ed., *Technology Transfer*, 137, 153, and "Teknologioverførsel eller hvor lærte de det?" *Fabrik og Bolig. Det industrielle miljø i Danmark* (1989), 3–21.

27. Bruland, "Skills, Learning and the International Diffusion of Technology," in *Technological Revolutions in Europe: Historical Perspectives*, ed. Maxine Berg and Kristine Bruland (Cheltenham: Edward Elgar, 1998), 179.

28. Chris Evans and Göran Rydén, "Kinship and the Transmission of Skills: Bar Iron Production in Britain and Sweden, 1500–1860," in Berg and Bruland, eds., *Technological Revolutions in Europe*, 198–199.

29. Lennart Jörberg, "The Diffusion of Technology and Industrial Change in Sweden during the Nineteenth Century," in Bruland, ed., *Technology Transfer*, 188.

30. Christensen, *Det Moderne Projekt*, 510–520.

31. See e.g., Bruland, *British Technology and European Industrialization*, 61–67.

32. Christensen, *Det Moderne Projekt*, 51–120.

33. Ibid., 518–519.

34. Ibid., 818.

35. Ibid., 519.

36. Rolf Adamson, "Borrowing and Adaptation of British Technology by the Swedish Iron Industry in the Early Nineteenth Century," in Bruland, ed., *Technology Transfer*, 96.

37. Ole Hyldtoft, *Teknologiske Forandringer i dansk industri, 1870–1896* (Odense: Odense Universitetsforlag, 1996), 171–172.

38. Poul Strømstad, "Artisan Travel and Technology Transfer to Denmark," in Bruland, ed., *Technology Transfer*, 137.

39. Christensen, *Det Moderne Projekt*, 517.

40. Ibid., 818.

41. Ibid., 519.

42. See Göran Ahlström, *Technological Development and Industrial Exhibitions 1850–1914: Sweden in an International Perspective* (Lund: Lund University Press, 1996).

43. Ibid.

44. Fritz Hodne, "Transfer Patterns of Technology: Theory and Evidence," in Bruland, ed., *Technology Transfer*, 171.

45. John R. Harris, "Law, Espionage, and the Transfer of Technology from Eighteenth-Century Britain," in *Technological Change: Methods and Themes in the History of Technology*, ed. Robert Fox (Amsterdam: Harwood Academic Publishers, 1996), 122–135.

46. See Christensen, *Det Moderne Projekt*, 268–284.

47. Rolv Petter Amdam, "Industrial Espionage and the Transfer of Technology to the Early Norwegian Glass Industry," in Bruland, ed., *Technology Transfer*, 75.

48. Poul Strömstad, "Artisan Travel and Technology Transfer to Denmark," in Bruland, ed., *Technology Transfer*, 144–147.

49. Lennart Jörberg, "The Diffusion of Technology and Industrial Change in Sweden during the Nineteenth Century," in Bruland, ed., *Technology Transfer*, 194.

50. Christensen, *Det Moderne Projekt*, 426, 517.

51. The term is from Bengt-Åke. Lundvall and Björn Johnson's analysis of current events in "The Learning Economy," *Journal of Industry Studies* 1 (1994), 23–42.

52. Christensen, *Det Moderne Projekt*, 507–508.

53. Ibid., 507–508, 516–522.

54. Ibid., 818–819.

55. Ibid., 518.

56. David J. Jeremy, "Damming the Flood: British Government Efforts to Check the Outflow of Technicians and Machinery, 1780–1843," *Business History Review* 51 (1977), 1–34.

57. This issue is raised in Bruland, "Skills, Learning and the International Diffusion of Technology," in Berg and Bruland, eds., *Technological Revolutions*, 183–184. But see also Loftur Guttormsson, "The Development of Popular Religious Literacy in the Seventeenth and Eighteenth Centuries," *SJH* 15 (1990), 7–35; Ingrid Markussen, "The

Development of Writing Ability in the Nordic Countries in the Eighteenth and Nineteenth Centuries," *SJH* 15 (1990); Lars Sandberg, "Ignorance, Poverty and Economic Backwardness in the Early Stages of European Industrialization: Variations on Alexander Gerschenkron's Great Theme," *Journal of European Economic History* 2 (1982), 686–688, and "The Case of the Impoverished Sophisticate: Human Capital and Swedish Economic Growth before World War I," *Journal of Economic History* 39 (1979), 230.

58. See Hans Jørgen Jørgensen, *Det Norske Tollvesens historie, Fra middelalder til 1814, Bind 1* (Oslo: Tolldirektoratet, 1969), 271–285, 296–299; Henrik Becker Christensen, *Proteksjonisme og reformer, 1660–1814, Dansk Toldhistorie, Bind 2* (Copenhagen: Toldhistorisk selskab, 1988), 467–473; Hans Christian Johansen, *Dansk økonomisk historie i årene efter 1784, Bind 1: Reformår 1784–88* (Aarhus: JSH, Universitetsforlaget i Aarhus, 1968), 400–404; Leo Müller, *The Merchant Houses of Stockholm c. 1640–1800, A Comparative Study of Early-Modern Entepreneurial Behavior* (Uppsala: Studia Historica Upsaliensia 188, 1998), 47; Staffan Høgberg, *Utrikeshandel och sjøfart på 1700-talet. Stapelvaror i svensk export og import 1738–1808* (Stockholm: Bonniers, 1969), 28–33; Eli F. Heckscher, *An Economic History of Sweden* (Cambridge, MA: Harvard University Press, 1963).

8

Crafting the Industrial Revolution: Artisan Families and the Calico Industry in Eighteenth-Century Spain

Marta V. Vicente

"If Catalonia were all of Spain, Spain would over shadow England," stated the Anglophile Spanish journalist Francisco Mariano Nipho in 1779.[1] A year later, he reiterated his admiration for Catalonia's industrial achievements by declaring it to be "a little England at the heart of Spain."[2] Nipho's exaggerated claim had a grain of truth. In less than fifty years, the growth of the cotton industry and its main product, calico cloth, had brought dramatic changes to the city's industry. From only eight manufactures in 1750, at the end of the century, Barcelona owed its "splendor and wealth" to its 150 calico concerns.[3] In fact, in the 1780s, Barcelona alone had more calico *fábricas*, or manufactures of printed calicoes, than Great Britain.[4] Visitors to Barcelona found calico manufactures on virtually every street corner. From large concerns employing up to three hundred workers to small enterprises with two or three family members, manufacturers sold everything from expensive taffeta to cheaper mixed calicoes of cotton and linen, fabrics of intricate designs or simple stripes and checks, and fabrics dyed in madder and indigo blue. At the turn of the century, contemporaries affirmed that these fábricas employed between ten and fifteen thousand men, women, and children. One probably exaggerated calculation brought the total to eighty thousand individuals in 1816. These thousands of workers made up to 7 million yards of cloth annually during the 1790s, fostering a culture of prosperity and conspicuous display of wealth among Catalans.[5] A number of calico entrepreneurs were granted titles of nobility; others, as chronicler Rafel d'Amat observed, "in a short time enriched themselves with the sale of those fashions."[6] Calico manufacturers who did not become nobles or wealthy entrepreneurs contented themselves with displaying their unique fabric designs from their shop doors.

Although impressive, is this evidence enough to affirm that the calico industry of Barcelona brought an industrial revolution to Spain? The cotton industry was not the only one to flourish in the eighteenth century. Production in the steel industry of Bilbao as well as in the royal wool "factories" in Castile also saw considerable growth.[7] But until the early nineteenth century, the growth of the calico industry was limited to Barcelona and a few other centers along the Mediterranean and Atlantic coasts, such as Málaga and Seville. Yet current scholarship would agree that after 1760, Barcelona's calico manufactures had already begun a process of industrialization marked by concentration of labor (in buildings with one hundred to three hundred workers), incorporation of new technology, and easy access to raw materials. These calico manufactures produced fashionable fabrics at affordable prices and sold them to thousands of consumers in Spain and the Americas.[8] In this chapter, I argue that the changes that took place in small calico manufactures became crucial to the growth of the Spanish calico industry. Large manufactures were able to outsource parts of the production process to smaller enterprises. This collaboration between small and large fábricas allowed the calico industry to survive its most difficult years. The study of the work of the family and production in small manufactures also illustrates how modest increments in production and technological advance in one industry could have an impact on large-scale ventures and how the adaptation of traditional practices resulted in long-term implications with a qualitative departure from past industrial forms.

In order to understand the impact of family forms of work and the relationship between small and large manufactures in the industrial process, it is necessary to broaden the usual scope of investigation from individual manufactures to a citywide structure. This citywide structure also needs to be connected to Catalonia's importance as an industrialized region— what Jordi Nadal labeled "the factory of Spain."[9] Catalonia's role, and especially Barcelona's, as "the factory of Spain" had a long history going back to the Middle Ages, when Barcelona's artisanal production was pivotal in the industrial development of the city as a major cloth-producing and commercial center. Among the eight principal centers of the Catalan woolen industry around 1400, Barcelona was the largest, with a production capacity of some forty thousand bolts of cloth a year.[10] During the early modern period, however, Barcelona declined to the rank of a peripheral player in the Mediterranean and Atlantic trades, fighting what increasingly seemed a losing battle to maintain its stature as the main industrial and commercial center of Spain.

In the early eighteenth century, Catalonia began an astonishing recovery in both agriculture and industry. As if fulfilling the predictions of political economist Narcís Feliu de la Peña in 1683, Catalonia became the legendary phoenix reborn from its own ashes.[11] Amid the eighteenth-century economic and demographic recovery of Spain, Barcelona reclaimed its earlier position as one of the major commercial and financial centers in Spain and the western Mediterranean. Spain's population grew from 7.5 million at the beginning of the century to 10.5 million in 1797, with the greatest density on the Mediterranean coast and in Andalusia. Such growth at the periphery was partly a result of the development of trade to the American colonies. This change allowed Barcelona to become increasingly influential in the economic life of the Empire.[12] As Albert García Espuche has demonstrated, Barcelona's growth can be understood only by taking into account the development of the city as part of the general reordering of merchants' market strategies that featured a decentralized industrial and commercial division of work to form the core of a "crown radiating" to surrounding secondary commercial and industrial centers.[13] Bartolomé Yun and Jaume Torras have further illustrated how this reordering became effective: Catalans used their local networks not only to trade with the rest of Spain but also to establish separate communities of merchants and manufacturers.[14] It is this early modern transition that makes Barcelona's industrial development unique and justifies focusing on the calico industry in order to tell the story of the Industrial Revolution in Spain.

ARTISANS, FAMILY, AND THE SPANISH CALICO INDUSTRY: 1730s–1780s

By the 1730s and 1740s, families of entrepreneurs established the first Spanish cotton manufactures in Barcelona. The primary challenge these families encountered was the technical difficulties of the calico trade. Of the three main parts of calico making—spinning the thread, weaving the cloth, and printing the calicoes with dyes—the critical part was printing the cloth with engraved blocks. The expert on dyeing was a fabricant, the only person who knew the secret of "making the colors," that is, making the mixture of mordant and thickener that ultimately determined the different colors of the cloth and made dyes permanent. Barcelona's entrepreneurs had long purchased good-quality cotton thread from Maltese merchants, and the city's artisans wove, bleached, and dyed various cotton fabrics. Yet as Bernard Glòria attested, since at least the late 1720s, manufacturers like him had been unsuccessful in their attempts to master the technique of calico printing.[15] Thus from the 1730s to the 1750s, Catalan calico manufacturers had to hire foreign experts from France, Germany, and Switzerland to produce the distinctive patterns that characterized calico fabric. It was not until the 1760s that a growing number of Catalan master dyers and journeymen learned calico-printing techniques.[16]

Growth in the industry during the 1760s and 1770s went hand-in-hand with increasing access to the technological knowledge of printing and engraving within the community of artisans. In fact, by 1766, knowledge about how to print calicoes had extended so widely that the owners of the largest manufactures advised the Board of Trade to enact ordinances to regulate the trade.[17] In 1768, the Royal Board of Trade (the institution that regulated the Spanish cotton industry) listed twenty-nine individuals who possessed printing skills; between 1780 and 1784, the official count reached fifty-four. The real numbers were probably much higher, since many individuals never became certified fabricants.[18] Aspiring calico entrepreneurs saved money by purchasing second-hand tools or by using terraces on the roofs of apartment buildings to dry calicoes instead of buying or renting a separate building.[19] These remodeled workshops for calico printing differed little from traditional artisanal workshops. Yet contemporaries still labeled these small enterprises "factories" because they printed calicoes, an activity unregulated by any guild.

The growth of the industry relied not only on the initiative of artisans to learn the techniques of the calico trade to establish their own manufactures, but also on the human and social capital represented by the family. If calico entrepreneurs' actions, regardless of the size of their investment and enterprise, rested on artisanal traditions, at the core of these traditions was the work of the family. Thus families provided the industry with human capital—individual skills—as well as social capital—social networks based on principles of trust and mutual reciprocity—with which entrepreneurs were able to run their factories successfully. It was also in the ideal of the family that entrepreneurs as well as authorities found models of economic behavior to apply to the new industry.

The challenges of a new industry and the demands of manufacturing a product with variable demand—in size and in fashion—made entrepreneurs in the calico industry particularly vulnerable to bankruptcy. To survive, owners of calico manufactures relied on their social capital: the networks of relatives and acquaintances whose affective links encouraged a high level of commitment to sustaining the business. This reliance on social capital was particularly crucial for small businesses. His older brother's production of cloth kept the veil maker Francisco Dordal's calico business afloat in his modest apartment on carrer Mònec in Barcelona's Sant Pere neighborhood. Although small, Francisco's output was characterized by a wide variety of designs, another factor that probably accounts for the survival of his business.[20] Josep Dordal's enterprise was smaller still; with only two looms, he wove white cotton cloth in his rented apartment, also in the Sant Pere neighborhood. He also sold his white cotton cloth to owners of larger fábricas. He had worked for some of them—like Mateu Farra—as a weaver.[21]

The familial component of these businesses affected their likelihood of survival, although there are no reliable data as to whether family-owned enterprises lasted longer than those with no relatives involved. Yet the number of factories that had, in one way or another, links with relatives in the business appears impressive. Fathers and sons, husbands and wives, widows and sons, and siblings owned both large and small calico enterprises. When the factory's name does not already underscore its family origins—"Rafel Arxer and Sons Co.," "Fábrica de Ramón and Teresa Clossa," or "Fábrica Widow Vicente & Co.—the extant archival documentation tends to demonstrate it. Manuel González owned a factory that manufactured printed silks, but it was Jaume Pons, his brother-in-law, who managed it.[22] The calico fábrica of Martín Riera was partly managed by his wife, Rosa. Factory owner and linen weaver Mateu Farra had his son, brother-in-law, and later even his grandson working for him.[23] In 1796, the fábrica of Joan and Antonia Costa opened in an apartment the couple rented on the recently paved carrer del Conde del Asalto. The Costas had their small calico manufacture on the second floor of a four-story building; the first floor of the same building was inhabited by another calico manufacturer, Francisco Babil and his wife, Teresa.[24] Bernard Llorens also lived with his family on the first floor of his house and had his factory on the second floor.[25]

In their small manufactures where they lived with their families, artisans also drew and engraved calico designs on cheap wooden molds. Engraver Pablo Vidal, a master turner, engraved molds for his friend Miquel Llorens, owner of a small calico factory. Vidal engraved Llorens's designs on the molds. He worked, as his wife said, "in the workshop of his own house."[26] Likewise, in the 1790s, Joan Ricart and Josefa Pou married and established a fábrica on carrer Riera d'en Prim. When Joan died shortly afterward, Josefa took over the business, which prospered into the first years of the next century.[27]

The familial context in which these artisans labored extended beyond the close family and the limits of the home. In 1792, engraver Josep Cabesa engraved calico blocks in his small workshop on carrer Mònec, not far from where the Dordal brothers had their businesses. Cabesa sometimes engraved designs brought to him by owners of larger factories, but on other occasions, he engraved his own designs onto the blocks. To market this aspect of his trade, his daughter's father-in-law, who lived with the Cabesas, sold in the streets the drawings that Josep made.[28]

The value of this social capital and the role of family networks also applied to owners of large fábricas in the daily running of their concerns. Factory owners had their own relatives participate in the making of calicoes.[29] In the Fábrica Sirés, for instance, Joan Baptista Sirés's brother, Pau, lived and worked in the factory building. Later their nephew Pere Antón, a certified fabricant, worked there. Sirés's sister, Antonia, who lived in Cardedeu, about thirty miles north of Barcelona, hired weavers to make cotton cloth for the factory. Their brother Antón delivered calicoes to shopkeepers around Catalonia. When Antón died in 1771, the youngest brother, Llorens, transported Sirés's calicoes. Another relative, located in Vinaroz, Valencia, became the company's principal broker in the region. In 1798, Pau Sirés, the brother who had been living in the Sirés factory, went to help their relative in Valencia by working as a merchant's assistant; he was charged with all business correspondence.[30]

This dense network of family relations at the managerial level was replicated on the factory floor. Sirés, as well as other manufacturers, hired entire families of workers, a pattern characteristic of other European calico manufacturers and also of textile factories throughout the eighteenth and nineteenth centuries.[31] In the Sirés factory, at least 25 percent of the workers had relatives laboring in the large building. Having entire families working on the factory premises allowed employers to better replicate some of the familial practices of work in artisanal shops: the male artisan worked at the main activities of the trade—the tasks that resulted in the most valuable products—while his wife, daughters, and young sons usually took up duties that either supplied or prepared the materials necessary to complete the main part of the trade. They warped and wound the thread and set it on the loom. This form of organizing work was aided by the way large fábricas replicated the artisan workshop, where each working room was a separate unit that resembled a family workshop. This structure allowed each room to have workers attending to a single task: winding thread, warping and weaving the cloth, or printing calicoes with engraved blocks. Such organization of the work process created a sense of familiarity, which perhaps reduced turnover in times of high demand. Nevertheless, family and factory work sometimes resulted in conflicting loyalties. Owners of large, relatively impersonal shops always ran the risk of having their workers leave without notice to go back to a family business or to work in a more family-like factory. In 1785, the Royal Spinning Company

(Real Compañía de Hilados), formed by a group of twenty-five owners of large factories, complained that calico workers preferred to leave the large fábricas to go to smaller ones, where they supposedly felt "more relaxed" and free from constraints.[32]

Despite such tensions, the family also served as the model for the imposition of industrial labor discipline. For instance, the report of the Royal Spinning Company of 1785 looked to the family as the template for the calico industry's correction of workers' disrespect. The owners of large manufactures firmly believed that only a paternalistic hierarchy, in which workers treated owners with filial respect and obedience, would save the industry from moral decadence. Even when employers delegated part of their power to a small group of foremen, they believed that all workers depended on the father-owner. This hierarchical division of work and the subordination of workers were necessary, the *patrons* believed, to organize work in the factory productively. The foremen ensured that workers arrived on time, were present at work, and labored according to the manufacture's daily needs.[33] Still, despite the foremen's control of the workers' daily performance, it was the owner's authority that formed "a class of enduring workers." Thus, the Royal Spinning Company maintained that workers "having entered the fábrica beardless, go gray working in it, loyal and faithful because of the love of their employers, who provide them with a continuous and secure mode of subsistence."[34] Of course, regardless of what the Royal Company thought, workers kept leaving their employers with the hope of establishing their own calico factories. In fact, throughout the industry, but particularly in large enterprises, 60 to 70 percent of employees stayed less than a year.[35]

Finally, the family framed the organization of production by both mighty and minor manufacturers. Entrepreneurs formed companies, also known as "commercial houses" (*casas comerciales*), a type of commercial association traditional among Catalan entrepreneurs in the early modern period.[36] Whether forming small or large companies, when partners drafted a contract for the establishment of a factory, they entered a "company and society," two terms with familial connotations. Legal texts sometimes referred to marriage as the "conjugal society" or the "matrimonial company."[37] In small manufactures, with only two or three partners and where most of the workers were close relatives, the trust and obedience expected of family members were especially necessary for the survival of the business. But even in a large factory, with five partners and dozens of workers not closely related, the core of the business remained familial. When, in 1770, the Fábrica Sirés merged with Alegre & Gibert Co., the resulting Sirés & Co., with its large factory, still functioned as a small family business. Joan Baptista Sirés remained the head of the factory. He contacted clients and dealt with brokers and merchants from whom the concern purchased dyes, raw cotton, and yarn on credit. He also kept the account books with the assistance of a journeyman.[38] By relying on a core of only a few partners, often related, large and small manufacturers alike established an industry that demonstrated a tremendous ability to make products in great demand. Such partners relied on the trust and obedience of the members of their commercial houses. Such loyalty was particularly expected from partners and clients who were also family members.

Large and small manufactures both based their organization of work and the smooth running of their businesses on the patriarchal hierarchy that the family represented. However, as the century advanced, it became obvious that besides their reliance on the model of the family, substantial and slight enterprises had more to share. Each type of factory specialized in one part of the production process or produced different types of calicoes; together these manufactures were able to overcome the difficulties that the long years of war imposed on the industry.

Calico Manufactures and the Specter of War: 1790s–1810s

At the turn of the century, despite the industry's increasing importance, many obstacles remained for Barcelona's artisans seeking to establish small fábricas. According to James K. J. Thomson, owners of large manufactures regarded their smaller competitors as "both a threat to the maintenance of control over their labor force and a challenge to their very existence as large, concentrated manufactures."[39] Members of the Board of Trade, who were merchants with interests in large fábricas, shared these fears. Small manufactures, sometimes informally directed by women, produced mostly low- and medium-quality calicoes (*indianes*) for the Spanish market. In contrast, the larger fábricas tended to specialize in the production of high-quality calicoes (*pintats*) chiefly for export to the lucrative Spanish-American market.[40] Despite these differences, owners of large manufactures resented the intrusion of small fábricas into their market. This tension had a long history. For instance, an article on the Magarola factory in Nipho's *Diario curioso, erudito, económico y comercial* describes how, in 1758, master linen weavers hurried to manufacture the kind of cotton cloth the Fábrica Magarola produced in order to sell it outside Barcelona.[41]

Yet the initial hostility between great and humble manufactures turned into cooperation during the 1770s and 1780s. Indeed, despite their protests, the owners of larger factories benefited from the presence of small manufactures. The smaller enterprises allowed the larger to grow without their owners having to make further investments in technology or having to hire more workers. In 1762, Bernardo Ward, a royal officer and adviser to King Ferdinand VI (1746–1759), noted that by having part of their workforce labor in their own homes, factory owners could save on maintenance expenses. Such costs included mainly "constructing houses [for the workers] and buying and keeping the looms and other tools." Living at home, Ward continued, meant that a worker could work at any time and with the help of his family.[42] Ward thus foresaw the advantages of what became common in Barcelona during the 1780s and 1790s: production in small factories based on family labor.[43] Moreover, small factories worked as subcontractors for larger ones, bleaching or printing small amounts of calicoes. Owners of large factories, such as Francisco Maleras, greatly benefited from the production of cloth in small factories. Maleras saved on salaries and raw materials by having some of his cotton cloth dyed and printed in Juan Gatell's unregulated factory.[44]

During the early 1790s, the climax of the calico industry and the increasing collaboration between small and large fábricas coincided with the expansion of trade between Barcelona and the American colonies. Between 1784 and 1786, Barcelona saw the establishment of 112 calico manufactures. Between 1792 and 1794, their number increased to 150.[45] The "Regulation of Free Trade" decree of 1778 had allowed merchants and entrepreneurs to export calicoes through the main ports of the Iberian Peninsula to twenty-two destinations in the colonies. Eleven years later, Venezuela and the large and populous viceroyalty of New Spain (Mexico), initially excluded from the 1778 decree, joined the list of regions that enjoyed direct trade with most Spanish ports.[46] In the fourteen years between 1778 and 1792, calico exports from Spain to the American colonies increased an impressive 500 percent. In 1792 alone, Barcelona exported 3.3 million yards of calicoes, an increase of 100 percent from the previous year.[47] That same year, the traffic of exports from Barcelona to New Spain reached its peak.

Besides increasing the number of yards directed to a thriving colonial market, entrepreneurs also tried to master the mechanical spinning of thread by perfecting the Catalan version of the spinning jenny, the Bergadana, introduced in 1786.[48] Small fábricas, which first introduced the jenny in Spain, had enabled large ones to experiment with technical innovations such as new spinning techniques.[49] Thus, in the 1770s and 1780s, substantial factories commonly outsourced the weaving of white cotton cloth and printing of calicoes to smaller production units, but in the 1790s, small factories also specialized in selling spun thread to larger ones.

This form of subcontracting was not unique to the printing industry. For instance, throughout the 1770s and 1780s, silk maker Francisca Comadura built a successful enterprise by relying on the help of her female relatives to find skillful silk spinners in Barcelona who worked as her subcontractors. Comadura's female supervisors provided these women with raw silk. In this way, Comadura obtained the silk she needed to make lace goods well known in Barcelona "for their excellence and beauty."[50] Yet the practice of subcontracting in the calico industry was distinctive: it involved the great majority of calico manufacturers in the city and took place primarily in the city, not in nearby villages.

Artisanal manufacturing became even more important during the war with England (1796–1808). During the wartime era, families working in small factories spun, wove, bleached, and even printed inexpensive calicoes. Similarly, during the 1790s, the Fábrica Pujadas, with twenty-one looms, regularly purchased plain cotton cloth from small factories in Barcelona; it then printed the cloth and sold the calicoes throughout Spain.[51] The Spanish market, dominated by Catalan towns, absorbed 60 percent of the factory's sales.[52] Pujadas's manager, Francisco Grau, sold spun cotton at "a fixed price" to small fábricas, which made cloth from it. Silk weaver Francisco Pons owned one of the small factories. He earned his living by working at home with his own tools, making "all kinds of cotton cloths to sell to large manufactures." His small factory of seven looms was located in the apartment where he lived with his family in the textile-manufacturing neighborhood of Sant Pere.[53]

Subcontracting allowed large manufactures to save on production costs by reducing their workforce. Consequently, where the average large factory employed 108 workers in 1784, by 1800, that number had dropped to 67.[54] The average number of workers per factory continued to shrink during the first half of the nineteenth century.[55] There was nonetheless a limit to how long large and small fábricas in Barcelona could adapt to continuing wartime conditions. The war against England lasted from 1796 until March 1802, and after a nearly three-year truce, it resumed between December 1804 and 1808. Between 1804 and 1808, bankruptcies of large factories and of merchants trading in calicoes rose 35 percent.[56] Total demand for printed calicoes declined, while prices of imported dyes, such as indigo, increased. The crisis equally affected considerable and modest manufactures, although small enterprises showed more flexibility. But the worst was yet to come. On July 13, 1808, French troops entered Barcelona. Between 1808 and 1814, France's occupation of Spain caused further economic dislocation.[57] Still, despite the acute economic crisis that depressed Spanish domestic demand, Barcelona's industry managed to survive the war years, though it was much reduced. Large factories specialized in printing cheap calicoes on low- or medium-quality cotton cloth purchased from small artisanal factories.[58]

In the 1820s, small factories managed to produce and sell cheap textiles, permitting the revival of the Barcelona cotton industry. Soon after the end of the war in 1814, the city's industrial activity recovered to the levels of the 1790s. The number of cotton factories in the city increased from 230 in 1802–1804 to 338 in 1823.[59] Realizing that small factories had become the rule, the government no longer enforced the royal decree of 1767, which had set a minimum of twelve looms for recognition as a legal factory.[60]

Conclusion

In 1832, Catalonia witnessed the establishment of its first steam fábrica, owned by Bonaplata, Vilaregut, Rull & Co. and called "The Steam" (El Vapor).[61] El Raval, one of the neighborhoods that saw the first calico manufactures in the 1760s and 1770s, became the site for the many spinning fábricas that relied on American cotton to make thread. But most of the industry—spinning and weaving factories alike—had left the city to occupy towns outside Barcelona, or, more commonly, to go to rural Catalonia, near rivers to make use of hydraulic power.

In spite of the calico industry's health in the 1830s, the years of war at the beginning of the century continued to reverberate. War was the main cause behind reduced levels of investment, the delays in technological innovation, and the slowing pace of production in general. The importance of the war as a disruptive factor to industrialization leaves us with unanswered questions. What complex forms, alliances, and paths to industrialization did the war hinder? Could a prolonged period of peace have yielded a larger concentration of workers in these manufactures and quickened the process of technical advancement? Or did the war instead favor the collaboration between petty and large manufactures and the survival of the small factory?[62] Significantly, these connections, characteristic of the beginning of war in the 1790s, endured throughout the nineteenth century.

In the first half of the nineteenth century, the number and importance of modest calico manufactures in Barcelona grew. During the 1820s, small manufactures that specialized in spinning and weaving cotton continued to supply thread and cotton cloth to a reduced number of larger manufactures that printed calicoes. As late as 1841, the small factory was still the norm in Barcelona. That year, the average number of workers per factory was only thirty-three, compared with sixty-seven in 1800.[63] Until the 1860s, small- and medium-size factories were still producing mostly low- and medium-quality calicoes for the Spanish market. These products did not bring in the high profits of previous decades, but at least they allowed large manufactures to survive when the markets for their high-quality products collapsed. In fact, some economic historians believe that thanks to these humble enterprises, by the mid-nineteenth century, the Barcelona cotton industry had started a more sustained process of Spanish industrialization that would continue into the twentieth century. The combination of large and small manufactures producing inexpensive commodities represented a balance that laid the basis for the Industrial Revolution in Spain.[64]

Until the 1830s, however, Spanish calico printing was largely limited to Barcelona. A recent study on Spanish industrialization has confirmed that by 1784, Barcelona was the main Spanish center of calico production, with eighty fábricas. Sixteen more fábricas were spread throughout Catalonia, and the rest of Spain had only seven more, five of them located near the main ports in Andalusia.[65] Although production in the city was impressive, it could not compete with other European countries, such as France and England, where the industry had spread widely. For instance, in 1775, Barcelona's manufactures produced 90,000 bolts of calico annually, while Normandy's forty calico factories alone yielded 152,000 calicoes in 1785.[66] And the 8.6 million kilograms of raw cotton that Catalan manufactures consumed annually between 1816 and 1820 may seem considerable, but they were less than a third of British consumption of 29.1 million kilograms during the same period.[67] In sum, calico output represented only 4 percent of the total industrial production in Spain, while calico manufactures employed barely 2.5 percent of the Spanish working population.[68]

At the outset of the nineteenth century, Castile and Andalusia focused on the production of wool cloth, while Valencia excelled in the manufacture of fine silk. These two industries, particularly the woolen industry, had a much larger quantitative impact than the calico industry in the total textile production of Spain. By the early 1830s, however,

Spaniards were again excited by the possibility that steam power would increase calico production even further, bringing renewed hopes of wealth and prosperity. Thus, throughout the 1830s, owners proudly placed their steam engines on their factory seals. In 1834, the economic writer Manuel María Gutiérrez stated that the Fábrica Bonaplata & Co. had indeed made "a true manufacturing revolution in Catalonia" by establishing the first steam factory.[69] More than a decade later, in 1849, Juan Illas y Vidal labeled the 1832 achievement "a true industrial revolution."[70] In this newly euphoric atmosphere, the Spanish-American market, despite Spain's recent loss of most of its colonies, played an important part. Merchants and entrepreneurs maintained links with their families and contacts in America, and never lost hope that regular traffic between the two sides of the Atlantic could be reestablished at prewar levels.[71]

The story of the efforts of the early manufacturers in the Spanish calico industry is much more than the tale of an interlude between the preindustrial years and the Industrial Revolution of the nineteenth century. In fact, the study of eighteenth-century small manufacturing and the role of the family allows a reconsideration of the place of Spain in the Industrial Revolution. It has long been argued that the Spanish empire, with its stifling mercantilist policies and rigid institutions that hindered economic growth, seemed irremediably impotent before foreign competitors and rampant smuggling.[72] In such an interpretation, there is little place for the study of ordinary individuals and even less for their families. The few exceptional men who became the protagonists of historians' narratives were ministers and wealthy merchants, who symbolized the individualistic character of capitalist growth. Yet when we study not only the entrepreneurial drive of Catalan families, but also the story of these families themselves, the story of the Spanish Industrial Revolution stands revealed as the result of constant readapting to new challenges as a means of opening up new possibilities. Once we look at the successful stories of families who manufactured and sold calicoes in the Iberian Peninsula and to the American colonies, the economic "failure" of Spain needs to be reconsidered. From the perspective of these families, a controlling bureaucracy that fossilized economic growth through monopolies and commercial restrictions did not limit the Spanish empire of the late eighteenth century. As revealed by their participation in the calico boom in the eighteenth century, families had great potential for flexible adaptation to the changes required. In fact, individuals, with the aid of their families, were able to develop successful trade strategies that either predated or inspired governmental policies. The new opportunities for riches in the manufacture and sale of consumer goods attracted the competing interests of Spaniards, criollos, and foreigners, each of whom sought to claim those profits for themselves alone. Through a microhistorical approach, historians can explain how individual decisions generated new sources of wealth in the Spanish empire and shaped the road toward the Industrial Revolution.

Acknowledgments

I thank Luis Corteguera, Carolyn Nelson, Giorgio Riello, and Leslie Tuttle, as well as all the participants of the Dibner Institute seminar, "Reconceptualizing the Industrial Revolution," in particular, Ian Inkster and the organizers, Jeff Horn, Leonard Rosenband, and Merritt Roe Smith, for their comments.

Notes

1. "Carta III sobre la exquisita política de los Ingleses en el modo de animar las ciencias, comercio, marina, y artes," in *Estafeta de Londres y extracto del correo general de Europa* (Madrid: Miguel Escribano, 1779), 1:67.

2. Quoted in Sidney Pollard, *Peaceful Conquest: The Industrialization of Europe, 1760–1970* (Oxford: Oxford University Press, 1981), 206.

3. James K. J. Thomson, "The Catalan Calico-Printing Industry Compared Internationally," *Anuari de la Societat Catalana d'Economia* 7 (1989), 77.

4. Ibid. Barcelona had 74 "factories" in 1780 and 150 in 1790, whereas there were 111 British "factories" in 1785.

5. See Alejandro Sánchez, "La era de la manufactura algodonera en Barcelona, 1736–1839," *Estudios de Historia Social* 48/49 (1989), 91. The number of eighty thousand workers comes from Antonio Buenaventura Gassó, *España con industria fuerte y rica* (Barcelona: Antonio Brusi, 1816), 62. The impressive annual output was three times higher than production in "factories" in Paris and Neuchâtel, with an annual output of 2.3 million and 2.0 million yards respectively, as noted in Thomson, "The Catalan Calico-Printing Industry," 80.

6. Rafel d'Amat, "Calaix de sastre en que se explicarà tot quant va succeint en Barcelona i veïnat desde mitg any 1769," housed in Arxiu Històric de la Ciutat de Barcelona (henceforth AHCB) Manuscrits, A-202, entry for May 21, 1798.

7. Luis María Bilbao and Emiliano Fernández de Pinedo, "Auge y crisis de la siderometalurgia tradicional en el País Vasco (1700–1850)," in *La economía Española al final del Antiguo Régimen*, ed. Pedro Tedde (Madrid: Alianza Editorial, 1982), 133–228; Agustín González Enciso, *Estado e industria en el siglo XVIII: La fábrica de Guadalajara* (Madrid: Fundación Universitaria Española, 1996).

8. Jordi Nadal, *El fracaso de la Revolución Industrial en España, 1814–1913* (Barcelona: Ariel, 1975); J. K. J. Thomson, *A Distinctive Industrialization: Cotton in Barcelona, 1728–1832* (Cambridge: Cambridge University Press, 1992); Jordi Nadal and Albert Carreras, eds., *Pautas Regionales de la Industrialización Española (siglos XIX y XX)* (Barcelona: Ariel, 1990).

9. Jordi Nadal, "Cataluña, la fábrica de España. La formación de la industria moderna en Cataluña," in *Moler, tejer y fundir. Estudios de historia industrial*, ed. Jordi Nadal (Barcelona: Ariel, 1992), 84–154.

10. Claude Carrère, *Barcelona, centre économique à l'époque des difficultés, 1380–1462*, vol. 1 (Paris: La Haye, Mouton et Cie, 1967), 500–505.

11. Narcís Feliu de la Peña, *Fénix de Cataluña. Compendio de sus antiguas grandezas y medio para renovarlas* (Barcelona: Editorial Base, 1975 [1683]).

12. David R. Ringrose, *Spain, Europe, and the "Spanish Miracle," 1700–1900* (Cambridge: Cambridge University Press, 1996), 188–206.

13. Albert García Espuche, *Un siglo decisivo: Barcelona y Cataluña, 1550–1640* (Madrid: Alianza Editorial, 1998), 285–341.

14. Jaume Torras Elias, "The Old and the New: Marketing Networks and Textile Growth in Eighteenth-Century Spain," in *Markets and Manufactures in Early Industrial Europe*, ed. Maxine Berg (New York: Routledge, 1991), 93–133; Bartolomé Yun-Casalilla, "City and Countryside in Spain: Changing Structures, Changing Relationships, 1450–1850," in *Early Modern History and the Social Sciences: Testing the Limits of Braudel's Mediterranean*, ed. John A. Marino (Kirksville, MO: Truman State University Press, 2002), 35–70.

15. Arxiu de la Corona d'Aragó (henceforth ACA) Real Patrimoni Batllia Moderna, vol. 234, fols. 441–452.

16. Ramón Grau and Marina López, "Empresari i capitalista a la manufactura catalana del segle XVIII: Introducció a l'estudi de les fàbriques d'indianes," *Recerques* 4 (1975), 37.

17. Biblioteca de Catalunya (hereafter BC) JC, lligall 53, capsa 71, n. 3, 1767, "Ordenanzas para el régimen y gobierno de las fábricas de indianas, cotonadas y blavetes." The ordinances were drafted in 1766 but did not receive royal approval until 1767.

18. BC JC, lligall 53, capsa 71, n. 21, fol. 2.

19. See the case of the "factory" of Josep Rius, AHCB Consellers Obreria C-XIV-37 (July 19, 1778).

20. ACA Real Audiència (henceforth RA) Consulat de Comerç Plets, no. 7259.

21. ACA RA Plets Civils no. 1196 (1806). For Josep's previous experience in the trade, see ACA RA Plets Civils, no. 14379. He was a witness during the trial against Mateu Farra, for whom Josep Dordal worked in 1775 for six months.

22. BC JC, lligall 54, n. 17 (December 12, 1773).

23. ACA RA Plets Civils no. 14379.

24. Arxius de Sants Just i Pastor List of communions for 1796.

25. AHCB Arxiu del Veguer (henceforth AV), XXXVII-1295, 1784. Llorens enlarged his "factory" in 1785, according to Thomson, *Distinctive Industrialization*, 188.

26. AHCB AV, XXXVII-1476, 1799.

27. AHCB Consellers Obreria C-IXV (1795).

28. ACA RA Civil Lawsuits n.1794.

29. In the Sirés "factory," about twenty families lived in the factory building. Other calico factories in Barcelona provided living quarters in the factory building for workers and their families. For the case of the Magarola factory, see "Historia de la fábrica de Magarola," in *Diario curioso, erudito, económico y comercial*, 1787, nos. 188, 190, 191, 192.

30. Jacinto paid Pau Sirés, their relative in Valencia, an annual salary of 150 *lliures*.

31. Stanley Chapman and Serge Chassagne, *European Textile Printers in the Eighteenth Century: A Study of Peel and Oberkampf* (London: Heinemann Educational Books, 1981); Tamara K. Hareven, *Family Time and Industrial Time: The Relationship between the Family and Work in a New England Industrial Community* (Cambridge: Cambridge University Press, 1982); Marta Vicente, "Artisans and Work in a Barcelona Cotton 'Factory' (1770–1816)," *International Review of Social History* 45 (2000), 1–23.

32. BC JC, lligall 51, capsa 68, n. 16, July 23, 1785.

33. This was the practice and philosophy of "factory" owner Juan Costa y Merla. See AHCB AV XXXVII-1494 (1789).

34. BC JC, llig. 51 caixa 68 no. 16 (1785 report of the *Real Compañía de Hilados*).

35. Pedro María Antón, "Salarios en las fábricas de indianas de Barcelona en el último tercio del siglo XVIII" (licentiate thesis, University of Barcelona, 1972).

36. Isabel Lobato Franco, *Compañias y negocios en la Cataluña preindustrial (Barcelona 1650–1720)* (Seville: Publicaciones de la Universidad de Sevilla, 1995).

37. Archivo General de Protocolos de Puerto Rico Notary Gregorio Sandoval (San Juan): 456-A; Paloma Fernández, *El rostro familiar de la metrópoli. Redes de parentesco y lazos mercantiles en Cádiz, 1700–1812* (Madrid: Editorial Siglo XXI, 1997), 128.

38. AHCB FC B-260 account books for 1783.

39. Thomson, *Distinctive Industrialization*, 289.

40. To produce these calicoes, large factories used high-quality white cloth imported from northern European countries.

41. January 4, 1787. A copy of the Diario is housed in BC Arxiu Gómina 68/4.

42. Ward, however, was thinking about a putting-out system in the countryside, with city merchants hiring peasant families to manufacture wool cloth. Bernardo Ward, *Proyecto económico en que se proponen varias providencias dirigidas a promover los intereses de España* (Madrid: Joaquin Ibarra, 1779).

43. This strategy was used by other calico "factory" owners in Europe. See Pierre Caspard, "The Calico Painters of Estavayer: Employers' Strategies toward the Market for Women's Labor," in *European Women and Preindustrial Craft*, ed. Daryl Hafter (Bloomington: Indiana University Press, 1995), 133.

44. AHCB AV, XXXVII-1278. Between 1768 and 1770, Maleras sent some of his cloth to Gatell's factory. In 1779, Maleras petitioned the Board of Trade to consider his factory regulated, BC JC, lligall 53, n. 24, January 21, 1779.

45. Sánchez, "Era de la manufactura," 90.

46. John Fisher, "El comercio entre España e Hispanoamérica, 1797–1820," *Estudios de Historia Económica* 27 (1993), 16.

47. Antonio García-Baquero González, "La industria algodonera catalana y el libre comercio. Otra reconsideración," *Manuscrits* 9 (1991), 33.

48. In a recent work, James Thomson examines artisanal participation in the implementation of the spinning jenny in Barcelona during the 1780s. James K. J. Thomson, "Transferring the Spinning Jenny to Barcelona: An Apprenticeship in the Technology of the Industrial Revolution," *Textile History* 34 (2003), 21–46.

49. BC JC, lligall 23, n. 14. During the 1790s, small "factories" put the spinning jenny to regular use. See Thomson, *Distinctive Industrialization*, 248–249.

50. ACA RA, civil lawsuits, n.14380

51. AHCB AV, Notary Miquel Mir i Llorens, manual 1796, lawsuits, n. 1426. Located in carrer Cuch, the Pujadas "factory," which operated from 1785 to 1797, was a medium-size regulated factory employing about forty-five workers.

52. Merçè Prats, "La manufactura cotonera a Catalunya: L'exemple de la fàbrica Magí Pujadas i Cia," (licentiate thesis, University of Barcelona, 1976).

53. AHCB AV XXXVII-1242, 1796.

54. Sánchez, "Era de la manufactura," 90. The small size of Barcelona's average "factory" is comparable to those in other countries in late eighteenth- and early-nineteenth-century Europe; see Maxine Berg, "Small Producer Capitalism in Eighteenth-Century England," *Business History* 35 (1993), 17–39.

55. Enriqueta Camps, *La formación del mercado de trabajo industrial en la Cataluña del siglo XIX* (Madrid: Publicaciones del Ministerio de Trabajo, 1995), 150–160.

56. Josep María Delgado, "La industria algodonera catalana (1776–96) y el mercado americano: Una reconsideración," *Manuscrits* 7 (1988), 151–169.

57. Francisco de Zamora, *Diarios de los viajes hechos en Cataluña*, ed. Ramón Boixareu (Barcelona: Curial, 1973 [1788]), 21; Thomson, *Distinctive Industrialization*, 268.

58. Sánchez, "Era de la manufactura," 89–94.

59. Ibid., 103.

60. It is possible that the owners of large factories, some of whom were members of the Board of Trade, influenced the latter's decision to legalize the small "factories."

61. Sánchez, "Era de la manufactura," 103.

62. The impact of the war on the industrialization process in Spain is a problematic historical issue. Gabriel Tortella examined the subject in *Los orígenes del capitalismo en España. Banca, industria y ferrocariles en el siglo XIX* (Madrid: Tecnos, 1973).

63. Sánchez, "Era de la manufactura," 104.

64. Jordi Nadal, "A Century of Industrialization in Spain, 1833–1930," in *The Economic Modernization of Spain, 1830–1930,* ed. Nicolás Sánchez-Albornoz (New York: New York University Press, 1987), 63–89; Pierre Vilar, "La Catalunya industrial: Reflexions sobre una arrencada i sobre un destí," *Recerques* 3 (1974), 7–22.

65. Jordi Nadal, *Atlas de la Industrialización de España* (Barcelona: Crítica, 2003), appendix I.1.2.4. I want to thank Bernart Hernández for helping me to access the appendixes of this work, available only on CD-ROM.

66. Chapman and Chassagne, *European Textile Printers*, 8.

67. Nadal, *Fracaso de la Revolución Industrial*, 207.

68. Juan Plaza Prieto, *Estructura económica de España en el siglo XVIII* (Madrid: Confederación Española de Cajas de Ahorros, 1975), 331.

69. Manuel María Gutierréz, *Comercio libre o funesta teoría de la libertad económica absoluta* (Madrid: Marcelino Calero, 1834), 130.

70. Juan Illas y Vidal, *Memoria sobre los perjuicios que ocasionaría en España así a la agricultura como a la industria y comercio, la adopción del sistema del libre cambio* (Barcelona: J. M. De Grau, 1850), 50.

71. Sánchez, "Era de la manufactura," 101.

72. Accordingly, the inability of Spain to open its markets to free competition in the eighteenth century would delay its industrialization until the early nineteenth century, James Laforce, *The Development of the Spanish Textile Industry, 1750–1800* (Berkeley: University of California Press, 1965), 182. See also Nadal, *Fracaso de la Revolución Industrial*, 188–245. On the other hand, James Thomson has demonstrated that government intervention in manufacturing during the eighteenth century contributed to increasing the demand for calicoes. See his *Distinctive Industrialization.*

9

Taking Stock of the Industrial Revolution in America

Merritt Roe Smith and Robert Martello

In December 1791, Treasury Secretary Alexander Hamilton delivered his *Report on Manufactures* to the Second Congress of the United States. In it, he argued that:

> not only the wealth, but the independence and security of a country, appears to be materially connected with the prosperity of manufactures. Every nation, with a view to those great objects, ought to endeavor to possess within itself all the essentials of national supply. These comprise the means of subsistence, habitation, clothing, and defence. The possession of these is necessary to the perfection of the body politic, to the safety as well as to the welfare of the society.[1]

As a government official writing at the beginning of the new federal Republic, Hamilton saw the need to accelerate American manufacturing. Perhaps more important, he understood the associated costs and benefits of such a course. While he recognized the complexity and importance of the subject—notably the interwoven economic, political, material, and even ideological implications of manufacturing—he was unaware of the truly staggering impact that the new nation's turn toward manufacturing would ultimately have.

This chapter explores the nature and relevance of the Industrial Revolution in the United States through a historiographic review of changing scholarship on the subject. It proceeds from an understanding that history is an open-ended interpretative discipline. Because the literature on the subject is so large, choices had to be made about what to include (and exclude) from this discussion. Such choices reflect our evaluations of the literature and our scholarly proclivities as historians of technology. There are doubtless other approaches to the subject, but to our surprise, we could find no previous publication that assessed America's Industrial Revolution from a historiographic perspective. This chapter is a beginning.

We view the American Industrial Revolution primarily as an economic transition that involved the replacement of craft methods with mechanized methods of production, the organization of work into larger and more specialized units, the more rigorous and pervasive management of labor, the construction of a national transportation system, and the growth of markets. America's so-called First Industrial Revolution occurred roughly between 1790 and 1860, although these starting and ending points obviously spilled over into earlier and later periods. The lion's share of historical scholarship on the United States focuses on the early take-off phase of the Industrial Revolution, including the factors contributing to economic growth as well as the anticipated and unanticipated consequences, beneficial and harmful, of such growth. Our focus on the pre-1860 period streamlines and contains this study without sacrificing any of the major themes pertaining to the process of industrialization.

Americanists have been thinking and writing about the Industrial Revolution since the nineteenth century, with a particular intensification of historical analysis during the past thirty years. The sheer density of scholarship on the subject testifies to its importance. Due to the fact that America's political and industrial revolutions overlapped, the industrialization of the new republic significantly shaped the development of its government, society, politics, and environment while, in turn, being shaped by them. Our current understanding of the Industrial Revolution owes much to the variety of disciplinary perspectives used to illuminate it, including economics, business and organizational history, social and labor history, the history of technology, and industrial archeology. In addition to reviewing the literature on America's Industrial Revolution, we highlight major scholarly trends during the past thirty years and point to major questions and issues, often raised by the most recent studies, that need to be addressed more fully.

Countless books and articles address the subject of American industrialization, some more explicitly and directly than others. As early as 1841, for example, Henry Howe published his filiopietistic *Memoirs of the Most Eminent American Mechanics*, and at the outset of the Civil War, J. Leander Bishop issued the first of three volumes in his landmark *History of American Manufactures*.[2] But much more was to come. After the war, an outpouring of popular books trumpeted the triumph of commerce and manufacturing with titles like *One Hundred Years' Progress, The Great Industries of the United States,* and *Triumphs and Wonders of the 19th Century*.[3] Next to the nation's independence and political unity, the celebration of America's technical "genius" and industrial progress mattered most to writers of the Gilded Age. These themes would conflate into a powerful nationalist economic ideology that influenced public policy throughout the twentieth century.

A full treatment of the origin and enduring popularity of such themes lies beyond the scope of this chapter. Suffice it to say that the advent of publications celebrating American industrial progress coincided with the centennial of American independence in 1876 and continued through the great Columbian Exposition in 1892–1893 and, beyond that, to Chicago's Century of Progress Exhibition in 1933 and New York's futuristic World's Fair six years later. Clearly a climate of boosterism and nationalist feeling prompted such publications. An important component of the American self was being constructed that identified American industry and inventiveness not only with the explosive growth of the world's largest industrial economy but also with the genius of American politics, especially democracy. Issues arose only when the ideas of progress and democracy conflicted, as they did in late-nineteenth-century clashes between labor and management.

While early celebratory histories were often the products of enterprising publishing houses intent on penetrating broad popular markets, more thoroughly researched studies began to appear in 1880s. In this regard, the U.S. Census of 1880 proved to be a landmark. Writing for the Census Office, Charles H. Fitch, Carroll D. Wright, and James M. Swank produced impressive reports on interchangeable manufacturing, the factory system, and iron and steel production.[4] After completing their respective census assignments, Fitch went on to publish a popular essay on interchangeable manufacturing, "The Rise of a Mechanical Ideal," while Wright and Swank produced well-received books on their respective subjects. Of the three, Wright adopted the broadest perspective, publishing the Chautauqua-sponsored and widely circulated *The Industrial Evolution of the United States*.[5] Wright not only traced the growth of the factory system, a process of which he did not entirely approve, but also provided one of the earliest accounts of the labor movement in the United States. Like their contemporaries, however, Wright, Fitch, and Swank clung to a linear perspective in their treatments of industrial development. For them, industrial progress was America's most important and triumphant product because it confirmed and legitimized the nation's most hallowed political values and institutions.[6]

Professionally trained historians did not begin to devote sustained attention to American industrialization until after World War I. Following in the footsteps of Charles A. Beard and other Progressive historians whose economic interpretations of American history represented a new departure for the discipline, a younger generation of historians came to the fore who specifically identified themselves as "economic historians." Among the best and brightest of this new generation were Louis C. Hunter, Vera Shlakman, Caroline F. Ware, and George Rogers Taylor, all of whom completed doctoral dissertations in the late 1920s and had contacts with an emerging business history program at Harvard University headed by Edwin F. Gay and Arthur H. Cole. Each of these scholars subsequently published foundational works—Hunter on steamboats and steamboating, Shlakman on metalworking in an early factory town, Ware on cotton textiles and the factory system, and Taylor on transportation and the emerging national economy—that paved the way for more than a generation of scholarship on the Industrial Revolution in America.[7] Their books became obligatory reading for anyone interested in the interrelated processes of technological innovation and industrialization in America.

With the foundations laid by scholars like Hunter, Shlakman, Ware, and Taylor, scholarly inquiry into American industrialization entered a new phase during the early 1960s when several books shifted attention to the role of technology in American economic growth. From a methodological and stylistic standpoint, the new work was more analytical and thematic than the older narrative and empirical tradition. It also posed a different set of questions, mostly concerned with the sources of economic growth, and in doing so inaugurated a new approach to economic history.

One of the new works was H. J. Habakkuk's *American and British Technology in the Nineteenth Century*, a wide-ranging comparative study that sought to explain why the United States, and particularly New England, made such rapid strides in mechanizing production (especially in metalworking and woodworking trades) by the time of London's 1851 Crystal Palace Exhibition. In searching for an answer, Habakkuk addressed rates of investment, trade cycles, demand factors, and the general level of economic development in both countries, including agriculture. But his most striking and original point disputed the significance of long-standing arguments that cited "labor scarcity" as the primary cause of America's rapid mechanization of production prior to the 1850s.[8]

While suggestive, Habakkuk's analysis fell short of the mark in explaining what induced American manufacturers to mechanize their operations and, in effect, challenge British leadership by 1851. However, it engaged the attention of a number of economists who set out to critique and refine the themes he advanced.[9] Foremost among them was Nathan Rosenberg, whose seminal articles and books on the American system of manufactures (and how it spread from the originating arms industry to other technically related industries) revealed what was innovative and different about the new American technology and why it disseminated so rapidly. In the process, Rosenberg and his collaborators questioned the effectiveness of neoclassical production theory and the way it characterized technological change, pointing instead to the need for more inductive, empirical historical research on the subject. For Rosenberg, a cluster of interrelated institutional, political, and social factors combined with economic factors to prompt the emergence of the American system.[10] At the same time, scholars in the newly inaugurated field of the history of technology began to produce books and articles on the transfer of technology from Europe to America, thus fleshing out the connections between the Old World and the New and, in the process, shedding light on how key technological choices were made during the early years of American industrialization. This body of work reinforced other industry-specific findings that the American system emerged not because of labor scarcity but rather to create demand-induced products for the American market.[11]

A significant historiographic shift occurred as Habakkuk's themes were being debated. Prior to the 1960s, economic history, while written by trained economists as well as by historians, relied primarily on a narrative style of presentation. Thereafter, the field became increasingly populated by economists who employed quantitative methods of analysis, popularly known as "cliometrics." A leading exponent of the "new economic history" was Robert William Fogel, whose *Railroads and American Economic Growth* employed counterfactual questions and econometric analysis to attack the longstanding belief that railroads were "indispensable" to American economic development, arguing instead that canals could have been equally effective in building the nation's transportation system.[12]

The impact of Fogel's thesis was memorable, if somewhat fleeting. To be sure, his work played an important role in spotlighting the new economic history. What is more, it succeeded in showing that no single industry should be seen as dominating an economy. But in the long run, it did little to lessen the significance historians assigned to the railroad in building the U.S. national economy. Nowhere was this point better demonstrated than in the work of Alfred D. Chandler, Jr.

Chandler eschewed the economic growth debate that engrossed econometric historians and instead asked a different set of questions about the rise of modern management, a critical yet understudied component of the Industrial Revolution. In *The Railroads* and even more forcefully in *The Visible Hand*, his Pulitzer Prize–winning book of 1977, Chandler showed that railroads were not only the nation's first big business but also a new phenomenon in American history. Because of their large size, heavy capitalization, and dispersal over long distances, railroads faced enormous problems of scale never before encountered. Railroad officials had to learn how to plan, coordinate, and control vast operations, and they did so by developing bureaucratically structured administrative systems that relied on new multiunit supervisory structures and modern cost-accounting methods to solve pressing day-to-day and long-term managerial problems. In effect, Chandler argued that railroads invented modern corporate management and, in doing so, substituted the "visible hand" of management for Adam Smith's "invisible hand" of the market, thus rationalizing choices and reducing chance and contingency in the allocation of resources.[13] What is more, Chandler showed that the experience gained on railroads allowed railroad managers to move into other kinds of large businesses, transferring their know-how to other complex, diverse, and widely scattered enterprises. As a result of Chandler's pioneering work, modern management assumed a place alongside technology as a critical factor in the process of American industrialization. The two went hand-in-hand.[14]

With the possible exception of Thomas Kuhn, no other American historian exerted greater influence both within and outside the historical profession, in the past forty years. Within the discipline, Chandler's work was instrumental in promoting the "organizational synthesis" of American history, a school that underscored the importance of organizations and professions in American society, politics, and economic growth.[15] His work also reinvigorated the field of business history, influencing new work on the emergence of the American business system while serving as a counterforce to the excessive scholasticism and mathematization of the new economic history.[16] By redirecting attention to "the dynamic factors in the growth of the American economy and its business system"—notably the railroads, urban markets, mass production, electrification, automobility, and research and development—Chandler revealed the intimate connections that existed between the economy and the pursuit of technology and science in America.[17] Above all, Chandler's influence spread beyond academia. Long before the publication of *The Visible Hand*, Chandler's second book, *Strategy and Structure* (1962), became required reading in business schools and in the business world.[18] As Thomas McCraw aptly observed, "The maxim 'strategy precedes structure' became a byword of corporate management during the 1960s and 1970s, not only in the United States but all over the world, perhaps most notably in Japan.

The notions of upper, middle, and lower management . . . became vivid in the minds of executives in all capitalist economies."[19]

While the Chandlerian approach spread both within and outside academia, the scope and content of mainstream history experienced a sea change with the advent of the "new social history." Prompted by the political tensions of the 1960s—most notably the civil rights movement, the Vietnam War, and the feminist movement—the new history drew on the Annales school, Marxist historiography, and advanced social science methods to challenge postwar consensus interpretations of the past that emphasized American exceptionalism and the orderly progressive character of American history. Rather than chronicling the exploits of great men and elite institutions, new social historians sought to write history "from the bottom up" by studying ordinary people, their social mores, and the tensions they experienced as the country moved from an agrarian past into an urban-industrial future. The results were telling. "Like a flood," Alice Kessler-Harris observes, "the new social history shifted the course of the profession's mainstream, carving out new directions for exploration, raising a series of questions about the nature of the craft of history, and transforming our understanding of the past."[20]

One criticism new social historians leveled against Alfred Chandler's treatment of the managerial revolution was that it "left workers out," thus biasing public understanding of American industrialization toward management and other elitist forces of production. They responded by drawing on the penetrating insights into workplace cultures by E. P. Thompson, Sidney Pollard, and other British historians to produce a range of community studies that focused on demographic changes, institutional influences, family networks, cultural traditions, and occupational mobility in an effort to flesh out what life was really like under the factory system.[21] Among the most influential studies were those that focused on life and labor in early industrial communities. At issue was not only how workers and their families responded to industrialization, but also how they participated in and influenced the process. Emphases varied, of course, depending on subject matter and the author's particular interests. In their studies of the boot and shoe trade in Lynn, Massachusetts, for example, Alan Dawley, Paul Faler, William Mulligan, and Mary Blewett showed how early craft-oriented household manufactures gave way to a vast merchant-controlled putting-out system that gradually transmuted into a full-blown factory system. In the process, they revealed how shoemakers went from being independent craftsmen to dependent wage earners, a transition that boded ill for the household system and ultimately led to considerable labor-management strife in Lynn, including, in 1860, the first large and prolonged strike in American history.[22]

The harvest produced by these "new labor historians" (in effect, the labor history wing of the new social history) was rich and wide ranging. Beginning in the mid-1970s and continuing well into the 1990s, dozens of studies investigated everything from craft traditions and traditional work practices to women and children in the workplace to the role of religion, education, and politics in early industrial communities.[23] Virtually every publication addressed the nature of paternalism and the forms of resistance that occurred among skilled and unskilled laborers in different industries, regions, and time periods. What is more, the new labor history underscored the enduring significance of skill as an aspect of production and as a component of technological change.[24]

The large body of work on early industrial communities revealed a diverse set of responses to industrialization, not all of which were negative.[25] Industrial communities, sometimes even individual factory enclaves within communities, had their own particular habits and customs within which labor practices, innovations, and responses developed and found expression. As for the temper and direction of various responses to industrialism, much depended on firm size (the larger the firm, the more frequent serious forms of resistance), the nature of paternalism, labor traditions, worker identities (particularly with respect to their perceptions of autonomy and skill), the postindustrial prospects and trajectories of workers in different trades and settings, and the ethnic and religious composition of communities undergoing industrial transformation.

While the new labor history offered an illuminating and instructive perspective on the social processes of early industrialism, it did not escape criticism. Historians of technology pointed out, for example, that the new history tended to overlook technology, thus missing key insights into mechanization and how it affected the division of labor and the organization of work on the shop floor. In an otherwise positive review of Susan Hirsch's monograph on the industrialization of crafts in antebellum Newark, New Jersey, for example, MIT's Michael Folsom complained that Hirsch ignored technology. "Technology is merely a given," he noted, "something that just happens sooner or later to a craft," adding that "the social history of industrial peoples simply cannot be practiced without factoring in the history of the technologies upon which modern industry is based."[26] The Smithsonian's Brooke Hindle emphatically agreed. "The trouble," he wrote in a review of Jonathan Prude's *The*

Coming of Industrial Order, "is that it lacks the central attention to technology that the large picture requires." In an earlier review of Howard Rock's *Artisans of the New Republic*, Hindle was even more pointed. "The gulf remains between good political history, even when it deals with a technology community, and the history of technology. The integration of the two is not likely to be produced by political or social historians," he concluded. "This must be done by historians of technology."[27]

By the time Folsom and Hindle published their reviews, two books had already met their demands that technology be taken into account in the study of early industrial communities. In *Harpers Ferry Armory and the New Technology*, Merritt Roe Smith paid close attention to the shift from craft to machine production between 1815 and 1850, showing how resulting labor tensions and conflict were not only closely tied to mechanization and stricter management practices but also entangled in local and national politics.[28] Similarly in *Rockdale,* anthropologist Anthony F. C. Wallace detailed the mechanization of textile manufacturing along Chester Creek in southeastern Pennsylvania while providing a rich ethnohistorical account of the small mill villages that sprang up in the area during and after the War of 1812. For Wallace, as for Smith, the big question was "who *should* control the machines of the Industrial Revolution."[29] Both placed considerable emphasis on the complex network of business and family relationships that fostered, and at times opposed, change in the two communities. Although the business orientation and internal dynamics of Rockdale and Harpers Ferry differed, they nonetheless experienced similar problems in getting workers to subscribe to an industrial ethos. In Rockdale, mill owners sought to introduce "Christian capitalism," a creed that emphasized social harmony rather than social conflict, by building churches, establishing Sunday schools, and demanding adherence to rigid work rules. At Harpers Ferry, a more utilitarian approach to order and control emerged through the imposition of military superintendents and military management methods. While both communities experienced labor turmoil, Harpers Ferry proved the most resistant to industrial rule. Indeed, the violence associated with John Brown's famous 1859 raid on Harpers Ferry provided a fitting epitaph to the community's troubled history. Within two years, the national armory was looted and burned to the ground by invading Confederate forces.

As the number of community studies multiplied during the 1980s and beyond, research and publication on the Industrial Revolution in America branched out in a number of different directions. Some of the best work revisited, revised, and expanded on older subjects such as the transfer of technology, power systems, iron and steel production, the textile industry, the American system of manufactures, and the role of the military as a catalyst of change. Three of the most notable works were David J. Jeremy's rich comparative study of the "transatlantic industrial revolution" in textile technology, David Hounshell's expansive treatment of the transition from the American system to mass production, and Louis C. Hunter's encyclopedic multivolume *History of Industrial Power in the United States*.[30]

While these and other monographs significantly deepened understanding of early American industrialization, new ground was being broken elsewhere. In *Proprietary Capitalism*, for instance, Philip Scranton challenged Alfred Chandler's emphasis on big business by arguing that small and medium-size proprietary enterprises, not just large enterprises, played a crucial role in American industrialization. Such firms possessed the flexibility to shift product lines much faster than the large corporate organizations touted by Chandler, a point that significantly reinforced a contemporary argument being made by political economists Michael Piore and Charles Sabel about the need for the United States to adopt a more flexible approach to industrial competitiveness in order to meet the challenges being posed by Japan and Germany in world markets.[31] Like Chandler, Scranton was in the vanguard of historians who influenced policy analysis.

Throughout the 1980s and 1990s, historians of technology produced an array of sophisticated case studies that blended the techniques and perspectives of technical, political, and cultural analysis. One notable contribution was Judith McGaw's *Most Wonderful Machine*, a superb work that cast fresh light on the little-known paper industry of western Massachusetts while establishing a standard of excellence for all subsequent studies of early industrial communities.[32] Likewise, Carolyn Cooper and Steven Usselman previewed their prize-winning monographs on invention, the patent system, and the politics of patent management in a special issue of the journal *Technology and Culture* in 1991.[33] Of special interest was Theodore Steinberg's *Nature Incorporated*, a pathbreaking book that revealed how large textile corporations owned by the "Boston associates" proceeded to buy up much of the Merrimack River's watershed, build dams, and control the flow of water to power-hungry textile mill towns from Manchester, New Hampshire, to Lawrence, Massachusetts. The process of damming and dumping waste had devastating effects on the river's wildlife, to say nothing of human health.[34] Equally important was Colleen Dunlavy's comparative study of railroad-system building in Prussia and the United States, a book that nicely demonstrated how archival research could be combined with political theory to yield fresh insights about the role of the state and private capital in the development of national transportation systems.[35] Even the larger socioeconomic implications of time and time-keeping received fresh treatment at the hands of David Landes, Michael O'Malley, Mark Smith, Carlene Stephens, and Ian Bartky.[36] These works, as well as others, underscored the point that the Industrial Revolution in America was more incremental than revolutionary in nature and followed many paths.

Setting themselves apart from the case study approach, a number of historians posed big picture cultural studies questions to the topic of American industrialization. The cultural history of technology is arguably best represented by David Nye's *American Technological Sublime*. Building on the earlier work of Leo Marx, John Kasson, and other scholars in the field of American studies, Nye examined public reactions to railroads, bridges, skyscrapers, and other monumental engineering projects to show how Americans reified technology and used it to reinforce the prevailing ideology of progress and, by extension, national greatness. Reinforcing, indeed influencing, Nye's work was Leo Marx and Susan Danly's visually rich investigation, *The Railroad in American Art*, a subject that Marx had first probed in his classic *The Machine in the Garden*. Vivid and compelling, both works complemented each other and established technology as an integral and powerful part of American culture.[37] Indeed, Marx went further to critique the indiscriminate use of "technology" by scholars as a catch-all term to describe all sorts of discrete developments with complex and varying outcomes.[38]

After years of struggling to establish itself as a serious scholarly enterprise, the field of industrial archeology reached maturity in the 1990s. The importance of material culture in elucidating and understanding processes of invention and industrialization dated back to the 1960s with the pioneering work of Robert Vogel, John H. White, and Edwin Battison–all, interestingly, curators of industrial technology at the Smithsonian's National Museum of History and Technology (now the National Museum of American History). Indeed, Vogel was instrumental in founding the Society for Industrial Archeology and fostering the work of the Historic American Engineering Record. Through the former's informative newsletter (and later its journal, *IA*) and the latter's detailed surveys and field reports, invaluable information on the physical remains of the Industrial Revolution was reported, recorded, and, often, saved from the wrecking ball. From a scholarly standpoint, a turning point occurred in 1994 with the publication of *The Texture of Industry* by Robert B. Gordon and Patrick M. Malone. Drawing on their backgrounds in metallography and hydrology, Gordon and Malone produced a broad-ranging study of American industrialization that used artifacts and other objects of material culture to enrich their analysis. Subsequent studies by Gordon, Malone, Gregory Galer, Sara Wermiel, and other scholars helped to establish the centrality of industrial archeology and material culture to the study of American industrialization.[39] In doing so, they also drew much-needed attention to the technological components of the emerging urban-industrial "built environment" both before and after the Civil War.[40]

Rounding out the new work of the 1980s and 1990s was Charles B. Dew's reconsideration of industrial slavery and its influence on Southern industrial development in antebellum America. Building on his earlier monograph about Richmond's Tredegar Iron Works and Ronald L. Lewis's instructive investigations of industrial slavery in Maryland and Virginia, Dew's *Bond of Iron* challenged earlier interpretations that emphasized the relentless brutality of industrial slavery while pointing to more complex give-and-take relationships between master and slave at William Weaver's Buffalo Forge and Etna Furnace in the Great Valley of Virginia.[41] Dew's probing inquiry revealed how Weaver allowed his slaves considerable leeway in the workplace and paid them for "overwork," even permitting one of his most skilled hands to open a savings account at a bank in nearby Lexington, Virginia. That he did so indicates the extent to which he relied on his slave labor force to produce some of the South's best iron.

Like many ironworks in the North, Weaver's business was a mixed enterprise that combined farming with iron making. But there the similarities ended. Indeed, Dew argues that the existence of slavery holds the key to understanding why the South, despite its growing enthusiasm for manufacturing, had fallen so far behind the North by the outbreak of the Civil War. "Unlike free artisans in the North," he observes, "slave ironworkers had little opportunity to travel widely, visit other industrial establishments, learn new techniques, witness technological innovations, and acquire scientific knowledge."[42] Lacking sufficient labor mobility, the South's slave-based economy failed to innovate. As Dew and, more recently, Angela Lakwete show, the South had a formidable industrial base by the 1860s, but it never came close to matching the industrial juggernaut that was building in the North.[43]

Given the large body of literature about the Industrial Revolution in America, it is curious that no general synthesis or survey of the subject had been published since George R. Taylor's *Transportation Revolution*.

Perhaps the success of Taylor's masterful effort intimidated others from making the attempt. Whatever the case, Taylor's book, as good as it was, had become dated by the 1980s.

The first to take up the challenge was Thomas C. Cochran, a distinguished business historian well known for his emphasis on the influence of cultural and social factors in the emergence of the American business system. In 1981, Cochran published a brief but stimulating interpretation of American industrialization, *Frontiers of Change.* While acknowledging the importance of foreign trade to American development prior to the War of 1812, Cochran, like Taylor, maintained that "the most important frontiers of change" involved faster transportation technologies, the rise of a small but innovative manufacturing sector, and the growth of hinterland markets connected to eastern production centers by roads, rails, and waterways.[44] In his view, merchants and merchant capital were critical to American industrialization. He also challenged a longstanding belief by pointing to the Middle Atlantic states rather than New England as the spearhead of change. Above all, he viewed American industrialization as much a product of sociocultural forces as market capitalism. His conclusion that "the technological revolution . . . had been welcomed" by American mechanics did not sit well with labor historians, who found quite the opposite to be true.[45] Nonetheless, the point was provocative, and Cochran's book attracted considerable attention.

By and large, Cochran found compatible intellectual companions in Brooke Hindle and Steven Lubar who, in 1986, published *Engines of Change: The American Industrial Revolution, 1790–1860.* For Hindle and Lubar, even more so than Cochran, technological innovation stood at the center of American industrialization. Having overseen the mounting of a major exhibit under the same title at the Smithsonian's National Museum of American History and doubtless influenced by Robert Vogel and other Smithsonian advocates of industrial archeology, they placed considerable emphasis on artifacts as evidence, arguing that "in some ways they constitute the book's primary message." Hindle and Lubar also stressed the importance of placing the artifacts of industry in larger context, thus situating the Industrial Revolution in America by pointing to its origins in Europe, particularly Great Britain. Interestingly, the centerpiece of the Smithsonian exhibit and one of the key artifacts discussed by the authors was the *John Bull,* a steam locomotive imported to the United States in 1831 and put to work on the Camden & Amboy Railroad in New Jersey. In many ways, it represented the ideal artifact because it was British built but subsequently modified by American mechanics (with a lead truck and cowcatcher) to better negotiate a different terrain.[46] What better way to illustrate the incremental nature of technological innovation?

For Hindle and Lubar, as for Cochran, the relative speed with which the United States industrialized owed much to the country's "vast natural resources," the active involvement of government, particularly the military, as a catalyst of change, and the "technological enthusiasm" with which Americans generally greeted new technology. They also acknowledged that industrialization had costs as well as benefits. Among the former were the tensions and strains workers experienced not so much with the new technologies themselves but rather with the organizations that owned and controlled them. Yet in the end, Hindle and Lubar considered the "exhilaration" with which inventors, engineers, capitalists, and the public embraced technological change as "the great internal force" that drove the Industrial Revolution in America. It was a theme that Hindle had developed some twenty years earlier in an important essay, "The Exhilaration of Early American Technology."[47] With *Engines of Change*, America's First Industrial Revolution became the purview primarily of historians of technology.

The large body of socially oriented historical work on American industrialization published during the 1970s and 1980s helped set the stage for an important departure in the kindred field of technologies studies. Led primarily by European sociologists interested in shifting scholarly discourse away from deterministic treatments of the impact of new technologies to a more open-ended and democratic emphasis on the social shaping of technology, the new school of thought adopted "social constructivism" as its moniker while underscoring the role that social environments and social processes played in the design, dissemination, and subsequent adaptation of new technologies.

Two edited volumes set forth the social constructivist program. In *The Social Shaping of Technology* (1985), Donald MacKenzie and Judy Wajcman drew on earlier historical work to show how social factors influenced the development and deployment of new technologies.[48] Two years later, Wiebe Bijker, Thomas Hughes, and Trevor Pinch expanded the approach to technological systems.[49] Several tenets of social constructivism had already emerged in some of the "new" labor histories that highlighted worker culture, traditions, and expectations as primary lenses for the study of industrialization and its implications. Yet, interestingly, while social constructivism provided fresh concepts and a new language with which to analyze technology as a social product and industrialization as a social process, relatively few students of the Industrial Revolution in America adopted an explicit social constructivist approach.[50] That said, there is no question that social constructivism made historians of all stripes more conscious of the social forces impinging on American industrialization and, more generally, technological innovation.[51]

More controversial from a historiographic perspective was the publication in 1991 of Charles Sellers's *The Market Revolution: Jacksonian America, 1815–1846*.[52] According to Sellers, the market revolution was a powerful force that gathered momentum after the War of 1812, with dire consequences for agrarian America. Propelled by demographic changes that brought increasing numbers of people, especially farmers, into the market in search of supplementary income and led primarily by merchant capitalists intent on enlarging their businesses and maximizing profits through the construction of mechanized factories and far-reaching transportation systems, the market revolution disrupted the traditional "moral economy" of communal face-to-face relationships while setting in motion political tensions and protests that eventually coalesced into the Democratic party with its states' rights platform and heroic leader, Andrew Jackson. "Scrambling to sustain the traditional family," Sellers argued, "the dislocated rural populace . . . was experiencing the transition to capitalist production that would presently overtake most Americans."[53] The result, he concluded, was a "capitalist hegemony over economy, politics, and culture" that "created ourselves and most of the world we know." The market had won.[54]

Sellers's thesis sparked a lively debate between Marxist scholars who applauded his class-conflict interpretation of the transition and more consensus-oriented historians who saw more complex ethnic and cultural forces at play.[55] Critics abounded. Mary Blewett, a leading labor historian, considered Seller's market revolution "barely visible" and "slightly offstage, relatively unexplored, its mechanics and power assumed rather than demonstrated."[56] Other scholars, led by Winifred Rothenberg, Cary Carson, Richard Bushman, and Timothy Breen, disputed Sellers's chronology by locating the origins of the market revolution and its "empire of goods" much earlier in the pre- or post-revolutionary periods.[57] Still others considered *The Market Revolution*'s argument overly simplified and inaccurate while questioning whether Jacksonian "democracy was born as a counterweight to capitalism."[58] Indeed, David R. Meyers challenged Sellers's thesis by emphasizing the agrarian origins of American industrialization as well as the synergies that existed between increasing farming productivity and industrial development.[59] Critics as well as supporters nonetheless concluded that *The Market Revolution* represented "a magisterial synthesis of social and political history."[60] Flawed as it was, the concept had a future, particularly among mainstream historians who felt indebted to Sellers for providing them with an organizing principle that connected diverse social, cultural, and political developments with the dynamic "capitalist transformation" that galvanized nineteenth-century America.

With its sweeping view of the changeover from an agrarian to an urban-industrial society, *The Market Revolution*'s influence spread during the 1990s. Indeed, what originated as a provocative thesis about the origins of Jacksonian democracy quickly became a paradigm for the entire nineteenth century. Yet as attractive as the market revolution concept was, it signaled a significant shift of perspective from a supply-side to a demand-side explanation of American development. More implicitly than explicitly, the new view moved away from longstanding economic and technological forces of production toward a consumer-driven model of change that emphasized the centrality of politics and social turmoil while black-boxing complex processes of technological change.

The shift had important historiographical ramifications. For one thing, the market revolution began to supplant the Industrial Revolution as the primary organizing concept in discussions of modernization. Daniel Walker Howe spoke for many historians when he noted that "the market revolution was broader [and, by implication, more inclusive] than the industrial revolution."[61] The latter was often subordinated to the former with the result that supply-side innovations in the great transformation became increasingly blurred, if not totally opaque. In effect, the workings of technology and management were moved to the periphery, often becoming lost in the larger scheme of things. Much to the disappointment of labor historians, as technology receded from view, working people and critical labor processes were also pushed to the sidelines and made all but invisible. In their place stood merchants, politicians, and middle-class social reformers—all engaged in a high-stakes gambit for control of the new society. As the internal dynamics of change receded from view, so did the role of the state as an organizing force and catalyst of change. In sum, the market revolution, with its emphasis on commerce, reform, and power politics, began to replace the Industrial Revolution, with its emphasis on technology, management, and labor, as the nineteenth century's primary engine of change.

But scholarly traditions tend to recede rather than totally disappear. While the market paradigm held sway, business and technological historians continued to publish important work on the transition to industrialism. In a noteworthy example, Matthew W. Roth's *The Platt Brothers and Company: Small Business in American Manufacturing* detailed how Connecticut farmers like the Platts moved from a subsistence-based moral economy to a market-based manufacturing economy over the course of two generations. Prior to the 1820s, Alfred Platt combined farming with various rural manufacturing pursuits (grist milling, saw milling, flax milling, and nail making). A turning point occurred in 1822 when a local merchant named Aaron Benedict persuaded Platt to undertake the manufacture of brass buttons, a decision that quickly introduced him to the faster pace and pressures of selling goods in distant markets like New York City. By the end of the Civil War, the transition was complete. The Platts, urged on by local merchants with New York connections, were no longer farmers but full-time manufacturers with a large stable of workers and machinery. Their proprietary enterprise became one of the "Big Five" in Connecticut's highly profitable Naugatuck Valley brass industry.[62]

Other scholars working primarily in the history of technology have probed deeply and achieved a similar level of insight into how early manufacturing enterprises grew from humble origins to carve niches for themselves in expanding regional and national markets. One immediate conclusion is that the paths to industrialization varied. Gregory Galer's fresh perspective on the Ames Company of Easton, Massachusetts, reveals patterns similar to those outlined by Roth.[63] On the other hand, Robert Martello's research on Paul Revere's transition from a highly skilled silversmith to owner-manager of a foundry and rolling mill that produced everything from church bells to copper sheeting for naval warships indicates a different scenario of technology transfers between trades and the confluence of craft and industrial practices. Particularly revealing is Martello's discovery that Revere, like Eli Whitney and other early arms contractors to the War Department, relied on contracts and monetary advances from the U.S. Navy to get his copper rolling and foundry business tooled up and running.[64] Both companies, like the Platt brothers, went on to become leading manufacturers of specialty consumer products: with the Ames, it was shovels; with Revere, it was copper cookware, both of which can still be found in retail stores.

Martello's argument about the importance of navy subsidies to Revere's initial success as a manufacturer serves to underscore an often overlooked point about the role of the state in early American industrialization. Numerous enterprises, from foundries and factories to canals, roads, and railroads, received crucial support from local, state, and federal governments prior to, during, and after the War of 1812. Yet, although a number of scholars—notably Carter Goodrich, Forest Hill, Louis Hartz, and Oscar and Mary Handlin—published foundational works on the role of the state during the 1940s and 1950s, the theme never gained much traction among historians.[65]

Thanks primarily to the work of historical sociologist Theda Skocpol and other social scientists collectively known as "the new institutionalists," a renewed interest in bringing the state back into discussions of political economy emerged during the late 1990s. Leading this resurgence among historians was Richard John, whose pathbreaking book on the antebellum postal system and subsequent articles on state entrepreneurship and its political implications in Jacksonian America aimed at correcting Charles Sellers's market revolution paradigm and, in the process, attracted considerable attention. Drawing on earlier work by Colleen Dunlavy, William H. Freehling, Allan Kulikoff, John Lauritz Larson, Patricia Limerick, and Merritt Roe Smith, John maintains that federal and state policies and initiatives were critical to the emergence of the United States as an industrial power. Indeed, he argues that "in the United States, no less than in France, Germany, or Great Britain, big government *preceded* big business," adding that "no private enterprise could match the organizational capabilities of the [U.S.] Post Office Department, the Treasury Department, or the War Department."[66] What is more, federal institutions not only preceded large private enterprise, they also provided organizational and managerial models on which private firms, including major railroads, initially relied and then extended.[67]

Supporting John's contention are numerous works that point to the innovative role played by the state, especially the military, in fostering new industries and underwriting the development of key technologies.[68] All of these publications reinforce the important but neglected "state in, state out" thesis of Carter Goodrich about the silent workings of the American political economy. Years ago Goodrich pointed out that in the case of internal improvements, state and federal governments frequently "got in" the business of underwriting or otherwise subsidizing and supporting risky business and technological ventures until they could prove themselves in the marketplace. Once a venture became profitable or otherwise economically viable, the government "got out" and the enterprise proceeded under private ownership. "The striking feature of the American experience," he concluded, was that "the public and private roles in promotion were almost always thought of as cooperative rather than competitive."[69]

Many examples of the state-in/state-out scenario can be cited. One of the most famous, and short-lived, was Samuel F. B. Morse's first telegraph line from Baltimore to Washington, which was funded by the federal government. As soon as its viability for instantaneous communication was demonstrated, telegraphy immediately became privatized (even though Morse initially wanted to sell his telegraph patent to the Post Office).[70] Other prominent examples are the engineering and construction of roads, canals, harbors, lighthouses, river levees, and railroads, as well as the development of interchangeable manufacturing methods, the essential precursor for mass production in the twentieth century.[71] In short, the state at all levels has always been an important player in the American economy even though the story has repeatedly been swept under the rug. Will the new institutionalists succeed where Goodrich and others of his generation failed? We will see.

As the preceding discussion indicates, a large and varied literature explicates the Industrial Revolution in America. From that literature, we know that at its core, the Industrial Revolution involved the substitution of machinery for traditional handicraft methods, with all the structural changes, managerial innovations, standardized practices, and labor turmoil that followed in its wake. We also know that industrialization was multifaceted and followed a more incremental than radical course of development. Building on trends dating back to pre-revolutionary times, it began in the early days of the republic, gathered momentum during and after the War of 1812, matured during the 1830s and 1840s, and reached an important juncture in 1851, when American entries won a number of prizes at the London Crystal Palace Exhibition. By then the United States was no longer a developing nation dependent on European technology. It had reached a point where mass production, mass transportation, and mass communication were possible. The extent of these developments would soon be demonstrated in the massively equipped armies and industrial methods that defined the Civil War as a new kind of war.

Yet as good as the literature is, a number of doors remain open for further research on the Industrial Revolution in America. We need more comparative studies like Colleen Dunlavy's and David Jeremy's on early railroads and the transatlantic textile industry. Following Theodore Steinberg's example, new research should continue to explore the environmental, medical, and public health implications of early industrialization and the ways such changes may, in turn, have shaped sociopolitical responses to industrialization. Shop floor workers' contributions to the design and development of new technologies, a subject about which little is known and even less written, requires further exploration. These new efforts might also probe the connection between labor innovation and labor protest in early industrial America. Equally important, we need to know more about decision making—specifically how and why various actors chose to substitute machinery for handicraft methods, build factories, and resort to steam power over waterpower.[72] We also need to better understand how concerns about political order, social control, technical uniformity, and conflicting ideologies complicated, and in some instances trumped, purely rational economic choices. All of these factors, along with many others, had significant long-term implications.

Doubtless the biggest question currently facing historians of the early Republic is the relationship of the Industrial Revolution to the market revolution. To date they have been studied separately, with the market revolution primarily the purview of political and social historians and the Industrial Revolution the domain of economic and technological historians. To our knowledge, no publication comprehensively delineates the linkages and common boundaries between these topics and their related themes. This situation needs to change. First and foremost, historians of all stripes have to break through the specialist boundaries that confine and blind them, and work to produce more inclusive studies that reveal how and in what ways markets intersected with forces of production. From whom and where did the initial impetus come? Enterprising merchants seeking larger markets and greater profits? Craftsmen and mechanics eager to expand their operations? Government officials interested in securing regional advantages or national self-sufficiency, or both? "Soldier-technologists" committed to engineering new technologies and systems in the interest of national expansion? Preachers and teachers interested in public education and the "improvement of the age"? A mix of these and other actors on the historical stage?

In short, we need to identify the individuals, groups, and institutions that initiated the great transformation to modernity and determine when, where, how, and why they did what they did. To be sure, a number of recent publications have provided valuable clues, but all of them are limited by specialist constraints. Our current understanding of the early national period and its bearing on the emergence of modern America will remain segmented, limited, and not very meaningful until we break free of self-imposed disciplinary constraints and adopt a more flexible and inclusive boundary-crossing approach.

NOTES

1. Alexander Hamilton, *Report on Manufactures* in *The Debates and Proceedings in the Congress of the United States,* 2nd Congress (December 1791), 1004. For discussions of the origins and significance of Hamilton's report, see Jacob E. Cooke, ed., *The Reports of Alexander Hamilton* (New York: Harper Torchbooks, 1964), esp. xx–xiii, 115–205; Jacob E. Cooke, *Tench Coxe and the Early Republic* (Chapel Hill: University of North Carolina Press, 1978); Lawrence A. Peskin, *Manufacturing Revolution: The Intellectual Origins of Early American Industry* (Baltimore, MD: Johns Hopkins University Press, 2003).

2. Henry Howe, *Memoirs of the Most Eminent American Mechanic* (New York: Alexander V. Blake, 1841); J. Leander Bishop, *History of American Manufactures* (New York: Augustus M. Kelley, 1966 [1861–68]).

3. For entry to the literature on American progress, see Charles L. Flint et al., *One Hundred Years' Progress of the United States by Eminent Literary Men* (Hartford, CT: L. Stebbins, 1871); James Parton et al., *Sketches of Men of Progress* (New York: Hartford Publishing Co., 1870–1871); Richard M. Devens, *Our First Century* (Springfield, MA: C. A. Nichols, 1871); Horace Greeley et al., *The Great Industries of the United States* (Hartford, CT: J. B. Burr & Hyde, 1872); *Industrial America; or, Manufactures and Inventions of the United States . . . A Biographical and Descriptive Exposition of National Progress* (New York: Atlantic Publishing and Engraving Co., 1876); Benson Lossing, *The American Centenary: A History of the Progress of the Republic of the United States during the First One Hundred Years of Its* Existence (Philadelphia: Porter & Coates, 1876); Theodore D. Woolsey et al., *The First Century of the Republic: A Review of American Progress* (New York: Harper & Brothers, 1876). For entry into similar literature published in later years, see James P. Boyd, *Triumphs and Wonders of the 19th Century: The True Mirror of a Phenomenal Era* (Chicago: C. W. Stanton Co., 1899); Trumbull White, ed., *Our Wonderful Progress* (Springfield, MA: Hampden Publishing Co., 1902); Henry Chase Hill, ed., *The Wonder Book of Knowledge: The Marvels of Modern Industry and Invention . . .* (Philadelphia: John C. Winston Co., 1917, 1927); Waldemar Kaempffert, ed., *Modern Wonder Workers: A Popular History of American Invention* (New York: Blue Ribbon Books, 1924). On the idea of progress and its influence, see John B. Bury, *The Idea of Progress* (New York: Macmillan, 1932); Charles A. Beard, "The Idea of Progress," in *A Century of Progress,* ed. Charles A. Beard (New York: Harper & Brothers, 1933); Arthur A. Ekirch, *The Idea of Progress in America, 1815–1860* (New York: Columbia University Press, 1944); Leo Marx and Bruce Mazlish, eds., *Progress: Fact or Illusion?* (Ann Arbor: University of Michigan Press, 1996).

4. Charles H. Fitch, "Manufactures of Interchangeable Mechanism," Carroll D. Wright, "Report on the Factory System," and James M. Swank, "Iron and Steel"—all in the *Tenth Census of the United States (1880): Manufactures,* vol. 2 (Washington, DC: U.S. Government Printing Office, 1883).

5. Charles H. Fitch, "The Rise of a Mechanical Ideal," *Magazine of American History* 11 (June 1884), 516–527; Carroll D. Wright, *The Industrial Evolution of the United States* (Meadville, PA: Chautauqua Century Press, 1895); James M. Swank, *History of*

the Manufacture of Iron in All Ages from 1588 to 1885 (Philadelphia: J. M. Swank, 1884). Also see William R. Bagnall, *Samuel Slater and the Early Development of the Cotton Manufacture in the United States* (Middletown, CT: J. S. Stewart, 1890), and *The Textile Industries of the United States, 1639–1810* (Cambridge, MA: Riverside Press, 1893).

6. For insight into the relationship of technology, industry, and political ideology in the early Republic, see John S. Kasson, *Civilizing the Machine* (New York: Viking, 1976).

7. Louis C. Hunter's *Steamboats on Western Rivers: An Economic and Technological History* (Cambridge, MA: Harvard University Press, 1949) is considered one of the earliest and most perceptive examples of the contextual history of technology. Hunter previewed his steamboat study with two earlier works on the economic history of the Ohio Valley: "The Influence of the Market on Technique in the Iron Industry in Western Pennsylvania up to 1860," *Journal of Economic and Business History* 1 (Feb. 1929), 241–281, and a brief monograph entitled *Studies in the Economic History of the Ohio Valley: Seasonal Aspects of Industry and Commerce before the Age of Big Business* (Northampton, MA: Smith College Studies in History, vol. 19, 1933–1934). For an excellent review of Hunter's work, see John K. Brown, "Louis C. Hunter, Steamboats on the Western Rivers," *Technology and Culture* 44:4 (2003), 786–793. Vera Shlakman's *Economic History of a Factory Town: A Study of Chicopee, Massachusetts,* which first appeared in the Smith College Studies in History (1934–1935), was reissued in 1969 by Octagon Books. Caroline F. Ware's *The Early New England Cotton Manufacture: A Study in Industrial Beginnings* (Boston: Houghton Mifflin, 1931) advanced "the cultural approach to History," giving equal treatment to mill workers (many of them women) as well as inventors and entrepreneurs. She went on to publish *The Cultural Approach to History* (1940) and is considered an early advocate of "history from the bottom up." George Rogers Taylor's *The Transportation Revolution, 1815–1860* (New York: Rinehart & Co., 1951) provided the first detailed scholarly synthesis of the Industrial Revolution in America, much of it based on original research. It remains one of the most frequently cited works on the subject. A helpful synthesis of this literature is provided by Stuart Bruchey, *The Roots of American Economic Growth, 1607–1861* (New York: Harper & Row, 1965).

8. Hrothgar J. Habakkuk, *American and British Technology in the Nineteenth Century* (Cambridge: Cambridge University Press, 1962).

9. See, for example, Peter Temin, "Labor Scarcity and the Problem of American Industrial Efficiency in the 1850s," *Journal of Economic History* 26 (Sept. 1966), 277–298; Temin, "Notes on Labor Scarcity in America," *Journal of Interdisciplinary History* 1 (Winter 1971), 251–264; Robert William Fogel, "The Specification Problem in Economic History," *Journal of Economic History* 27 (Sept. 1967), 283–308.

10. See the following works by Nathan Rosenberg: "Technological Change in the Machine Tool Industry, 1840–1910," *Journal of Economic History* 23 (Dec. 1963), 414–443; "Anglo-American Wage Differences in the 1820s," *Journal of Economic History* 27 (June 1967), 221–229; ed., *The American System of Manufactures* (Edinburgh: Edinburgh University Press, 1969); and by Edward Ames and Nathan Rosenberg, "Changing Technological Leadership and Industrial Growth," *Economic Journal* 73 (Mar. 1963),

13–31, and "The Enfield Arsenal in Theory and History," *Economic Journal* 78 (Dec. 1968), 827–842.

11. See Brooke Hindle, *Technology in Early America: Needs and Opportunities for Study* (Chapel Hill: University of North Carolina Press, 1966), 18–21; Eugene S. Ferguson, "On the Origin and Development of American Mechanical 'Know-How,'" *Midcontinent American Studies Journal* 3 (1962), 3–16; Norman B. Wilkinson, "Brandywine Borrowings from European Technology," *Technology and Culture* 4 (1963), 1–13; Carroll W. Pursell, Jr., "Thomas Digges and William Pearce: An Example of the Transit of Technology," *William and Mary Quarterly* 21, 3rd ser. (1964), 551–560, and *Early Stationary Steam Engines in America: A Study in the Migration of a Technology* (Washington, DC: Smithsonian Institution Press, 1969); Nathan Rosenberg, "Economic Development and the Transfer of Technology," *Technology and Culture* 11 (Oct. 1970), 550–575; Rosenberg, "Factors Affecting the Diffusion of Technology," *Explorations in Economic History* 10 (Fall 1972), 3–34; Darwin H. Stapleton, *The Transfer of Early Industrial Technologies to America* (Philadelphia: American Philosophical Society, 1987); David J. Jeremy, *Transatlantic Industrial Revolution: The Diffusion of Textile Technologies Between Britain and America, 1790–1830s* (Cambridge, MA: MIT Press, 1981).

12. Robert William Fogel, *Railroads and American Economic Growth* (Baltimore, MD: Johns Hopkins University Press, 1964). Also see Albert Fishlow, *Railroads and the Transformation of the Ante-Bellum Economy* (Cambridge, MA: Harvard University Press, 1965). For a succinct but devastating critique of the Fogel thesis, see William Cronon, *Nature's Metropolis* (New York: Norton, 1991), 407, n. 79. For useful overviews of the new economic history, see Douglass C. North, ed., *Growth and Welfare in the American Past* (Englewood Cliffs, NJ: Prentice Hall, 1966); Lance E. Davis et al., *American Economic Growth: An Economist's History of the United States* (New York: Harper & Row, 1972). For a useful review of cliometrics, see Naomi Lamoreaux, "Economic History and the Cliometric Revolution," in *Imagined Histories: Americans Interpret the Past*, ed. Anthony Molho and Gordon S. Wood (Princeton, NJ: Princeton University Press, 1998), 59–84.

13. Albert D. Chandler, Jr., ed., *The Railroads: The Nation's First Big Business* (New York: Harcourt, 1965), and *The Visible Hand: The Managerial Revolution in American Business* (Cambridge, MA: Harvard University Press, 1977).

14. For perceptive assessments of Chandler's contributions to business history, see Glenn Porter, *The Rise of Big Business, 1860–1920*, 2nd ed. (Wheeling, IL: Harlan Davidson, 1992 [1973]), and "Technology and Business in the American Economy," in *An Emerging Independent American Economy, 1815–1875* (Tarrytown, NY: Sleepy Hollow Press, 1980), 1–28; Thomas K. McCraw, "Introduction: The Intellectual Odyssey of Alfred D. Chandler, Jr.," in *The Essential Alfred Chandler: Essays toward a Historical Theory of Big Business*, ed. McCraw (Boston: Harvard Business School, 1988), 1–22; Richard R. John, "Elaborations, Revisions, Dissents: Alfred D. Chandler, Jr.'s *The Visible Hand* after Twenty Years," *Business History Review* 71 (Summer 1997), 151–200; David A. Hounshell, "Hughesian History of Technology and Chandlerian Business History: Parallels, Departures, and Critics," *History and Technology* 12 (1995), 205–224.

15. See Louis Galambos, "The Emerging Organizational Synthesis in Modern American History," *Business History Review* 44 (Autumn 1970), 279–290, and "Technology, Political Economy, and Professionalization: Central Themes of the Organizational Synthesis," *Business History Review* 57 (Winter 1983), 471–490. For Galambos's most recent discussion of the subject, see "Recasting the Organizational Synthesis: Structure and Process in the Twentieth and Twenty-First Centuries," *Business History Review* 79 (2005), 1–38. For an alternative view that assigns greater significance to politics and culture in the development of American industrial capitalism, see Douglas Miller, *The Birth of Modern America, 1820–1850* (New York: Pegasus, 1970); Kasson, *Civilizing the Machine*; Edward Pessen, *Jacksonian America: Society, Personality, and Politics* (Homewood, IL: Dorsey Press, 1978).

16. On Chandler's influence, see McCraw, "Intellectual Odyssey," 1–22. Also see John and Hounshell, cited above (n. 14).

17. McCraw, "Intellectual Odyssey," 12.

18. Albert D. Chandler, Jr., *Strategy and Structure* (Cambridge, MA: MIT Press, 1962).

19. McCraw, "Intellectual Odyssey," 13.

20. Alice Kessler-Harris, "Social History," in *The New American History*, ed. Eric Foner (Philadelphia: Temple University Press, 1990), 163–164. Also see Peter Novick, *That Noble Dream: 'The Objectivity Question' and the American Historical Profession* (New York: Cambridge University Press, 1988), esp. chap. 13.

21. A number of Thompson's most important essays, including "Time, Work-Discipline, and Industrial Capitalism" (1967) and "The Moral Economy of the English Crowd in the Eighteenth Century" (1971), are collected in E. P. Thompson, *Customs in Common: Studies in Traditional Popular Culture* (New York: New Press, 1991). Also see his *The Making of the English Working Class* (New York: Vintage Books, 1963); Sidney Pollard, *The Genesis of Modern Management* (Baltimore, MD: Penguin Books, 1965) and "Factory Discipline in the Industrial Revolution," *Economic History Review* 16 (1963), 254–271; Eric Hobsbawm, *Labouring Men* (London: Wiedenfield and Nicolson, 1964); and Eric Hobsbawm and George Rudé, *Captain Swing* (New York: Pantheon, 1968). For an early example of the "new labor history" that clearly acknowledges British influence, see Herbert Gutman, "Work, Culture, and Society in Industrializing America, 1815–1919," *American Historical Review* 78:3 (June 1973), 531–588, and *Work, Culture, and Society in Industrializing America* (New York: Vintage Books, 1977). David Brody provides a helpful assessment of the "new labor history," including its British influences, in "Workers and Work in America: The New Labor History," in *Ordinary People and Everyday Life: Perspectives on the New Social History*, ed. James B. Gardner and George R. Adams (Nashville, TN: American Association for State and Local History, 1983), 139–159.

22. Alan Dawley, *Class and Community: The Industrial Revolution in Lynn* (Cambridge, MA: Harvard University Press, 1976); Paul Faler, "Cultural Aspects of the Industrial Revolution: Lynn, Massachusetts, Shoemakers and Industrial Morality, 1826–1860," *Labor History* 15 (1974), 367–394, and *Mechanics and Manufacturers in the Early Industrial Revolution* (Albany: State University of New York Press, 1981); William H. Mulligan,

Jr., "Mechanizing the Gentle Craft: From Family to Factory Training in Lynn, Massachusetts, 1800–1920," in *Essays from the Lowell Conference on Industrial History*, ed. Robert Weible (Lowell, MA: Lowell Conference on Industrial History, 1981); Mary H. Blewett, *Men, Women, and Work: Class, Gender, and Protest in the New England Shoe Industry, 1780–1914* (Urbana: University of Illinois Press, 1988).

23. In addition to the work of Gutman, Faler, Dawley, and Blewett (cited above), other well-known examples of the new labor history are Daniel J. Walkowitz, *Worker City, Company Town: Iron and Cotton-Worker Protest in Troy and Cohoes, New York, 1855–1884* (New York: Oxford University Press, 1978); David Montgomery, *Workers' Control in America* (New York: Cambridge University Press, 1979); Thomas Dublin, *Women at Work: The Transformation of Work and Community in Lowell, Massachusetts, 1826–1860* (New York: Columbia University Press, 1979); Howard B. Rock, *Artisans of the New Republic* (New York: New York University Press, 1979); Bruce Laurie, *Working People of Philadelphia, 1800–1850* (Philadelphia: Temple University Press, 1980); Michael H. Frisch and Daniel J. Walkowitz, eds., *Working-Class America* (Urbana: University of Illinois Press, 1983); Jonathan Prude, *The Coming of Industrial Order: Town and Factory Life in Rural Massachusetts, 1810–1860* (New York: Cambridge University Press, 1983); Philip Scranton, *Proprietary Capitalism: The Textile Manufacture at Philadelphia, 1800–1885* (New York: Cambridge University Press, 1983); Walter Licht, *Working for the Railroad: The Organization of Work in the Nineteenth Century* (Princeton, NJ: Princeton University Press, 1983); Sean Wilentz, *Chants Democratic: New York City and the Rise of the American Working Class, 1785–1851* (New York: Oxford University Press, 1984); Steven J. Ross, *Workers on the Edge: Work, Leisure, and Politics in Industrializing Cincinnati, 1788–1890* (New York: Columbia University Press, 1985); William J. Rorabaugh, *The Craft Apprentice: From Franklin to the Machine Age in America* (New York: Oxford University Press, 1986); Cynthia J. Shelton, *The Mills of Manayunk: Industrialization and Social Conflict in the Philadelphia Region, 1787–1851* (Baltimore, MD: Johns Hopkins University Press, 1986); Herbert G. Gutman and Donald H. Bell, eds., *The New England Working Class and the New Labor History* (Urbana: University of Illinois Press, 1987); David A. Zonderman, *Aspirations and Anxieties: New England Workers and the Mechanized Factory System, 1815–1850* (New York: Oxford University Press, 1992); Laurence F. Gross, *The Course of Industrial Decline: The Boott Cotton Mills of Lowell, Massachusetts, 1835–1955* (Baltimore, MD: Johns Hopkins University Press, 1993); Mary H. Blewett, *Constant Turmoil: The Politics of Industrial Life in Nineteenth-Century New England* (Amherst: University of Massachusetts Press, 2000). For an excellent synthesis of this literature, see Bruce Laurie, *Artisans into Workers: Labor in Nineteenth-Century America* (New York: Noonday Press, 1989).

24. See, for example, Montgomery, *Workers' Control*, and Wilentz, *Chants Democratic*.

25. See, for example, Gerald G. Eggert, *Harrisburg Industrializes: The Coming of Factories to an American Community* (University Park: Penn State University Press, 1993); Barbara M. Tucker, *Samuel Slater and the Origins of the American Textile Industry, 1790–1860* (Ithaca, NY: Cornell University Press, 1984). Also see Anthony F. C. Wallace, *Rockdale: The Growth of an American Village in the Early Industrial Revolution* (New York: Knopf, 1978), and Judith McGaw, *Most Wonderful Machine: Mechanization*

and Social Change in Berkshire Paper Making, 1801–1885 (Princeton, NJ: Princeton University Press, 1987).

26. Michael Folsom, review of Susan Hirsch, *Technology and Culture* 21:2 (April 1980), 261–262.

27. See Brooke Hindle, review of Jonathan Prude, *Technology and Culture* 26:4 (Oct. 1985), 855; Hindle review of Howard Rock, *Technology and Culture* 21:3 (July 1980), 508. For Hindle's further observations on the new social history, see "'The Exhilaration of Early American Technology': A New Look?" in *The History of American Technology: Exhilaration or Discontent?* ed. David A. Hounshell (Wilmington, DE: Hagley Museum & Library, 1984), 7–17, and "A Retrospective View of Science, Technology, and Material Culture in Early American History," *William and Mary Quarterly* 41, 3rd ser. (1984), 422–435. Also see Richard Bushman's review of John J. McCusker and Russell R. Menard, *The Economy of British North America* in *Technology and Culture* 28:2 (1987), 357–359. On the relationship of labor history to the history of technology, see Philip Scranton, "None-Too-Porous Boundaries: Labor History and the History of Technology," *Technology and Culture* 29:4 (Oct. 1988), 722–743.

28. Merritt Roe Smith, *Harpers Ferry Armory and the New Technology* (Ithaca, NY: Cornell University Press, 1977). For a related study with a similar thesis, see James J. Farley, *Making Arms in the Machine Age: Philadelphia's Frankford Arsenal, 1816–1870* (University Park: Penn State University Press, 1994).

29. Wallace, *Rockdale*, 240.

30. Jeremy, *Transatlantic Industrial Revolution*; David A. Hounshell, *From the American System to Mass Production* (Baltimore, MD: Johns Hopkins University Press, 1984); Louis C. Hunter, *History of Industrial Power in the United States*, 2 vols. (Charlottesville: University Press of Virginia, 1979, 1985). Other notable works on the transfer of technology include Stapleton, *Transfer of Technology*, and Joseph Bradley, *Guns for the Tsar: American Technology and the Small Arms Industry in Nineteenth-Century Russia* (De Kalb: Northern Illinois University Press, 1990). For other important works on power systems, see Patrick M. Malone, *Waterpower in Lowell: Engineering and Industry in Nineteenth-Century America* (Baltimore, MD: Johns Hopkins University Press, 2009), and "Canals and Industry: Engineering in Lowell, 1821–1880," in *The Continuing Revolution: A History of Lowell, Massachusetts,* ed. Robert Weible (Lowell, MA: Lowell Historical Society, 1991), 137–157; Richard L. Hills, "Steam and Waterpower: Differences in Transatlantic Approach," in *The World of the Industrial Revolution*, ed. Robert Weible (North Andover, MA: Museum of American Textile History, 1986), 35–53; Terry S. Reynolds, "The Emergence of the Breast Wheel and Its Adoption in the United States," in *The World of the Industrial Revolution*, ed. Robert Weible, 55–88. On iron and steel, see W. David Lewis, *Iron and Steel in America* (Greenville, DE: Hagley Museum, 1976); Robert B. Gordon, "The 'Kelly' Converter," *Technology and Culture* 33: 4 (Oct. 1992), 769–779, and *American Iron* (Baltimore, MD: Johns Hopkins University Press, 1996); John N. Ingham, *Making Iron and Steel* (Columbus: Ohio State University Press, 1991); Charles B. Dew, *Bond of Iron: Master and Slave at Buffalo Forge* (New York: Norton, 1994); Thomas J. Misa, *A Nation of Steel: The*

Making of Modern America, 1865–1925 (Baltimore, MD: Johns Hopkins University Press, 1995); Gregory Galer, Robert Gordon, and Frances Kemish, "Connecticut's Ames Iron Works," *Transactions: Connecticut Academy of Arts and Sciences* 54 (Dec. 1998), 83–194. On the American system, see Rosenberg, *American System of Manufactures*; Smith, *Harpers Ferry Armory*; Otto Mayr and Robert C. Post, eds., *Yankee Enterprise: The Rise of the American System of Manufactures* (Washington, DC: Smithsonian Institution Press, 1981); Donald R. Hoke, *Ingenious Yankees: The Rise of the American System of Manufactures in the Private Sector* (New York: Columbia University Press, 1990); Carolyn C. Cooper, *Shaping Invention: Thomas Blanchard's Machinery and Patent Management in Nineteenth-Century America* (New York: Columbia University Press, 1991). On the role of the military in industrialization, see William McNeill, *The Pursuit of Power* (Chicago: University of Chicago Press, 1983); Merritt Roe Smith, ed., *Military Enterprise and Technological Change* (Cambridge, MA: MIT Press, 1985); Todd Shallat, "Building Waterways, 1802–1866: Science and the U.S. Army in Early Public Works," *Technology and Culture* 31:1 (1990), 18–50, and *Structures in the Stream: Water, Science, and the Rise of the U.S. Army Corps of Engineers* (Austin: University of Texas Press, 1994); Barton C. Hacker, "Engineering a New Order: Military Institutions, Technical Education, and the Rise of the Industrial State," *Technology and Culture* 34:1 (Jan. 1993), 1–27, and "Military Institutions, Weapons, and Social Change: Toward a New History of Military Technology," *Technology and Culture* 35:4 (Oct. 1994), 768–834; Sara E. Wermiel, *Army Engineers' Contributions to the Development of Iron Construction in the Nineteenth Century* (Kansas City, MO: Public Works Historical Society, 2002).

31. Philip Scranton, *Proprietary Capitalism: The Textile Manufacture at Philadelphia, 1800–1885* (New York: Cambridge University Press, 1983), and *Endless Novelty: Specialty Production and American Industrialization, 1865–1925* (Princeton, NJ: Princeton University Press, 1997); Charles Sabel and Michael Piore, *The Second Industrial Divide* (New York: Basic Books, 1983). Another model study is John K. Brown, *The Baldwin Locomotive Works, 1831–1915: A Study in American Industrial Practice* (Baltimore, MD: Johns Hopkins University Press, 1995).

32. McGaw, *Most Wonderful Machine.*

33. Carolyn Cooper, "Social Construction of Invention through Patent Management: Thomas Blanchard's Woodworking Machinery," *Technology and Culture* 32 (1991), 960–998, led to her *Shaping Invention.* Steven Usselman, "Patents Purloined: Railroads, Inventors, and the Diffusion of Innovation in Nineteenth-Century America," *Technology and Culture* 32:4 (Oct. 1991), 1047–1075, resulted in *Regulating Railroad Innovation: Business, Technology, and Politics in America, 1840–1920* (Cambridge: Cambridge University Press, 2002).

34. Theodore Steinberg, *Nature Incorporated: Industrialization and the Waters of New England* (Cambridge: Cambridge University Press, 1991). Of related interest is Dolores Greenburg, "Energy, Power, and Perceptions of Social Change in the Early Nineteenth Century," *American Historical Review* 95 (June 1990), 693–714.

35. Colleen Dunlavy, *Politics and Industrialization: Early Railroads in the United States and Prussia* (Princeton, NJ: Princeton University Press, 1994).

36. David S. Landes, *Revolution in Time* (Cambridge, MA: Harvard University Press, 1983); Michael O'Malley, *Keeping Watch: A History of American Time* (New York: Viking, 1990); Mark M. Smith, *Mastered by the Clock: Time, Slavery, and Freedom in the American South* (Chapel Hill: University of North Carolina Press, 1997). Articles by Ian Bartky and Carlene Stephens in *Technology and Culture* (January 1989) previewed the former's *Selling True Time: Nineteenth-Century Timekeeping in America* (Stanford: Stanford University Press, 2000), and the latter's *On Time: How America Has Learned to Live by the Clock* (Washington, DC: Smithsonian Institution Press, 2002), which accompanied a major Smithsonian exhibit at the National Museum of American History.

37. David Nye, *American Technological Sublime* (Cambridge, MA: MIT Press, 1994); Susan Danly and Leo Marx, eds., *The Railroad in American Art* (Cambridge, MA: MIT Press, 1987); Leo Marx, *The Machine in the Garden* (New York: Oxford University Press, 1964). Also see Nye, *America as Second Creation* (Cambridge, MA: MIT Press, 2003); Kasson, *Civilizing the Machine;* and Laura Rigal, *The American Manufactory: Art, Labor, and the World of Things in the Early Republic* (Princeton, NJ: Princeton University Press, 1998).

38. Leo Marx, "*Technology*: The Emergence of a Hazardous Concept," *Social Research* 64:3 (Fall 1997), 965–988.

39. Robert B. Gordon and Patrick M. Malone, *The Texture of Industry: An Archeological View of the Industrialization of North America* (New York: Oxford University Press, 1994). Also see Patrick M. Malone and Robert Weibel, *Lowell Water Power System: Pawtucket Gatehouse Hydraulic Turbine* (Lowell, MA: American Society of Mechanical Engineers, 1985) and *Waterpower in Lowell*; Galer, Gordon, and Kemmish, "Connecticut's Ames Iron Works"; and Sara Wermiel, *The Fireproof Building: Technology and Public Safety in the Nineteenth-Century American City* (Baltimore, MD: Johns Hopkins University Press, 2000), for exemplary studies that incorporate the use of artifacts as evidence.

40. Perhaps the best example of how the built environment influenced not only urban but larger regional development during this period is Cronon, *Nature's Metropolis*.

41. Dew, *Bond of Iron* and *Ironmaker to the Confederacy: Joseph R. Anderson and the Tredegar Iron Works* (New Haven, CT: Yale University Press, 1966); Ronald L. Lewis, *Coal, Iron, and Slaves: Industrial Slavery in Maryland and Virginia, 1715–1865* (Westport, CT: Greenwood Press, 1979).

42. Dew, *Bond of Iron*, 107.

43. Angela Lakwete, *Inventing the Cotton Gin: Machine and Myth in Antebellum America* (Baltimore, MD: Johns Hopkins University Press, 2003). Also see Curtis J. Evans, *The Conquest of Labor: Daniel Pratt and Southern Industrialization* (Baton Rouge: Louisiana State University Press, 2001); Robert S. Davis, Jr., *Cotton, Fire, and Dreams: The Robert Findlay Iron Works and Heavy Industry in Macon, Georgia, 1839–1912* (Macon, GA: Mercer University Press, 1998). For a recent exploration of industrial slavery in the Revolutionary and early national periods, see John Bezis-Selfa, *Forging America: Iron-*

workers, Adventurers, and the Industrious Revolution (Ithaca, NY: Cornell University Press, 2004).

44. Thomas C. Cochran, *Frontiers of Change: Early Industrialism in America* (New York: Oxford University Press, 1981), 39. See also Glenn Porter and Harold C. Livesay, *Merchants and Manufacturers* (Baltimore, MD: Johns Hopkins University Press, 1971), and Diane Lindstrom, *Economic Development in the Philadelphia Region, 1810–1850* (New York: Columbia University Press, 1978).

45. Cochran, *Frontiers of Change*, 144–145.

46. Brooke Hindle and Steven Lubar, *Engines of Change: The American Industrial Revolution, 1790–1860* (Washington, DC: Smithsonian Institution Press, 1986), 7, 125–151.

47. Ibid., 277–278; Brooke Hindle, *Technology in Early America: Needs and Opportunities for Study* (Chapel Hill: University of North Carolina Press, 1966), 3–28. See also Judith McGaw, ed., *Early American Technology: Making and Doing Things from the Colonial Era to 1850* (Chapel Hill: University of North Carolina Press, 1994).

48. Donald MacKenzie and Judy Wajcman, eds., *The Social Shaping of Technology* (Philadelphia: Open University Press, 1985).

49. Wiebe Bijker, Thomas Hughes, and Trevor Pinch, *The Social Construction of Technological Systems* (Cambridge, MA: MIT Press, 1987).

50. Significant examples of the social constructivist approach applied to American industrialization are Cooper's "Social Construction of Invention through Patent Management" and *Shaping Invention*; Misa, *A Nation of Steel*; and Ruth S. Cowan, *More Work for Mother: The Ironies of Household Technology from the Open Hearth to the Microwave* (New York: Basic Books, 1983).

51. Social constructivism did not escape criticism. See Langdon Winner, "On Opening the Black Box and Finding It Empty: Social Constructivism and the Philosophy of Technology," *Science, Technology and Human Values* 18 (1993), 362–378, and Edward W. Constant II, "Reliable Knowledge and Unreliable Stuff: On the Practical Role of Rational Beliefs," *Technology and Culture* 40 (1999), 324–357. Constant's article elicited a spirited exchange with Philip Scranton and John Law and Vicky Singleton in *Technology and Culture* 41:4 (2000), 752–782. Another critical exchange occurred among Nick Clayton, Wiebe Bijker, Trevor Pinch, and Bruce Epperson in *Technology and Culture* 43:2 (2002), 351–373.

52. Charles Sellers, *The Market Revolution: Jacksonian America, 1815–1846* (New York: Oxford University Press, 1991).

53. Ibid., 19.

54. Ibid., 5.

55. See, for example, Richard E. Ellis, Mary H. Blewett, Joel H. Silbey, Major L. Wilson, Harry L. Watson, Amy Bridges, and Charles Sellers, "A Symposium on

Charles Sellers, *The Market Revolution: Jacksonian America, 1815–1846,*" *Journal of the Early Republic,* 12: 4 (1992), 445–476; Melvyn Stokes and Stephen Conway, eds., *The Market Revolution in America: Social, Political, and Religious Expressions, 1800–1880* (Charlottesville: University Press of Virginia, 1996); and "Special Issue on Capitalism in the Early Republic," *Journal of the Early Republic* 16:2 (1996), 159–308.

56. Mary Blewett, "Society and Economic Change," *Journal of the Early Republic* 12:4 (1992), 451

57. Winifred Rothenberg, *From Market Place to a Market Economy: The Transformation of Rural Massachusetts, 1750–1850* (Chicago: University of Chicago Press, 1992); Cary Carson, Ronald Hoffman, and Peter J. Albert, eds., *Of Consuming Interests: The Style of Life in the Eighteenth Century* (Charlottesville: University of Virginia Press, 1994); Richard Bushman, "Markets and Composite Farms in Early America," *William and Mary Quarterly*, 3rd ser., 55 (July 1998), 351–374; Timothy H. Breen, *The Marketplace of Revolution: How Consumer Politics Shaped American Independence* (New York: Oxford University Press, 2004). Also see Maxine Berg, "The British Product Revolution in the Eighteenth Century" (chapter 3, this volume), as well as her *Luxury and Pleasure in Eighteenth-Century Britain* (New York: Oxford University Press, 2005) for supporting documentation.

58. Joel H. Silbey, "Hegemony of the Market Questioned," *Journal of the Early Republic* 12 (Winter 1992), 455; Major L. Wilson, "Conflict or Consensus?" *Journal of the Early Republic* 12 (Winter 1992), 461. Sellers's severest critic was William E. Gienapp, "The Myth of Class in Jacksonian America," *Journal of Public Policy History* 6:2 (1994), 232–259.

59. David R. Meyers, *The Roots of American Industrialization* (Baltimore, MD: Johns Hopkins University Press, 2003).

60. Wilson, "Conflict or Consensus," 464.

61. Howe, "The Market Revolution and the Shaping of Identity in Whig-Jacksonian America," in Stokes and Conway, *The Market Revolution in America*, 267. A critic from the outset, Howe would later strongly disavow Sellers's market revolution thesis in his Pulitzer Prize–winning book, *What Hath God Wrought: The Transformation of America, 1815–1848* (New York: Oxford University Press, 2007), which pointed instead to a twin "Communications/Transportation Revolution" as "a driving force in the history of the era" (p. 5). For a review of Howe's thesis, see Merritt Roe Smith, "America's Coming of Age," *Technology and Culture* 50 (Jan. 2009), 187–193.

62. Matthew W. Roth, *Platt Brothers and Company* (Hanover, NH: University Press of New England, 1994).

63. Gregory J. Galer, "Forging Ahead: The Ames Family of Easton, Massachusetts and Two Centuries of Industrial Enterprise, 1635–1861" (Ph.D. diss., MIT, 2002). Also Galer, Gordon, and Kemmish, "Connecticut's Ames Iron Works."

64. Robert Martello, "Paul Revere's Last Ride: The Road to Rolling Copper," *Journal of the Early Republic* 20:2 (Summer 2000), 219–240, and *Midnight Ride, Indus-*

trial Dawn: Paul Revere and the Rise of American Enterprise (Baltimore, MD: Johns Hopkins University Press, 2010). Also see Janet Siskind, *Rum & Axes: The Rise of a Connecticut Merchant Family, 1795–1850* (Ithaca, NY: Cornell University Press, 2002).

65. Carter Goodrich, *Government Promotion of American Canals and Railroads, 1800–1890* (New York: Columbia University Press, 1960), and *The Government and the Economy: 1783–1861* (Indianapolis: Bobbs-Merrill, 1967); Oscar and Mary Flug Handlin, *Commonwealth: A Study of the Role of Government in the American Economy: Massachusetts, 1774–1861* (New York: New York University Press, 1947); Louis Hartz, *Economic Policy and Democratic Thought: Pennsylvania, 1776–1860* (Cambridge, MA: Harvard University Press, 1948); Milton S. Heath, *Constructive Liberalism: The Role of the State in Economic Development in Georgia to 1860* (Cambridge, MA: Harvard University Press, 1954); James N. Primm, *Economic Policy in the Development of a Western State: Missouri, 1820–1860* (Cambridge, MA: Harvard University Press, 1954). All of the above focused primarily on local and state governments as agents of economic change. In contrast, see Forest G. Hill, *Roads, Rails, and Waterways: The Army Engineers and Early Transportation* (Norman: University of Oklahoma Press, 1957), and "Formative Relations of American Enterprise, Government and Science," *Political Science Quarterly* 75 (Sept. 1960), 400–419. On the army's involvement in opening and developing the trans-Mississippi West, see, for example, William H. Goetzmann, *Army Exploration in the American West, 1803–1863* (New Haven, CT: Yale University Press, 1959); Durwood Ball, *Army Regulars on the Western Frontier, 1848–1861* (Norman: University of Oklahoma Press, 2001); Laurence J. Malone, *Opening the West: Federal Internal Improvements before 1860* (Westport, CT: Greenwood Press, 1998); and Michael L. Tate, *The Frontier Army in the Settlement of the West* (Norman: University of Oklahoma Press, 1999). On the U.S. Navy's role, see Brendan P. Foley, "Fighting Engineers: The U.S. Navy and Mechanical Engineering, 1840–1905" (Ph.D. diss., MIT, 2003); Kurt Hackemer, *The U.S. Navy and the Origins of the Military-Industrial Complex, 1847–1883* (Annapolis, MD: Naval Institute Press, 2001); Edward W. Sloan III, *Benjamin Franklin Isherwood: Naval Engineer* (Annapolis, MD: U.S. Naval Institute, 1965). Still very useful is A. Hunter Dupree, *Science in the Federal Government* (Baltimore, MD: Johns Hopkins University Press, 1987 [1957]). For a revealing account, see David Ciepley, "Why the State Was Dropped in the First Place: A Prequel to Skocpol's 'Bringing the State Back In,'" *Critical Review* 14: 2–3 (2000), 157–213.

66. Richard R. John, "Governmental Institutions as Agents of Change: Rethinking American Political Development in the Early Republic, 1787–1835," *Studies in American Political Development* 11 (Fall 1997), 347–380, and "Affairs of Office: The Executive Departments, the Election of 1828, and the Making of the Democratic Party," in *The Democratic Experiment: New Directions in American Political History*, ed. Meg Jacobs, William J. Novak, and Julian E. Zelizer (Princeton, NJ: Princeton University Press, 2003), 50–84, 56. Also pertinent is Richard John, ed., *Ruling Passions: Political Economy in Nineteenth-Century America* (University Park: Pennsylvania State University Press, 2006).

67. See Charles O'Connell, Jr., "The Corps of Engineers and the Rise of Modern Management, 1827–1856," in Smith, ed., *Military Enterprise and Technological Change*,

87–116; Keith W. Hoskin and Richard H. Macve, "The Genesis of Accountability: The West Point Connections," *Accounting Organizations and Society* 13:1 (1988), 37–73, "Cost Accounting and the Genesis of Managerialism: The Springfield Armory Episode," in *Proceedings of the Second Interdisciplinary Perspectives on Accounting Conference* (Manchester, UK, 1988), and "Reappraising the Genesis of Managerialism: A Reexamination of the Role of Accounting at the Springfield Armory, 1815–1845," *Accounting, Auditing and Accountability Journal* 7 (1994), 4–29.

68. See also Mayr and Post, eds., *The American System of Manufactures*; Merritt Roe Smith, "Army Ordnance and the American System," in Smith, ed., *Military Enterprise*, 39–86; Robert G. Angevine, *The Railroad and the State: War, Technology, and Politics in Nineteenth-Century America* (Stanford, CA: Stanford University Press, 2004). Two works on the American political economy that bolster John's arguments are John Lauritz Larson, *Internal Improvement: National Public Works and the Promise of Popular Government in the Early United States* (Chapel Hill: University of North Carolina Press, 2001), and Peskin, *Manufacturing Revolution*. Also noteworthy are John R. Nelson, Jr., *Liberty and Property: Political Economy and Policymaking in the New Nation, 1789–1812* (Baltimore, MD: Johns Hopkins University Press, 1987), and James L. Huston, "Economic Landscapes Yet to Be Discovered: The Early American Republic and Historians' Unsubtle Adoption of Political Economy," *Journal of the Early Republic* 24 (Summer 2004), 219–231.

69. Goodrich, *Government Promotion*, 291–97.

70. On early telegraphy, see Robert L. Thompson, *Wiring the Continent* (Princeton, NJ: Princeton University Press, 1947), and Menaheim Blondheim, *News over the Wire: The Telegraph and the Flow of Public Information, 1844–1897* (Cambridge, MA: Harvard University Press, 1994). On the privatization of telegraphy, see Richard R. John, "Private Enterprise a Public Good? Communications Deregulation as a National Political Issue, 1839–1851," in *Beyond the Founders*, ed. Jeffrey L. Pasley and David Waldstreicher (Chapel Hill: University of North Carolina Press, 2004), 328–354.

71. On lighthouses and iron construction methods, see Wermiel, *Army Engineers' Contribution*; on harbors, rivers, and canals, see Shallat, *Structures in the Stream;* on railroads, Angevine, *The Railroad and the State;* on interchangeable manufacturing and its larger impact, Smith, "Army Ordnance and the American System"; Hounshell, *From the American System to Mass Production*.

72. An excellent treatment of decision making is Usselman, *Regulating Railroad Innovation*.

10

The Many Transitions of Ebenezer Stedman: A Biographical and Cross-National Approach to the Industrial Revolution

Leonard N. Rosenband

In 1829, Ebenezer Stedman began the difficult task of restoring a derelict paper mill. The old workshop in Georgetown, Kentucky, the first to produce paper in the "Great West," had been the victim of a series of ineffectual and alcoholic tenants. To reconstruct his newly leased mill, Stedman, just twenty-one years old, relied on practices and instruments inherited from the papermakers of Massachusetts, Great Britain, and Europe. He soon prospered and "began to feel" that he was "some Boddy."[1] He rented a room near the courthouse, where he bartered with blacks for hemp tow, the coarse, less desirable fibers that served as the raw material for his paper. He endured a regimen of long hours, made longer by the rotten timbers and literal pitfalls of the Georgetown site: "The old Mill Being near worn out i had to be Continual Repairing."[2] Still, he was sure that he was on the high road to plenty, "that Some day, i wold Stick a feather in my Cap, that I was E. H. Stedman."[3] He was a Jacksonian "go-getter," so aroused by the prospects of getting ahead that he could not let go, even when it was past time to do so.[4] Much later, he realized it was all "a delusion, a Bright Light of Some Metor."[5] He had negotiated the passage from handicraft to mechanized papermaking successfully, but lost his fortune to the ever-present hazards of his trade. He went from shirtsleeves to shirtsleeves in one generation.

Although Stedman lamented his "unprofitable life," he produced two accounts of it.[6] The first, barebones version depicted the location and techniques of early Kentucky papermaking. In this undated, pamphlet-length letter, he presented the cold facts of his trade, from the fiery demise of neighboring mills to the backbreaking duties of the layboy.[7] Typically, he explained his downfall with admirable concision: "Fire, war, intrest Eat me out."[8] But Stedman's second telling of his works and days (and nights in the mill), an epistolary autobiography addressed to his daughter, Sophronia, was an extended meditation on independence, industriousness, and the price of profit.[9] Largely chronological, Stedman's contemplative odyssey is often dark, but he had the storyteller's knack of shifting hues with asides about the utility of strong drink, the trickster's joy, and the tastiness of fried squirrel. His artful construction of twice-told tales was surely the residue of practice. Written during his exile in Live Oak County, Texas, the second self-portrait was dated 1878, Stedman's seventieth year. Like its precursor, Stedman's expanded autobiography centered on the 1820s and 1830s. Both were coming-of-age accounts: the time of his time included his marriage, his embrace of the papermaking machine, and his rise to the rank of mill owner. Above all, however, Stedman's story, as he told it, was about work or, more precisely, a life in which work was not a stage but a condition.[10]

As the design of this book confirms, economic historians and historians of technology increasingly recognize the value of exploring large-scale processes comparatively.[11] Living under the trade winds or equatorial heat no longer constitutes a sufficient climate for such comparison. Now scholars juggle variables from capital formation to cultural habits, technological styles to patterns of consumption, institutional incentives to the accessibility of mineral deposits. Generally the framework of analysis remains the nation-state, although Pat Hudson, Sidney Pollard, and Kenneth Pomeranz made the region the locus for effective comparison.[12] More recently, I have turned to one trade, papermaking, as a framework for comparative consideration.[13] My intention, as is inevitably the purpose of comparative studies, has been to highlight both equivalence and distinction, such as the surprising advantages of backwardness enjoyed by eighteenth-century English papermakers. My contribution to our collection offers comparative history in a somewhat different key. It tightens the focus to one man's journey within the contours and choices of this cross-national craft. It reveals the resonance of the everyday and the intergenerational in his transatlantic art.[14] It connects chance encounters and the familiar troubles of papermaking with its broadly diffused technological developments and shifting market opportunities. Thus, the high-flying transition to advanced industrial capitalism is brought to ground, to the promise and pitfalls of paper production in the bluegrass. There, Stedman's work ethic and prescriptive knowledge, his longings and regrets, paved one pathway of transformation within his transnational trade.

Stedman's experience with early industrial capitalism and his assessment of it form the core of this chapter. His was a life of labor, whether he was in pursuit of profits or wages. For fifteen years, the bulk of Stedman's working life was tied to the Georgetown mill and its cycles of production. When the springs that fed the mill dried, as they did every May, he had to look for hours and wages elsewhere. He took what he could get, including several summer stays in carding mills and an extended turn at the potter's wheel. But he always returned to the Georgetown mill if the most recent in a string of tenants had not run it into the ground. Still, his time on the road matured him. He learned to mold, forge, and turn the lathe, as well as the secrets of the steam engine, in a Lexington woolen mill. In an Ohio paper mill, he experienced the ribaldry of his shopmates and recognized that his voyage through the craft would be more rowdy than tranquil, a discovery that dovetailed neatly with his own trickster's gleam. He even encountered an English paper worker who had tramped through several states and "Hundreds of papermills."[15] The years that centered on Georgetown, the heart of his autobiography, schooled him in the durable national and transnational custom and technique of his trade. But this period of his career also provided Stedman with access to the evolving tools, machines, and sources of power within his trade and others, establishing him as a potential improver despite his usually thin pockets.

Essentially, hand papermaking consisted of three stages in the Anglo-American mills of 1810: the reduction of castoff linen into pulp, the creation of the sheet, and the preparation of the reams for ink and shipment.[16] Female hands started the production process. They separated white rags from gray, removed caked dirt, and cut away matted patches. An experienced hand oversaw the Hollander beater, or "rag engine," a rotating cylinder fitted with knives and fixed to a bedplate, which tore the linen into filaments. The vatman transformed this material into a sheet of paper. First, he dipped his mold, a rectangular wire mesh bounded by a wooden frame, into a vat of warm, watery pulp. Then this craftsman lifted the mold, and with water streaming through the sieve, shook it in a customary pattern to ensure that the fibers of the infant sheet "shut." (He repeated these motions two thousand or three thousand times each day.) Next, the vatman passed the mold, with a fresh sheet clinging to it, to the coucher. This artisan's main tool was a stack of hairy felts, which he rested on a small easel. He needed steady hands and good timing, since he flipped six or seven sheets per minute from wire to felt. Once his pile of felts, each bearing a moist sheet of paper, reached the customary height known as a post, it was submitted to the press. (An English statute mandated twenty posts per day in 1796, the same output dictated by custom in the French province of Angoumois.[17]) The layboy then separated the paper from the felts, a delicate task "suitable only for people who have practised it from an early age and not for uneducated, inexperienced country-folk."[18] Stedman was proud that "It was said That i was the Best lay Boy that Could Be Found."[19] But working with newly minted sheets "Barely dry Enough to Bare their weight, so Easly to tare" required both a sensitive touch and an iron spine: "When The days work was done, my Back woold achake So that manny a night i could not Sleep."[20] Not surprisingly, when Stedman got the layman's stool, he claimed that "My Life of Slavery" had commenced.[21]

More pressing of the paper followed, and the sheets were draped over cords to dry. The sizerman gathered this paper and plunged handfuls into an emulsion of hides, hoofs, tripe, and alum. This gelatin bath filled the paper's pores, thereby preventing ink blots. A deft vatman learned much from the color and consistency of his pulp. But the sizerman tested his work with his tongue: if it left a balanced impression on a sheet, akin to a fan or a butterfly's wings, and the paper had "crackle," the size was good. Finally, women sorted the paper, excised stained and clotted swatches, and assisted the loftsman in wrapping the reams. Rich in artisanal lore, marked by a venerable division of labor, and far more intricate than this brief description suggests, this process took weeks to turn hemp and linen scraps into paper.

Still, changes were in the works. Early in the nineteenth century in England, the mechanical agitator, or hog, a relatively recent invention, prevented the pulp from settling in the vats and slowing production. At the same time, American manufacturers were experimenting with the addition of size to the linen during beating in order to avoid the step of tub sizing. And a French prototype of the papermaking machine, patented in 1799, had crossed the Channel, where the industrial engineer Bryan Donkin had embarked on the tasks of refining and enlarging it. Rising demand as well as rising costs of production, prompted by scarce rags and calculating journeymen, were the twin impulses compelling the restless modification of papermaking's transnational technology.

The French astronomer and technical writer Lalande remarked in 1761 that paper had become an "everyday merchandise."[22] The rise of the novel, the rants of revolutionaries, and the demand for fashionable wallpaper, among many events and enticements, continued to feed this appetite throughout the Atlantic world. So did the expanding desire of the fin-de-siècle state to count, regulate, explain, and document. But the purchase of paper was no everyday affair. As early as the 1760s, New York advertisers offered seventeen sorts for sale.[23] This was an epistolary era, when "friends and neighbors" greeted and assailed each other with "little notes, invitations to dinner, thank-you notes, begging notes, scolding notes, and notes for no reason at all."[24] The medium itself was a critical part of the message, since the firmness and tint of the paper spoke volumes about rank and the worth of a relationship. Consumers cautiously evaluated the properties of the stationery and the paper in the books they bought. They shared an arcane vocabulary of quality, rubbed the paper between their fingers, and lifted it up to the light for a clear look at its knit and blemishes. Not every producer reached the printers' and scribblers' standards, but the wide markets for low-grade wrapping, newsprint, and pasteboard were also inviting.

Whether in France, England, or Kentucky, hand papermaking had always been both capital and skill intensive. The trade was composed largely of petty producers, many of whom sweated at the disposal of stationers and printers, and skilled, wage-earning workers. A conventional mill contained one or two units of production and twelve to twenty hands. Manufacturers often had trouble disciplining the footloose journeymen, and many, barely up from the workers' ranks, indulged in a nip with them. The familiarity of these small enterprises, however, did not bar either side from advancing its interests. To cushion the blows of downturns and capture favorable moments, both masters and men organized to regulate markets. In England, manufacturers leagued to control the price of labor, paper, and raw materials and to act in concert against a government always eager to increase the excise burden on their reams.[25] Around Philadelphia, the producers set standards for the weight and measurements of their paper to prevent price competition among themselves and disputes with their customers.[26] In both the Old World and the New, employers and craftsmen adhered to the "day's work," a production quota.[27] Yet the entrepreneurs did not hold all the cards: the journeymen of England forged a national trade union in 1800, the Original Society of Papermakers. Their skills ensured a great deal of leverage, and they were quick to "leave for wages or customs."[28] Vatfuls of perishing pulp and the insistence of printers and stationers on prompt delivery buckled even the most stubborn manufacturer's knees. Less is known about the collective powers of America's early paper workers. But just as their French and English counterparts did, the vat crews at Zenas Crane's Berkshire, Massachusetts, mill continued to provide tramping journeymen with a quire of broken paper and a dram when the master could not provide them with jobs.[29]

Even the mechanized production of paper, which began in England during the first decade of the nineteenth century, "required almost incessant attention from its operators."[30] The mortar of machine-age papermaking was the tacit knowledge about rags, finishes, and shortcuts at the fingertips of mill hands and mill owners alike. For men like Stedman, who held his early mechanized efforts together with shoestrings and an artisan's know-how, this meant ceaseless toil. He was aware of class, fancied conspicuous display, and railed against those who would dilute skill into tasks. But the categories of artisan and entrepreneur are too static and too freighted with interpretive weight for the proper telling of Stedman's tale. After all, he had to sort through a bed of rags constructed by rats for the raw material at his first mill. If anything, his was a work consciousness.

Work was the unifying thread of Stedman's memories. Recalling an early turn as a skilled hand in the Georgetown mill, he concluded that "Never was a slave kept as Constant at work Day & Night as i was."[31] His departure from his marriage bed, told with ledger-like precision, was doubly chilling because Stedman and his brother, Sam, owned the mill: "The wedding Company dispersed about nine o'clock. At 10 we ware in Bed. I got up & dressed & went to the mill at two o'clock in the morning; went Back to [his father-in-law's home] twelve o clock the next night; up By day Brake & of to the Mill. I Stopped the mill at 12 o clock. This was Saturday night."[32] Later, almost wistfully, he remembered: "Manny nights this winter your dear Mother wood Come to the Mill & Stay with me till 12 o'clock at night."[33] He also considered drink in terms of work: "Paper Makers thought they Could not work without whiskey. They had to have their hands & arms in the watter & without whiskey they Said they woold take Cold. You will not be Surprised that i thought so too."[34] He was even an industrious trickster and pulled his pranks with the things of his trade. Despite the title of boss, Stedman convinced James Martin, a "most Reliable man," "to walk down on the ice that had accumulated over [a mill] wheel." Martin "Slipped & Slid into the tale Race as i Knew he wood & Before he got to the Mill his Clothe ware Stiff Frozen & Such a laugh i had."[35] But in the end, Stedman believed that he had not played enough.

In his two portraits of his maturation as a man and as a manufacturer, Stedman dwelled on the material objects of his trade and the routine difficulties of toiling with them. He bemoaned the "Six years i Slept on a pile of Rags with My Clothes on, & often with the ice Frozen to My pantaloons up to My neese An inch thick."[36] He rested content when some tramping landed him in a mill with an indoor firepot. He was troubled when a dry stream or a worn millwheel drove him to the road and time "passed Since i Left home with So Sad a heart Beleiving that I never wood se home again."[37] (In contrast, constructing his first paper-making machine was a lark: "The thing was to me So Simple," he boasted.[38]) Whenever he took over a broken-down mill—and these were the only sites available to him—he had to replace spent timbers, vats, and rag beaters. He did the onerous tasks of restoration by hand and with the help of several others. Accordingly, the wages of his labor and the profits of his manufacture blended in the twilight of his days. He had known "Hay day times" and the flush of "acumilatin Evry day."[39] He attributed them to high demand for printing paper, cheap stacks of cast-off linen, and "a large Family of Hands," the "helpers that mad all the property we Ever acumulated" in both hand and mechanized papermaking.[40] Thus was wealth making legitimated and the machine naturalized.

A German established papermaking in British North America in 1690. During the colonial period, American producers counted on imported tools and men who had mastered their craft abroad. Even after the Revolution, these papermakers turned east for the wire to string their molds. Lacking skilled men in the Revolutionary era, one manufacturer, Hugh McLean of Milton, Massachusetts, entrusted his mill to James, an "indented Negro Man." McLean had instructed James in his "Art," and the black man "has been able to transact the Affairs of the Paper Mill for a long time past without other Assistance." He was indispensable to McLean, who "for a long time past has been unable to procure any workmen who understand the Papermaking Business." But James ran away, and "without him [McLean's] Paper Mill must stand entirely Still."[41] (Had McLean's property been in France, James, without roots in a paper-working family, would have found it difficult to land a spot.) During the 1790s, McLean signed several youths to apprenticeship contracts.[42] Nevertheless, at the turn of the century, the mills of eastern Massachusetts continued to rely on English, Irish, and Scots migrants for skilled work.[43] These travelers had carried the craft's familiar formulas as well as its technique across the sea. Ebenezer Stedman's father, also called Ebenezer and also a papermaker, completed his apprenticeship in their company. Born in 1776, or so his son thought, the elder Ebenezer was a fourth-generation American, the offspring of a poor carpenter with a thunderous temper. Ebenezer senior was "verry industerious," his son bragged.[44] Confirming this claim, after he "Commenced work for Himself," Ebenezer *père* purchased a paper mill. "He made money," his son remembered, but the day before he made the last payment on the mill, it went up in flames. "Evry thing Burnt, no insurance," his son despaired.[45] The father resurfaced as the foreman in a shop near Boston, but Ebenezer *fils* recognized, in capital letters, the deep wounds of "The Troublesome Times. Father had lost all of his mill Property."[46]

In the spring of 1815, Ebenezer senior ventured to Kentucky, alone. He had been invited by two Boston merchants, James and Thomas Prentiss, to supervise the construction of the papermaking section of an integrated cotton, woolen, and paper mill in Lexington. Powered by an unreliable steam engine that generated 40 horsepower, the five-story factory was built on an equally unreliable foundation of, ironically, paper money. At that point, banks were free to issue their own notes, which served as negotiable currency, but they often lacked the specie reserves to back these instruments. States eager to sponsor internal improvement simply incorporated fresh banks and their cheap money; in just one legislative session, Kentucky's governors licensed forty-six of these institutions. When the charter of the Prentisses' Lexington bank expired in 1818, they failed to meet their obligations and their little empire collapsed, including the mill.[47] Ever the craftsman, the younger Stedman noted that it had furnished good paper.

For Ebenezer senior, this second misfortune became a defining disaster. His son believed that he had journeyed to Kentucky "to Retreive his Loss By Fire."[48] He lived expansively during the Lexington mill's salad days, spending liberally since "Money was plenty."[49] Once the family arrived, they enjoyed a well-furnished house with "plenty of negro Survants."[50] Young Ebenezer claimed that his father "was looked upon as one that was to build up the Manafacturing interest of Lexington."[51] It all disappeared, as did the money that the elder Stedman had placed in other investments, along with the man who looked after it. Worse yet, he had advertised "plenty of work At Good wages" to his papermaking comrades in Massachusetts, and a small colony had ventured to Lexington. Now they looked to him "in Some measure for inployment" and he felt responsible for them. He sought "relief in Strong Drink."[52] Henceforth, his son lamented, "with all his industry" his failures multiplied "Because Drink Beclouded his Judgment in matters of Business."[53]

In 1818, Stedman moved his family to Georgetown, with many of the Massachusetts paper workers and their families in tow. A moneyed man, William Richardson, had arranged terms for him to rent and then buy the pioneer mill there. The elder Stedman lacked capital, but his son remained convinced "that he ought to have made money Very Fast." The site was promising: eight months each year of flowing spring water, new vats, new beaters, and a new waterwheel. Although the end of wartime tariffs flooded America with fine English paper and low-priced French, Spanish, and Italian reams, Kentucky's few papermakers, Stedman claimed, could count on cheap rags and "good Demand."[54] His father missed this chance as well and entered a spiral of decline. With his son Leander, he turned out wrapping paper at a small mill built in the shell of a gunpowder works, but abandoned this enterprise about three years later. He then worked and wandered in Ohio, sweated with slaves and his namesake in a wretched mill, made sandpaper peddled by young Ebenezer and even labored in his shops, and perished, his son raged, in 1844 or 1845, without the funds for his own coffin.[55] He died an impoverished journeyman.

Historians such as James Henretta and Herbert Gutman have emphasized the traditional, nonacquisitive, nonaccumulative habits and notions stowed in the baggage of the European artisans and small farmers who made their way to colonial and industrializing America. Whether Henretta's preindustrial "*mentalité*" or Gutman's premodern "subculture," these conceptions do not fit neatly into papermaking's European luggage.[56] It was packed by manufacturers attuned to local and distant markets and eager to maximize the advantage of seasonal water and new devices, and journeymen eager to squeeze the last sou or pence of reward and sure of their means to do so. A sharp youth like Ebenezer senior might trade his molds for an account book in this craft, casting off the mutuality among men for the mutuality among masters. (His British and French counterparts would have taken this journey less frequently and found it more hazardous.) After all, the elder Stedman, like his son, dreamed of becoming "some Boddy" through wealth making, and he acted the part before his riches were secure. His ambitions remind us of Joyce Appleby's image of the acquisitive inheritors of the American Revolution.[57] Even after the fiery loss of his New England shops, his son still "heard him Say that he intended to own a mill Some Day."[58] Yet whereas Appleby's new men struggled against the "traditional economic practices" of their fathers, Stedman's father struggled with new economic practices and the enduring, transnational burdens of his trade.[59] He was brought down by an old nemesis of his craft, fire, as well as a fresh one, paper currency (except, that is, for those who produced its substance). Like every papermaker, New World or Old, he knew his trade held the promise of profit—and in paper-starved Kentucky, windfall gain—but he withdrew his trust in it. Thus it was fatigue with the twists of papermaking and his resultant flight to drink, according to his son, that led the industrious Stedman to squander his last big opportunity. The younger Stedman would drive harder than his irresolute father; he would persevere through the trade's tests and purchase the mill his father had desired. Here was yet another stream of the work consciousness that fed his capitalist practice.[60]

Stedman presented his own tale as a series of polaric pairings: bawdy and disciplined, fortunate and luckless, dependent and independent. It was a self-conscious telling, for he knew he had bridged working for another and for himself, and he constantly referred to his ambivalence about riches. Always, however, there was the work itself, with its inescapable rhythms: "I worked day & night as the watter For the mill did not last more than Six month."[61] With his storyteller's instincts, he began his letter to Sophronia about the hardships and tools of hand papermaking with an account of the great Kentucky squirrel hunt of 1820. An "Emigration" of the rodents from the other side of the Ohio River was destroying the cornfields. So many men pursued the creatures "Thare Could not Be a pound of Shot purchased in Geotown." Out in the countryside, "We Boys had Fun after them." But quickly he got down to work: "From January till May i woold have to Get up in the morning at Two oclock."[62] He made a fire to warm the vat, awakened the skilled veterans, and later blew "a horn to notify all the men in the mill to Come to help press the post."[63] As a layboy, his "day's work" consisted of freeing 2,520 sheets of paper from the felts and washing the cloths. He then dragged into town for a jug of whiskey "for the men to drink next Day."[64] For all this, he received regular wages and "overwork Money," an uncommon incentive in most trades, but a cherished reward in French and English paper mills.[65] In fact, from firestarter to potboy, Stedman had just depicted the layman's day across the sea.

After his father's failed tenure in the Georgetown mill, the vats were taken over by a Pennsylvanian named Denormandie. Ebenezer, his father, and his brother Leander sweated in these shops, which were rich in resentment: "I was working For a man that Loved money more than anny thing Else. He had no scruples How he aquired it. I dont know that he woold Steal, But he woold Cheat Evry Boddy that worked for him, or Delt with him."[66] In the summer of 1822, the mill "Stopt for Watter."[67] Ebenezer senior and his son left for Ohio, where the lad was pleased to find "a nice Room as a Fire Room to make my Fire & have all my wood Dry."[68] He also took comfort in the master's table, another time-honored trade practice on both shores of the Atlantic.[69] He was instructed to eat as much as he wanted so long as he finished the pile on the plate. And he got an education in his craft brothers' penchant to turn every misstep and chance opportunity into humor and more than twice-told tales. When Ebenezer junior discovered a feverish man passed out near his fire, he raced to tell the veteran hands that "a crazy man" had fallen near the flame and was burning up. The old hands "Joked me about it, for a long time." He learned that hard liquor and hard work were not incompatible; by the time William Webb, "an Englishman, A fine workman," had finished his workday, "he was in A fine Humor."[70] Soon Stedman crowed that he "Could Drink all Day & not get Drunk."[71] He also witnessed the sexual pranks of his fellow craftsmen, "the war Dance of Delight."[72] He got his own tricks in, but recoiled from the fiercest interplay of sex, drink, and horseplay among the workers. When one journeyman slept with a servant, his wife pressed his shopmates to tar and feather her. Only the firm stance of an armed paper worker, who kept his brothers from entering the girl's boardinghouse, saved her. So Stedman's initiation in the trade had prepared him for a hard-working, rough-and-tumble journey. And these were the passages he chose to mark, far more than the moment he turned to the machine.

A more worldly Ebenezer rejoined Leander and their father in the Georgetown mill. Denormandie, who made money "verry Fast," had died, and one of his half-brothers now rented the site. It made little difference to young Stedman: "The Same Rotiene of work this winter as in year[s] past, in work Day & night was my portion."[73] But a vacant property had attracted his father and Leander. It had been a powder mill and still possessed "a good tite watter wheel," enough "to Run the little Rag Engine" they mounted.[74] (Refitting silent mills, paper or otherwise, was a commonplace in the trade.) Young Ebenezer was temporarily freed from the shackles of the Georgetown mill and gratefully named his bed of fine flax gleanings "tow harvest."[75] His father's new enterprise lasted only two or three papermaking seasons, and after interludes in other employments, Ebenezer, with his brothers John and Leander, spent yet another season in the aging Georgetown mill. "Nothing of importance happend worth Relation Through this winter," he muttered. The work did not grind to a halt until the familiar warmth of May.[76]

Stedman toiled around the Georgetown mill's vats in the winter and spring of 1827, but that summer he moved to the nearby mill at Great Crossing. Once a gristmill, this ramshackle establishment had been fitted with the machinery from the Prentisses' great debacle. With Frank McDonald, a white journeyman, he had found work in shops where the "Ballance" of the hands were black.[77] As a boy, Stedman had seen a black man in Dorchester: "They told me it was the Devil & he frightened me most into fits," he recalled.[78] Ever mindful of racial boundaries, he recounted a "live Mass of Men Women & Children, Negros & Horses" at a parade.[79] Capable of recognizing black effort—later he would describe his wife's servant as a "most industrious old Negro"—Stedman still shared the prevalent sense of African-American limitations: a free black body-snatcher for Lexington's medical college was "As Smart as Negros Get to Be."[80] The black hands around him at Great Crossing had been trained by a white foreman, whom the owners dismissed to cut costs. This partnership "let the Mill Run it Self," save for a daily inspection by one of its members.[81] Even so, papermaking's wide-ranging custom persisted there. Work began before light, as it did in Surrey and the Auvergne, with the horn blowing the hands in for breakfast at eight. Both the craft's order and racial distinctions were maintained at table. Stedman, now a vatman, and McDonald, "a Regular papermaker" who had served a formal apprenticeship, would naturally have prime seats.[82] That said, "As Frank and Myself ware the ondley whites we Both took the head of the table." Once there, they "had the Honor to preside over" the rough-hewn oak planks. While each slave ate from a wooden bowl, Stedman and McDonald merited pewter plates. Granted, the metal dishes were "Full of dents"; but, as Stedman observed, "it was Necessary to Make A distinction Between White men & negros." Nevertheless, he was attentive to the skill of the black workers who shared his art, noting that the cook's husband "was Considered the Best negro paper Maker In the Mill."[83]

Regardless of color, the men consumed, when they dared, less-than-appetizing food. Breakfast consisted of fried hog's jowl and cornbread baked without sifting the flour. Apparently the loaf was a foot thick. Worse yet, "In our tin Cups thare was Buttermilk thick and old Enough for a Duck To walk on."[84] Stedman never tried it. When he left the mill and the cold comforts of its boardinghouse, he weighed barely one hundred pounds. His first "'Journey work,'" his "First attempt to work for E. H. Stedman," had ended badly.[85] He expected cash wages, but had to settle for store goods. His first moments of independence, then, amounted to another lesson in his dependence. He returned to the mill with his father in 1828 "with the understanding that we woold live Better which we did."[86] He had yet to stand on his own feet.

The original Georgetown mill had fallen on hard times. Its rooms were dark and musty, the "Rag Engines & vats dry & Rotten," and the molds Stedman found there in 1829 were "Empty Frames like my Empty pocket." But Stedman "Saw how Easily i Could Make a Start If i had Anny thing to Start with."[87] He needed capital. William Richardson, who still owned the mill he had once retooled for Stedman's father, offered the son the property rent free until some coins fattened his purse.[88] Stedman then journeyed to Lexington for brass wire so his father could string the molds. "You Can Congecture The Castles that imagination

Built while Comming Home with the wire," he told Sophronia. His mother gave him old blankets to cut into felts on the promise of two new ones "when i made paper." He scavenged rats' nests for rags and collected bits of hemp tow and ended up with perhaps fifty pounds of stock. He fed this mass into the beater, and "the First Roar of the Engine on my Rat Stock was the Sweetest Mill Musick that i have Ever heard Before or Since." He realized that he had turned a corner, though his "first attempt To Commence Business for Myself" had taken shape "on the Bottom Floor of Poverty." Yet, he delighted, "thare appeared a Bright lining to the Dark undertaking."[89] With the traditional skills and machines of his trade, he was stepping beyond the rotten timbers of dependence.

He finished four reams of wrapping paper, which got him credit for linen rags and hemp scraps at the grocery, where blacks exchanged their tow gleanings for goods. Soon this barter underwrote a thousand pounds of hemp tow, and Stedman hired his father, brother John, and some of the Massachusetts migrants. They began to churn out five reams of wrapping paper "Evry day." Stedman purchased the emblem of industrial time discipline, an old English watch, "as i now Run the Mill day & night."[90] He had long known that time was money and thought his hours as a potter, which included plenty of hunting, fishing, and church meetings, had been wasted, since "I had no way to Make anny Money."[91] But he recognized the value of his endless nights in the steam-powered Lexington carding mill. Here, he understood, was the fulcrum of independence—and he could afford nothing else, as his father's missed transitions meant that "i had to row my own. Row & help all the Rest of the family."[92]

In 1830, he worked the mill winter and spring, until the stream dried, and "made Some Money."[93] He then took to the road and peddled the hat boxes his mother fashioned from his wrapping papers. Alone in this market, he got "a big price."[94] But he lacked the wanderlust of his trade and had parted from the journeymen who labored only to fatten their pockets enough to "Carry them From one Mill to another."[95] He was thrilled to return home and resolved to quit drumming. Besides, it was now late in the fall, and Stedman's mill had water again. That winter, he explained, "i made all the paper i Could."[96] (As Stedman testified, "the gingle of Silver has its Charms & often Creates a desire for more."[97]) He was thriving. During the papermaking autumn of 1832, he formed a partnership with his brother Sam, who had "Business qualites," was "Sharp in Trading," but "never do much in the mill."[98] Around New Year's Day 1833, the brothers took the substantial step of manufacturing printing paper, a cut above the ambit of wrapping.[99] At the same time, Ebenezer "concluded i was old Enough to have a wife," yet worried that he remained too poor for one. Nevertheless, he and Mary Steffee set a date. Meanwhile, Stedman ran the mill eighteen hours each day and "John Steffee Run the machine."[100] This was Stedman's first mention of the presence of the papermaking machine at the Georgetown mill, an event he failed to date precisely.[101] So much for mechanization taking command.

Perhaps Stedman did not think Sophronia was interested in the device, yet he had detailed the tools and techniques of hand papermaking to her. Perhaps he wanted to cloak his furtive duplication of a fellow producer's machine; still, he revealed any number of character flaws and even celebrated a near miss in securing ill-gotten gains as a peddler. Whatever the case, he took note of the instrument that retooled his trade in a letter about his wedding charivari and his abandonment of the wedding bed for the mill. It was industrious work that had permitted him to marry, and it was the same sweat equity that produced his machine. His night work in the carding mill had paid off, since he had the skills to build this device—almost effortlessly, he reported in his first autobiography. The machine was simply the extension of his hand work. No doubt he imagined that the device would insulate him from the pain and humiliation of his first journey work and from the scant rewards of the boarding house he knew as "Cold Comfort." But the machine, like hand papermaking and its instruments, was still subject to seasonal stream flows, still needed the attention of skilled men, and was still vulnerable to fire. And the device had its own demands: "What a Beginnin of life, day, and night, and night, and day," Stedman wrote about his first weeks as the working proprietor of a mechanized paper mill.[102]

Much had changed at the old mill: "The Machine made at Geotown worked well."[103] Whereas it took twelve hands to fashion six reams per day, the machine produced rolls measured in feet. According to Stedman, "Could Some of the old papermakers, & Printers, Se the papermill & printing Press of to Day, they woold think they Had waked up in the wrong world."[104] Summer production still slowed or stopped, but the Stedmans had new means to profit from the abundant water of autumn and winter. Much had also remained the same in Georgetown. "In the morning By day light," Ebenezer would steadfastly "Get up & go to the Mill, Eat My dinner & go to the Mill, Eat my Supper & go to the Mill."[105] He continued to trust in good drink as "a Stimulus of Strength."[106] Along with his manufacture of printing paper, he persisted in "Grinding hemp tow" and furnishing wrapping paper, especially when the water was muddy. And he remained lodged in a barter network that involved black hemp gleaners, his own hands, and grocers, though he also sold paper for cash. His dreams, too, were not yet vast: "In my devotion to the Business of the Mill i Cant Say that it was ambition to be Rich." His "polar star" was "a Home of My own" rather than the one he shared with the temperamental Sam.[107]

Still, the more he made, the more he worried. He described the pursuit of wealth as the chase after "Jack with his lantern." He brooded about the sacrifice of the "Small pleasures" of hours with his family and friends, particularly when the mill received "all my might & Strength, night & day, in the ice & snow of winter & the Scorching heat of Summer." But his burdensome labors were necessary, even just, since he "was Poor, had no home That i Could Call My own, Had No Rich Kin to Give me Annything." So, "I made the paper. Sam he sold it."[108] It was a division of labor Ebenezer would come to regret.

Much has been written about the relative contributions of merchant capitalists and artisan entrepreneurs in the transformation of American handicraft production. At issue is primacy in the creation and diffusion of technological change.[109] William Richardson had invested in Stedman's skill, work ethic, and enterprise, and now his old mill was mechanized. Stedman's next effort involved a different sort of symbiosis: he sent samples of his printing paper to Albert Hodges, the state printer and a newspaper publisher.[110] This bait lured Hodges to Georgetown, where he proposed that the Stedman brothers purchase a mill in Franklin County owned by Amos Kendall, who had left Kentucky and later became postmaster general of the United States. After a series of luckless tenants had run it into the ground, the site was ready for the Stedmans: it had been idle for some time and was rich in rotten timber. The mill house had been built in 1823, ten years earlier. It had served as a hog pen for three years, and the corncobs and manure were two feet deep. It contained a vat, a beater, a press, and a stock chest, "all Rotten."[111] The waterwheel was no better, and the second-story drying room, which had sheltered corn, oats, and hay, possessed few poles between which to string cords. The Stedmans would have to construct a new dam and a new head race and gates; they would have to install a new waterwheel and a papermaking machine. The tail race had to be mucked free of three feet of mud and then blasted two feet lower, since it had never been finished. It was "the most uninviting desolation that Two young men Ever had the nerve to undertake to Repair & Build up."[112]

So Hodges cut the brothers a deal: if they bought the property, he would cover the mortgage payments in exchange for the paper they furnished. The moneyed printer and the cash-poor artisan entrepreneurs had discovered their mutual interests. Sam, however, proved reluctant, despite Ebenezer's reminder that Georgetown's parched summer springs meant that they would never improve their circumstances. But Ebenezer had learned that his trade offered no orderly life cycle of stable comfort, much less accumulation. He had to strike at the main chance. To clinch the bargain, Ebenezer promised Sam risk-free terms: should he remain discontented after their new millwheel began to turn, Ebenezer would find someone to buy him out. With that, the brothers became proprietors.[113]

At first, Ebenezer claimed, the Stedmans' new neighbors were wary of them. Both Kendall and his tenants had defaulted on their pledges, and long-gone journeymen had skillfully vanished without paying their debts. Nevertheless, the brothers needed working capital. Once again, Ebenezer explained, their industriousness prevailed, and "in So short a time we Could Get Credit at anny place in town."[114] When Sam repaid a fifty-dollar loan promptly, his creditor even waived the interest. So the Stedmans cut and hauled logs, and as the summer progressed, they had their dam erected. They returned to Georgetown for the papermaking season, and the old mill performed its last duty for them. With the price of paper robust and castoff linen cheap, they settled the summer construction debt and moved to a hamlet on the main fork of Elkhorn Creek in May 1834. Here their artisanal experience nourished their pursuit of propertied, independent production—the aspiration deeded to them by Ebenezer senior, who eventually followed them to Franklin County. They hired a millwright to turn out their waterwheel and beater, but the Stedmans created a small blacksmith's shop in the mill yard where Ebenezer, primarily, built a papermaking machine. "By the time the Creek Rose in the fall," they were manufacturing paper. With favorable circumstances in both the paper and rag markets, they "Made money Fast." Ebenezer "Confined My Self to the Mill Night & day," while Sam "attended to all the out door Business." And then Ebenezer abruptly abandoned the tone of his letter, trading the excitement of construction and flush times for the dark hues of his future. How different his fate might have been, he howled, if he had "worked [his] brain insted of [his] Hands!"[115] He respected both the wages and the profits produced by industrious labor, but regretted his failure to master the entrepreneur's skill at securing them. Always in a rush, always eager to "Keep things Mooving," Sam pulled out of their partnership in 1852, before Ebenezer was ruined.[116] But

Ebenezer knew the work of papermaking; his "ambition was to Keep The papermill Mooving day & Night," and hence he stayed with it.[117] His work consciousness and, doubtless, his fear of replicating his father's failure kept him in harness when it was time to bolt. Consequently, "the waves of time have washed all on a Barren desert Shore."[118]

Ebenezer's remaining letters never strayed far from the theme of loss, but they continued to center on gain. The Stedmans had made much out of little and continued to do so. After Ebenezer added some whiskey and rainwater to a half-dozen molasses barrels, he proudly noted that a visitor to the mill yard exclaimed, "'Ill Be damd if they dont Make money out of old Rags and then out of their old Molasses Barrels they have Emptyed, and they are full of First Rate Vinigar.'"[119] The brothers had imposed their wills ("whare thare is a will there is a way") in a recalcitrant setting, one that was "wild, Both in formation & sosiety."[120] To tame their shops, they resorted to the custom of Georgetown, where Ebenezer once bartered with the grocer for goods to compensate his hands. Now the Stedmans possessed a well-stocked storeroom; as a result, "we Did not pay out But little money."[121] They counted on the workers' fondness for drink to extract greater effort for less cash: a dram of whiskey three times a day "Gave us more work to the hand than half more wages."[122] Put simply, they exploited the production culture of their trade and its material trappings, but distant from the mills of Philadelphia and Kent, their tactics did not ignite its oppositional culture.[123] Instead, they hired or trained a loyal, stable core of workers, including Frank McDonald, who stayed with them for five years. And Ebenezer found it easy to celebrate James Martin, who remained for fifteen or sixteen years: "In Cold or Heat Rain or Sunshine, i have Reason to Believe Jim worked for our intrest."[124] With such mutuality, the Stedmans and their hands, Ebenezer believed, might be spared the bleak circumstances of Massachusetts shoemaking. In that craft, there had once been "No overgrown wealthy Capitalist to Screw Down the wages of honest workmen & Cause them to Slight their work." Shoemakers had turned out "Good Substantial work" before they made only uppers and soles, working with patterns—"working In paper," Ebenezer mockingly labeled it.[125] The machine had transformed Ebenezer's craft, but he was sure the Stedmans still made good paper. The evidence was clear: they "Could not Supply the demand" of Kentucky's printers, and they were "making money verry Fast."[126]

To convince their skeptical Franklin County neighbors of their worth, to "Surmount all predices against us," the Stedmans turned to conspicuous display.[127] They painted the mill white and the shutters on the drying loft red and white, making things "Look New & Bright."[128] Sam purchased a buggy, and Ebenezer opened his purse for "a Silver Mounted Barouch," a four-wheeled carriage with a folding top over the back seat.[129] Both brothers bought slaves.[130] In no time, Ebenezer recalled, "Our Suxcess was the Common talk."[131] They settled with Hodges in three years, and Ebenezer supposed that printing paper alone had erased the debt.[132] Ever the trickster, Ebenezer was pleased that his younger brother Anderson, who labored in the mill's sizing room, had persuaded a green hand to undress in order to climb into the sizing tub. He was delighted to report in the same letter that "the two little Boys," Ebenezer and Sam Stedman, "ware men."[133]

Sam and Ebenezer had made it. They were the independent proprietors of a successful paper mill. They enjoyed esteem, easy credit, and eye-catching display. They had arrived and even "promised ourselfs a Rest from the Strain" that purchased the mill and their homes.[134] They had left boyish dependence, the lot of the journeyman—and their father. Even the bank considered them "Industerious Money Making Men."[135] But Ebenezer was looking back, and he knew the story ended in dependence. Perhaps he had always shared the anxieties that haunted the Jacksonian nouveau riche.[136] Still, if he had retained his fortune, would he regret that he had been "Such a Slave to Business"? Would he grieve that his pursuit of independence meant that he "Never Nursed" Sophronia? Or that he had taken his wife from her family in the move to Franklin County, provoking her tears, just as his father had made his mother weep when he brought her to Kentucky? All that is certain is Ebenezer's horror that he had spent "all of my life striving for that that has Evaporated in thin air."[137]

In 1837, the Stedmans installed a steam engine at the mill. Ebenezer marveled at the machine's appetite. Over the years, he claimed, it consumed fifty thousand cords of wood—at a rate of five cords or more each day—and hundreds of thousands of bushels of coal. But the Stedmans failed to take advantage of another breakthrough from the early Industrial Revolution, chlorine bleaching. Ebenezer was uncertain of his mastery of the substance, and at ten cents per pound, it was costly. So he turned to his trade's tacit knowledge: he exposed newly minted sheets to nighttime cold, just as European papermakers had done for centuries. He knew that when the sheets froze, they whitened.[138] Meanwhile, Stedmantown, the mill and its surrounding village, flourished. Ebenezer estimated that he produced $75,000 to $100,000 worth of paper during his first twenty-four years there.[139] (He did so with a dozen mill hands, the same number that once furnished six reams each day.) In 1845, the brothers sought "200 tons or any amount of cotton and linen rags to be bought at the highest market prices."[140] Evidently their days of bartering for hemp tow were over. But the persistent perils of papermaking now engulfed Ebenezer. In 1847, the Elkhorn rose, demolishing a series of dams and devastating the Stedmans' mill and machine shop. The brothers lacked insurance, but, as Ebenezer boasted, their credit was good.[141] They rebuilt, and in 1850 the value of the mill was $15,000, with another $12,000 worth of castoff linen in the yard.[142] Two years later, Sam made his timely exit from the partnership. A flood carried away Ebenezer's dam in 1854, and in 1856 the gristmill on his property burned, "with all The wheat, & flower with no insurance." He put up another in its place, "Better than the First." A year later, the weir washed downstream again. Ebenezer erected a new $4,000 dam, only to watch fire consume "4 large Stock of Rags" in 1859.[143] Still, he did not stir; he had engineered a small world of production that garnered him wealth, respect, and even a sense of manliness. Mulling over the distant trials of his father and memories of the Kentucky mills that had burned or "rotted down," he counted on misfortune.[144] He would work through the hard times, as he always had. Consequently, he reconstructed his mill in 1860 and installed "the Best of Machienry That Could Be Made."[145] He invested $40,000 in this retooling, which proved overwhelming when the war deprived him of essential labor and the Confederacy defaulted on an order for currency paper.[146] His creditors sold Stedman's machinery. "The place whare i have Spent the most of my life Belongs to others," he acknowledged.[147] He left Kentucky a bankrupt and died in 1885.

———

Stedman's proud lamentations about his long hours of labor and restless pursuit of fast money may seem to reflect a classic American paradox. But master papermakers throughout the Atlantic world, driven by pressures from printers and stationers and dependent on the rush of mountain streams and thin stocks of threadbare linen, worked their hands hard—and boiled over when these men played hard or negotiated by threatening to move on.[148] Most manufacturers also knew how to pad a ream of fine wares with inferior paper, whiten yellow sheets with quicklime, or imitate an elevated producer's watermark. So masters and men alike were well versed in the art of profiting from their demanding trade. Its market revolution had occurred centuries earlier. What was new for a lesser producer like Stedman was that he could accumulate, share in substantial, even refined consumption, and turn out vastly more reams of paper. He was not alone: thirty-seven of the first forty-two English papermaking machines were installed in mills that housed only two or three vats, that is, small enterprises.[149] The French inventor of this device imagined a settled trade, in which his machine had stilled the hands of overmighty journeymen. Many of Stedman's American and transatlantic counterparts adopted it to secure this outcome as well as swollen output.[150] But despite his silver barouche, he still had pulp under his fingernails, both from his earlier days around the vats and as the proprietor of a mechanized mill. Meanwhile, the Stedmans' relative isolation, their cagey paternalism, and the loyalty it engendered (and perhaps the prospect of hiring or training slave paper workers) had evidently enabled them to run their shops as they saw fit. Ebenezer, then, did not build his machines to erect social boundaries; he did so to hurdle them, to stick a feather in his cap. Small wonder he had so much trouble abandoning his last mill.

In a book devoted to the comparative consideration of national industrial trajectories, what does the tale of a single man, however detailed, add? That large-scale processes are also, inevitably, personal. Stedman's story brings nuance and shadow to the seemingly clear-cut, such as sharp periodizations of industrial change based on production aggregates and censuses of machines. Why, for example, did he follow in his father's footsteps and fail to insure his shops? Where did the elder Stedman's dreams and drives trail off and his son's begin? How did his time in a steam-powered carding mill and sleepless nights of hand papermaking figure in his rapid, even gleeful, construction of his own labor-saving machine? In our age, when expanding global trade has highlighted distinctions among producers, Stedman's story is a cautionary tale warning against the revival of neat, Rostovian stages in the history of industrialization. After all, he worked with advanced and antique tools, clutched and set aside the production custom of his craft, found inspiration in old notions of independence, and despite his denials, developed new habits of accumulation. He wrote movingly of both the moment "when that first sheet was dipt up By hand" in "the Great West" and the "Inventive jenius" whose machine rendered the dipping vat obsolete and spared the spines of countless layboys.[151]

Finally, Stedman's tale offers a bridge across the historiographical divide of scale, the chasm that separates the micro- and macrohistorians. Stedman's corner of Kentucky had always been part of his transatlantic trade and its development. Perhaps this is why most of the final page of his early autobiography centered on "a Rellick of 1792." It was a "fine" iron screw for the Georgetown mill's press. It weighed, Stedman guessed, about three hundred pounds. It was imported from England, but Stedman did not know "By what means it was transported To the wiles of Ky." He thought it might adorn the Louisville Public Library.[152] In fact, long after Stedmantown had been deserted, it rested in the muddy bank of the Elkhorn, a reminder of the deep, transatlantic sediments of Ebenezer's trade.[153] It vanished without a trace, unlike the journey it represented.

Notes

1. Ebenezer Hiram Stedman, *Bluegrass Craftsman: Being the Reminiscences of Ebenezer Hiram Stedman, Papermaker: 1808–1885*, ed. Frances Dugan and Jacqueline Bull (Lexington: University of Kentucky Press, 1959), 136. All capitalization, punctuation, and spelling are quoted as they appear in Stedman's original accounts.

2. Ibid., 130.

3. Ibid., 136.

4. On Jacksonian ambitiousness, see Marvin Meyers, *The Jacksonian Persuasion: Politics and Belief* (Stanford, CA: Stanford University Press, 1957), 76–107. See also Charles Sellers, *The Market Revolution: Jacksonian America, 1815–1846* (New York: Oxford University Press, 1991). For a recent, sweeping account of ambition, acquisitiveness, and failure in American history, see Scott Sandage, *Born Losers: A History of Failure in America* (Cambridge, MA: Harvard University Press, 2005).

5. Stedman, *Bluegrass Craftsman*, 136.

6. Ibid., 1. Stedman's first account appears as the appendix in *Bluegrass Craftsman*, 211–222.

7. The layboy separated the infant sheets from the felts.

8. Stedman, *Bluegrass Craftsman*, 221.

9. This lengthy, epistolary autobiography appears in ibid., 1–210.

10. For a fine account of the place and meaning of work in early American history, see Stephen Innes, "Introduction: Fulfilling John Smith's Vision: Work and Labor in Early America," in *Work and Labor in Early America*, ed. Stephen Innes (Chapel Hill: University of North Carolina Press, 1988), 3–47.

11. Perhaps the most visible recent effort is David Landes, *The Wealth and Poverty of Nations: Why Some Are So Rich and Some So Poor* (New York: Norton, 1998).

12. Pat Hudson, ed., *Regions and Industries: A Perspective on the Industrial Revolution in Britain* (Cambridge: Cambridge University Press, 1989); Sidney Pollard, *Peaceful Conquest: The Industrialization of Europe, 1760–1970* (Oxford: Oxford University Press,

1981); Kenneth Pomeranz, *The Great Divergence: China, Europe, and the Making of the Modern World Economy* (Princeton, NJ: Princeton University Press, 2000).

13. Leonard N. Rosenband, "The Competitive Cosmopolitanism of an Old Regime Craft," *French Historical Studies* 23 (Summer 2000), 455–476, and "Comparing Combination Acts: French and English Papermaking in the Age of Revolution," *Social History* 29 (May 2004), 165–185.

14. Although transatlantic economic history has recently received much attention, the focus has been on commerce rather than manufacture. Some valuable starting points are Bernard Bailyn, *Atlantic History: Concept and Contours* (Cambridge, MA: Harvard University Press, 2005); David Hancock, *Citizens of the World: London Merchants and the Integration of the British Atlantic Community, 1735–1785* (Cambridge: Cambridge University Press, 1995); David J. Jeremy, *Transatlantic Industrial Revolution: The Diffusion of Textile Technologies Between Britain and America, 1790–1830s* (Cambridge, MA: MIT Press, 1981).

15. Stedman, *Bluegrass Craftsman*, 44.

16. On American papermaking, see Judith McGaw, *Most Wonderful Machine: Mechanization and Social Change in Berkshire Paper Making, 1801–1885* (Princeton, NJ: Princeton University Press, 1987). On British papermaking, see D. C. Coleman, *The British Paper Industry, 1495–1860: A Study in Industrial Growth* (Oxford: Clarendon Press, 1958), and Gary Magee, *Productivity and Performance in the Paper Industry: Labour, Capital, and Technology in Britain and America, 1860–1914* (Cambridge: Cambridge University Press, 1997). On French papermaking, see Louis André, *Machines à papier: Innovation et transformations de l'industrie papetière en France, 1798–1860* (Paris: Editions de l'école des hautes études en sciences sociales, 1996), and Leonard N. Rosenband, *Papermaking in Eighteenth-Century France: Management, Labor, and Revolution at the Montgolfier Mill, 1761–1805* (Baltimore, MD: Johns Hopkins University Press, 2000).

17. The English Act is 36 George III, c. 111. On Angoumois, see Nicolas Desmarest, "Papier (Art de fabriquer le)," *Encyclopédie méthodique: Arts et métiers mécaniques 5* (Paris, 1788), 510. McGaw reports that Berkshire, Massachusetts, workers also "completed twenty posts of paper and then quit for the day." (*Most Wonderful*, 54). But John Bidwell, in his editor's introduction, maintains that "American journeymen definitely made less paper [per day] than their counterparts in England." See Bidwell, ed., *Early American Papermaking: Two Treatises on Manufacturing Techniques Reprinted from James Cutbush's American Artist's Manual (1814)* (New Castle, DE: Oak Knoll Books, 1990), 35.

18. Joseph-Jérôme Lefrançois de Lalande, *The Art of Papermaking*, trans. Richard Atkinson (Kilmurry, Ireland: Ashling Press, 1976), 41.

19. Stedman, *Bluegrass Craftsman*, 41.

20. Ibid., 39.

21. Ibid., 35.

22. Lalande, *Art*, 56.

23. Timothy H. Breen, *The Marketplace of Revolution: How Consumer Politics Shaped American Independence* (New York: Oxford University Press, 2004), 56–57.

24. Edmund Morgan, *The Genuine Article: A Historian Looks at Early America* (New York: Norton, 2004), 169–170.

25. Coleman, *British Paper Industry*, 268–287.

26. Bidwell, ed., *Early American Papermaking*, 29.

27. Coleman, *British Paper Industry*, 297–298; Bidwell, ed., *Early American Papermaking*, 28–36; Rosenband, *Papermaking*, 102–103.

28. Quoted in D. C. Coleman, "Combinations of Capital and of Labour in the English Paper Industry, 1789–1825," *Economica*, n.s. 21 (1954), 44.

29. McGaw, *Most Wonderful*, 53. The journeymen could sell or trade the paper to sustain their tramping.

30. Ibid., 98.

31. Stedman, *Bluegrass Craftsman*, 35.

32. Ibid., 143.

33. Ibid., 163.

34. Ibid., 40.

35. Ibid., 168.

36. Ibid., 40.

37. Ibid., 67.

38. Ibid., 220.

39. Ibid., 178

40. Ibid., 177, 196.

41. American Antiquarian Society (henceforth AAS), Worcester, MA. Tileston and Hollingsworth papers, Box 1, Folder 1, n.d.

42. AAS, Tileston and Hollingsworth Papers, Box 1, Folders 4 and 5.

43. McGaw, *Most Wonderful*, 40.

44. Stedman, *Bluegrass Craftsman*, 3.

45. Ibid., 4.

46. Ibid., 117.

47. For Stedman's account of this venture and its failure, see *Bluegrass Craftsman*, 8–26, 199. For a recent, informed version, see John Bidwell, "American Papermakers and

the Panic of 1819," in *A Potencie of Life: Books in Society*, ed. Nicolas Barker (London: British Library, 1993), 92, 100–101.

48. Stedman, *Bluegrass Craftsman*, 24.

49. Ibid., 14.

50. Ibid., 199.

51. Ibid., 14.

52. Ibid., 25.

53. Ibid., 26.

54. Ibid., 27.

55. Ibid., 168, 198.

56. James Henretta, *The Origins of American Capitalism: Collected Essays* (Boston: Northeastern University Press, 1991), chap. 3; Herbert Gutman, *Work, Culture, and Society in Industrializing America: Essays in American Working-Class and Social History* (New York: Knopf, 1976), 39–40.

57. Joyce Appleby, *Inheriting the Revolution: The First Generation of Americans* (Cambridge, MA: Belknap Press, 2000). See also Gordon Wood, *The Radicalism of the American Revolution* (New York: Knopf, 1992).

58. Stedman, *Bluegrass Craftsman*, 24.

59. Joyce Appleby, "The Vexed Story of Capitalism Told by American Historians," *Journal of the Early Republic* 21 (Spring 2001), 18. See also "Special Issue on Capitalism in the Early Republic," *Journal of the Early Republic* 16 (Summer 1996).

60. On intergenerational legacies and dynamics among enterprising men in the early Republic, see Barbara Tucker and Kenneth Tucker, Jr., "The Limits of *Homo Economicus*: An Appraisal of Early American Entrepreneurship," *Journal of the Early Republic* 24 (Summer 2004), 208–209, 212.

61. Stedman, *Bluegrass Craftsman*, 114. On labor in early national and Jacksonian Kentucky, see Craig Friend, ed., *The Buzzel About Kentuck: Settling the Promised Land* (Lexington: University of Kentucky Press, 1999), esp. 125–151, 175–193. See also Stephen Aron, *How the West Was Lost: The Transformation of Kentucky from Daniel Boone to Henry Clay* (Baltimore, MD: Johns Hopkins University Press, 1996).

62. Stedman, *Bluegrass Craftsman*, 35

63. Ibid., 38.

64. Ibid., 39.

65. Ibid., 40.

66. Ibid., 35.

67. Ibid., 41.

68. Ibid., 42.

69. See, for example, Rosenband, *Papermaking*, 89–92.

70. Stedman, *Bluegrass Craftsman*, 43.

71. Ibid., 47.

72. Ibid., 48.

73. Ibid., 70.

74. Ibid., 71.

75. Ibid., 72.

76. Ibid., 97.

77. Ibid., 101. On the role of slavery in industrial production, see Charles S. Dew, *Bond of Iron: Master and Slave at Buffalo Forge* (New York: Norton, 1994), and Robert Starobin, *Industrial Slavery in the Old South* (New York: Oxford University Press, 1970). Starobin cited several sources (238, n. 42), including Stedman's memoirs, to support his claim that "Carolina and Kentucky papermakers used bondsmen" (19).

78. Stedman, *Bluegrass Craftsman*, 12.

79. Ibid., 75.

80. Ibid., 89, 145.

81. Ibid., 101.

82. Ibid., 176.

83. Ibid., 102–103.

84. Ibid., 103.

85. Ibid., 102, 104.

86. Ibid., 104.

87. Ibid., 110.

88. Ibid., 109.

89. Ibid., 111–112.

90. Ibid., 112.

91. Ibid., 80.

92. Ibid., 113.

93. Ibid., 114.

94. Ibid., 115.

95. Ibid., 45.

96. Ibid., 130.

97. Ibid., 123.

98. Ibid., 138.

99. Ibid., 139.

100. Ibid.

101. In 1809, John Dickinson, an English paper manufacturer, invented and patented the cylinder version of the papermaking machine. In 1817, the Gilpin brothers of Delaware erected a device patterned after Dickinson's instrument. (It was not until 1827 that the Fourdrinier machine, the French device reconfigured by Bryan Donkin, was installed in a New York mill.) Stedman built a cylinder machine at Georgetown (*Bluegrass Craftsman*, 220).

102. Stedman, *Bluegrass Craftsman*, 200.

103. Ibid., 221.

104. Ibid., 3.

105. Ibid., 146.

106. Ibid., 181.

107. Ibid., 146–147.

108. Ibid.

109. On this issue, see the fine article by Gary Kornblith, "The Craftsman as Industrialist: Jonas Chickering and the Transformation of American Piano Making," *Business History Review* 59 (Autumn 1985), 349–368, esp. 350.

110. In his first, brief autobiography, Stedman placed Hodges's visit to the Georgetown mill in 1834 and noted that the machinery in the brothers' new mill began to furnish paper in 1835 (*Bluegrass Craftsman*, 221). But Stedman surrounded the earlier dates in his lengthy, later autobiography with a rich context (*Bluegrass Craftsman*, 148–169). In addition, in their editors' introduction to *Bluegrass Craftsman* (xii), Dugan and Bull also chose the earlier dating.

111. Stedman, *Bluegrass Craftsman*, 151.

112. Ibid., 152.

113. Ibid., 154–155.

114. Ibid., 157.

115. Ibid., 165–166.

116. Ibid., 177. For the date of Sam's withdrawal from the partnership, see Frances Dugan, "A History of Papermaking in Kentucky," *Papier Geschichte* 10 (February 1960), 7.

117. Stedman, *Bluegrass Craftsman*, 178. In his recent study, *Born Losers*, Sandage emphasized the cultural underpinnings of the fear of failure, the anxiety about ending up with an "identity in the red" (2). For Stedman, this concern was certainly a profound impulse; but so were his vivid memories of his father's collapse and the pain and dependence of handwork around the vats.

118. Stedman, *Bluegrass Craftsman*, 166.

119. Ibid., 203.

120. Ibid., 159.

121. Ibid., 175.

122. Ibid., 201.

123. Walter Licht emphasized the "relative antitraditionalism and unboundedness" of industry in the young American Republic. "Capital, labor, and ideas could flow toward new opportunities unimpeded for the most part by custom, law, group ties, and the need of elites to maintain social ways." See *Industrializing America: The Nineteenth Century* (Baltimore, MD: Johns Hopkins University Press, 1995), 40. But note too how Stedman treated custom as an opportunity to be turned to his advantage.

124. Stedman, *Bluegrass Craftsman*, 168 (for his years of service), 198.

125. Ibid., 23.

126. Ibid., 176.

127. Ibid., 202.

128. Ibid., 175.

129. Ibid., 179.

130. Ibid., 177.

131. Ibid., 202.

132. Ibid., 177.

133. Ibid., 169.

134. Ibid., 180.

135. Ibid., 177.

136. On these anxieties, see Meyers, *Jacksonian Persuasion*, and the valuable essay by Gary Kornblith, "Becoming Joseph T. Buckingham: The Struggle for Artisanal Independence in Early-Nineteenth-Century Boston," in *American Artisans: Crafting Social*

Identity, 1750–1850, ed. Howard Rock, Paul Gilje, and Robert Aster (Baltimore, MD: Johns Hopkins University Press, 1995), 123–134, esp. 133.

137. Stedman, *Bluegrass Craftsman*, 168–169.

138. Ibid., 207–208.

139. Ibid., 221.

140. Quoted in Dugan and Bull, Introduction to *Bluegrass Craftsman*, xii.

141. Dugan and Bull, Introduction to *Bluegrass Craftsman*, xii.

142. Dugan, "History," 6.

143. On the flood, see Dugan and Bull, Introduction to *Bluegrass Craftsman*, xiii. On the fire, see Stedman, *Bluegrass Craftsman*, 221.

144. Stedman, *Bluegrass Craftsman*, 222.

145. Ibid., 221.

146. On his wartime losses, see Stedman, *Bluegrass Craftsman*, 221, and Dugan, "History," 7.

147. Stedman, *Bluegrass Craftsman*, 221.

148. On the cultural impulses that led to the persistence of long hours in the machine and wood pulp era of French papermaking, see Michael Huberman, "Working Hours of the World Unite? New International Evidence of Worktime, 1870–1913," *Journal of Economic History* 64 (December 2004), 990.

149. Coleman, *British Paper Industry*, 197, table 15.

150. Bidwell, ed., *Early American Papermaking*, 36.

151. Ibid., 213.

152. Ibid., 222.

153. Ibid., n.24.

11

Reconceptualizing Russia's Industrial Revolution

Peter Gatrell

The long history of Russian industrialization was indelibly marked by two dramatic periods when investment, employment, and output increased at a rapid rate. The first phase belongs to the final years of the old regime and lasted from around 1885 to 1913. The second upsurge belongs to the period of early Stalinism, between 1928 and 1941. Comparisons between these two phases of industrialization have been rare in the historiography. A notable exception was Alexander Gerschenkron, who insisted on the need to relate economic and political factors to understand Russian economic backwardness and the strategies adopted for overcoming it, in the czarist and Soviet eras alike. In his view, the Russian state simultaneously constrained and fostered economic activity. The constraints included the maintenance of outmoded or inappropriate institutions, particularly in agriculture, and the dearth of skilled labor. The support mechanisms included the use of "substitutes" for prerequisites that were missing or in short supply; hence the important role played by czarist government fiscal and monetary policy and by state enterprise. Notwithstanding the challenges that have been made over the past half-century to aspects of his interpretation of European industrialization, Gerschenkron was right to draw attention to the state as part of the problem as well as the solution.[1]

To be sure, significant industrial and economic progress occurred outside these periods, with or without state assistance. One thinks of Petrine investment in ferrous metallurgy, armaments, and textiles in the early eighteenth century; the subsequent growth of the food-processing and paper industries; the boom in railroad construction that followed the Crimean War of 1854–1856; and the creation (albeit short-lived) of a modern small-arms industry. Nor should one overlook the less visible, unregulated, but economically important cottage (*kustar*) industries that flourished across the length and breadth of the country until they were snuffed out after the Revolution of 1917. Likewise, the two decades following World War II witnessed rapid economic growth. This coincided with a dramatic increase in the international profile and prestige of Soviet industry. However, evidence of a "revolution" in economic activity, accompanied by political turmoil, dramatic social upheaval, and cultural change, is to be found in the decades mentioned above, when total output increased by 3.4 percent (1.7 percent per capita) and 5.1 percent (3.9 percent per capita), respectively.[2]

Isolating these two decisive periods of rapid growth not only invites us to confront the magnitude and pattern of industrial expansion but also directs our attention to the political and institutional conditions under which industrialization took place. During the first period, the growth of output was characterized by the integration of the Russian economy into the world economy, a process that encouraged the inflow of foreign direct investment and the transfer of technology in mining and manufacturing. As a result, additional capacity was created in coal, oil, ferrous metallurgy, and some branches of engineering. Government economic policy supported a mixture of private and state-owned enterprise. Corporate forms of organization became more prominent in industry and in banking. The demand for labor led to a rapid expansion in nonagricultural employment, albeit with evidence of continuing ties to the relatively backward rural economy.[3]

Between 1929 and 1941, following the Bolshevik Revolution and the subsequent recovery of the war-ravaged economy, the Soviet government embarked simultaneously on an ambitious industrialization drive, a deliberate restructuring of society, and a cultural revolution. Now industrialization was associated with central economic planning and the mobilization of domestic resources for investment in newer as well as established branches of industry and energy. Collectivization of agriculture completed the picture. In contrast to the czarist period, there was no foreign direct investment and less reliance was placed on foreign technology and expertise. Industrialization also rested on more extreme conditions than under czarism, notably the extensive Stalinist recourse to forced labor. Planned industrialization was invested with enormous political significance. Society was expected (and, where necessary, compelled) to contribute to the achievement of industrial and technological strength and, ultimately, superiority over the USSR's capitalist rivals. Economic modernization would validate the Bolshevik project. So, indeed, it did in the eyes of many contemporaries.[4]

Similarities between late czarist and Soviet industrialization should not be overlooked. Examples abound from both eras of a lament for the precarious condition of Russian agriculture, with an equally strong endorsement of state intervention to restructure the basic institutions of peasant farming. In both periods, the growth of the labor force drew heavily on migration from low-wage peasant agriculture to construction, extractive, and manufacturing industries. Official backing for the economic transformation of Russia was an important impulse before and after 1917. In each instance, the government sought to establish the state's security on firmer economic foundations, against the background of an acknowledged disparity between Russia and its more advanced competitors. Bureaucratic debates about economic backwardness and its implications for Russia's future took an equally intense and bitter form in both czarist and Soviet Russia.[5]

Nor should it be thought that the Revolution of 1917 completely ruptured established institutions, practices, and relationships. The relationship between the technical intelligentsia, industrial managers, and the state was problematic in both periods. For example, under czarism the technical intelligentsia often criticized prevailing social and political arrangements, but questions of trustworthiness continued to cause unease to the Soviet leadership, particularly when engineers had been trained before 1917. Issues of professional autonomy were at stake under both political systems. To take another example, the discourse of conquest and mastery reached its apogee in the Stalinist era, coloring the form in which technological achievements were represented. But it did not originate with Stalin. In other words, we should relate ideas about technology to public attitudes, including the public espousal of the romance of technology, without assuming a sharp discontinuity across the revolutionary divide.

THE POLITICS AND ECONOMICS OF INDUSTRIALIZATION IN RUSSIA

Russian industrialization before 1914 was stimulated by a mixture of direct and indirect government intervention. Under Minister of Finance Sergei Witte, the czarist government embarked on a massive program of railroad building, including the construction of the Trans-Siberian Railroad, designed to open up markets in the Far East and Central Asia. Railroad construction in turn contributed to the expansion of heavy industry. This was not the sole element in government strategy. By adopting tariff protection and taking Russia onto the gold standard in 1897, the government created an environment favorable to direct foreign investment. Foreign specialists and financial journalists painted a picture of Russian technical weakness (e.g., in the "dormant" Urals iron industry, with its small-scale blast furnaces), coupled with opportunities for foreign investors to generate profits from the introduction of advanced technologies in metallurgy, mining, chemicals, and electrical equipment. Witte did his utmost to advertise these opportunities.[6]

The results were impressive: an annual industrial growth rate of 8 percent during the 1890s and around 6.3 percent between 1907 and 1913, combined with technological modernization in key branches of industry. Through the medium of capital movements and its export trade in commodities such as grain and oil, Russia became increasingly integrated into the international economy. Modern industries were established in the southern industrial region, where foreign engineers reorganized iron making and steelmaking along best international practice. Integrated steel production became the norm. Major foreign companies such as Singer and International Harvester established retail operations and extended credit to Russian consumers before dipping a toe into local manufacturing.[7]

This positive assessment requires modification. To be sure, the czarist Industrial Revolution provides evidence of import substitution in basic industries and key products such as rolling stock and agricultural equipment. But other sectors, such as machine tools, were in their infancy (although they received a boost during World War I). Furthermore, czarist industrialization tended to consolidate preexisting regional disparities rather than transform the pattern of industrial location. Thus, new investment in modern metalworking and textile factories was concentrated in Russia's northwest, a region that was already relatively highly developed in terms of educational attainment and income per head.[8]

Russia's industrial upsurge sparked controversy at the time, and its welfare consequences have been debated ever since. Some economic historians have questioned the primacy of the state and the efficacy of government intervention. Conservatives charged Witte with the "neglect" of agriculture and for failing to avert an "agrarian crisis." The landed elite felt threatened by the social and economic changes wrought by rapid industrialization. Witte's strategy gave rise to interministerial conflicts over decisions about tariff protection, fiscal policy, and labor policy. Gerschenkron defended the choice of strategy on the grounds that it enabled Russia to substitute for factors of production that were missing or in short supply, although he acknowledged the burden borne by the Russian peasantry.[9]

Industrialization did not cement the place of entrepreneurs in Russian society. Russia's mercantile and entrepreneurial elites were internally divided and organizationally weak, and their political leverage was correspondingly limited. The relations between private defense contractors and government departments remained tense. Employers who needed to raise large sums of money bemoaned the archaic and cumbersome system of company law for permitting arbitrary decision making by government officials. Some industrialists such as Putilov, Riabushinskii, and Tereshchenko achieved political prominence, and a handful of Moscow merchants played influential roles as cultural patrons, but the revolution swept them away.[10]

Other critiques came from liberals and radicals alike, who drew on a long tradition that denounced exploitation and "enslavement."[11] Liberals argued that Russian industrialization entrusted too much power to bureaucrats and employers; they called instead for greater civic freedom, which the Revolution of 1905 granted only to a limited extent. Even more challenging was the voice of Russian labor and those who spoke on its behalf. Organized labor protest—prominent in 1905 but crushed by repression and slump—revived on the eve of World War I. There were few signs of an accommodation between capital and labor. Employers resented the often ham-fisted intervention of officialdom in the workplace. Workers complained of humiliating treatment at the hands of foremen. The social system was profoundly unstable. During the revolutionary upheaval in 1917, the Bolshevik Party exploited workers' demands for democracy in the workplace and for an eight-hour day.[12]

The October Revolution significantly altered the political and social context for Russian industrialization. The Bolsheviks claimed to act in the name of the proletariat and poor peasantry, and they shed no tears when the terrorized industrial, financial, and landed elite fled abroad in the wake of measures to seize and redistribute privately held assets. Despairing, this elite was also disillusioned with the old regime, which they berated for its failure to introduce a secure and coherent economic and legal framework for private enterprise.[13]

When the civil war ended in 1921, Russia's damaged economic base required emergency rehabilitation. The Bolsheviks ruled a country now shorn of relatively advanced industrial centers in Poland and the Baltic region. Established economic ties were ruptured. Food and fuel supply collapsed, and the labor force hemorrhaged. Peasants, workers, and the radical sailors of Kronstadt took up arms against the new "commissarocracy," which forcibly extracted grain, imposed one-man management in industry, and curbed Soviet democracy. The siege mentality of the civil war persisted throughout the interwar period, profoundly affecting the course of Soviet industrialization and paving the way for mass terror.[14]

In 1921, Lenin imposed the New Economic Policy (NEP) which legalized private trade and encouraged the revival of economic relations between town and country. The Soviet economy under NEP extended the czarist experiment with electrification (recall Lenin's famous slogan, "Communism equals Soviet power plus electrification of the whole country") and introduced new products, such as oil drilling equipment, while consolidating others, such as aircraft. By 1926, industrial production had recovered to its prewar level. However, vital sectors such as iron and steel, complex agricultural machinery, and machine tools lagged. Radical elements within the party, as well as labor leaders, complained that NEP was failing to address the low technological level of Russian industry and construction compared to the greater dynamism in the developed capitalist West. Large-scale industry operated according to principles of cost accounting (*khozraschet*), which led some factories to jettison labor, turning unemployment into a politically explosive issue.[15]

Attempts were made to resurrect relations with foreign expertise and to gain renewed access to foreign technology. The most important such initiative was taken by the Russian engineer Iu. V. Lomonosov, who insisted on external assistance with the task of rapidly modernizing rolling stock and infrastructure. The Bolsheviks' repudiation of Russia's foreign debt and the civil war placed him in a difficult position, akin (it was said at the time) to "foraging by a besieged fortress." However, in 1920–1921, contracts were placed with suppliers in Sweden, Germany, and Britain. This controversial measure helped revitalize the ailing railroads, picturesquely described by one American engineer as "strings of matchboxes coupled with hairpins and drawn by samovars."[16]

Following the cautious strategy adopted under NEP, the First Five-Year Plan (FYP) in 1928 marked a fresh attempt to engineer rapid economic growth by means of concerted state intervention. With its ambitious targets for capital investment, increased labor productivity, and the expansion of output, the FYP reflected a clear redirection in Soviet life. Enthusiasts such as the economist Stanislav Strumilin, who espoused a "teleological" commitment to economic planning, triumphed over others who preferred an "organic" approach to growth. In this tense atmosphere—there was talk of "enemies" who threatened the "economic front"—it took considerable courage to proclaim the need for caution. As Strumilin put it in 1929, "specialists prefer to stand for high rates of growth rather than to sit [in jail] for low ones." Why then did the Communist Party commit itself to a new course? Apart from the distasteful encouragement that NEP appeared to give to "hostile" elements (i.e., private traders and kulaks), the existing economic system had not "solved" the problems of unemployment, technological backwardness, and a rising foreign trade deficit. A commitment to rapid industrial growth implied the absorption of unemployed labor, further import substitution, and the creation of a modern defense industry, something that the war scare in 1927 made imperative.[17]

Czarist officials sometimes referred to "His Excellency, the Harvest" as the factor governing economic affairs in prerevolutionary Russia. In Stalin's Russia, "His Excellency, the Plan" lay instead at the core of the economic system. Unlike the harvest, plans took a monthly, quarterly, and annual form, while for broad strategic purposes, the FYP dominated decision making. Gosplan and the economic ministries imposed plans on state-owned enterprises. Targets normally took the form of physical indicators, but were also expressed in terms of the gross value of output in "constant prices." Other elements of planned performance included targets for product assortment, cost reductions, labor productivity, and so forth. Quality considerations were secondary. Accompanying the targets were centrally allocated supplies. A large economic bureaucracy supported this hugely ambitious and unprecedented exercise in coordination, with officials intervening as needed to restore a degree of balance.[18]

Additional output reflected the priorities given to investment and government spending, notably on defense. The capital stock in industry trebled during the First FYP. Defense production increased twenty-eight-fold during the 1930s (far in excess of total industrial production), imposing a heavy burden, particularly after 1936. The Stalin era witnessed Russia's emergence as a modern military power. The hallmarks were new tank and aviation industries, supported in turn by steel, metalworking, fuel, chemicals, and rubber production. Qualitative improvements also took place. But pronounced difficulties remained. Inappropriate rolling stock bedeviled the performance of the Soviet railroads. The privileged defense sector was not immune to the inefficiency and waste prevalent in the economic system as a whole. Nor did Stalinist industrialization fundamentally transform the location of industry; notwithstanding famous ventures such as Magnitogorsk and the Ural-Kuzbass Combine, industry continued to be disproportionately concentrated in European Russia, where factories remained vulnerable to invasion from the West.[19]

The consequences were profound in terms of economic behavior. A complex interplay of interests among the party; planning agencies; economic ministries; republican, regional, and local authorities; and the enterprises (e.g., farms, factories) determined the detailed formulation and implementation of plans. In principle, Soviet planners dictated the targets, but at each level, subordinate agents within the system entered into complex strategies with their superiors to obtain the best possible set of instructions—in other words to negotiate a plan that was achievable. Since no superior had access to perfect information, subordinates were able to understate and conceal productive capacity. Similarly, plan fulfillment required astute and timely action by enterprises. No manager could afford to be exposed to failure to meet the targets, and horizontal networks and contacts flourished in these circumstances; managers engaged "pushers" (*tolkachi*) to obtain inputs over and above the planned allocation. Thus, the formal system of subordination disguised the fact that the principals did not have perfect information about the behavior of agents. In a sense, therefore, the system was sustained less by the hierarchical character of central economic planning than by the interaction of dictators and subordinates. Managers developed a range of skills, including those designed to outwit those in administrative command. As one scholar aptly says, this was "a system in constant motion, evolving in multiple and contradictory ways, driven by forces only partly controlled by central state and political authorities."[20] Attempts to impose control had unpredictable results for Soviet labor policy.

Russian Workers and Labor Productivity

The low levels of labor productivity in prerevolutionary Russian industry prompted considerable soul searching. One approach was to advocate rapid improvements in elementary education to provide the labor force with basic skills. Observers of the steel industry commented that the low quality of labor contributed to Russia's disadvantage vis-à-vis European producers. At the same time, there were marked regional disparities in labor productivity, which differed by a factor of six or more between the south Russian steel industry and its older counterpart in the Urals. Such differentials could not be explained purely by reference to education.[21] Some account also needs to be taken of deficiencies in the internal organization of the plants. As industrialization gathered momentum, established factories acquired equipment without rethinking the layout, which took on a haphazard and congested appearance. Another problem related to the uncertain and erratic nature of demand for manufactured goods. In engineering, firms were obliged to produce a bewildering array of products that hindered attempts to introduce mass-manufacturing methods.[22]

Lest it be thought that czarist construction projects after 1861 drew exclusively on free labor, we should recall that the Trans-Siberian Railroad required the labor of fourteen thousand convicts and exiles. They probably made up around one-fifth of the total labor force. Like their counterparts in the late Stalin period, unfree workers received a wage, minus deductions for their upkeep. The labor force on the railroads turned out to be particularly volatile, so perhaps the greatest contribution of the Trans-Siberian Railroad in czarist Russia was to deliver larger numbers of workers to the revolutionary cause.[23]

On the eve of World War I, attempts were made to improve labor productivity by intervening directly in the labor process and embracing the doctrine of scientific management. The attempted application of Taylorist principles in the engineering factories of St. Petersburg met with little success. Workers went on strike to protest the imposition of the system known as *amerikanka* and to protect traditional work routines that conferred a degree of shop floor autonomy. Hard-pressed employers, particularly those struggling to complete lucrative naval defense contracts, were tempted to concede defeat rather than risk lengthy stoppages.[24]

Soviet attempts to improve labor productivity also met with obstacles. Lenin was prepared to embrace the doctrine of Taylorism. In 1924, Stalin famously spoke of combining "American efficiency" with Bolshevik ideology. But their vision of "rational" factory organization (exemplified by the "scientific organization of work," or NOT, the Soviet acronym for *nauchnaia organizatsiia truda*) (scientific organization of labor) encountered stiff opposition from within the Communist Party, from trade unions, and from health and safety officials.[25] Subsequently, the First Five-Year Plan rested on a significant planned reduction in labor costs to finance increased investment. Attempts were made to improve the productivity of workers by means of moral and material inducements, such as "shock work," widening wage differentials (endorsed by Stalin in 1931), and creating differential access to rationed goods. But the mass influx of unskilled peasant migrant labor made it difficult to improve output per person. In the engineering industry, continuous-flow methods designed to economize on the scarce factor (skilled labor) did not achieve the desired result because management chose to broaden the product mix.[26]

Rapid technological change created pronounced difficulties for Soviet enterprises. Most investment was channeled toward basic production processes, with scant regard for auxiliary operations such as lifting and dispatching items within the factory. This contributed to bottlenecks and breakdowns in production. In order to avoid blame and penalties, enterprise managers typically employed labor on the more mundane tasks that could have been mechanized. This was not a purely Soviet problem;

czarist managers had done the same. But it meant that younger workers took relatively well-paid jobs as "beasts of burden" in otherwise dead-end jobs rather than take the lower wages offered to those with moderate skills. In the absence of proper attachments, young men simply held the pieces to be machined. Contrary to expectations, the ratio of workers to machines tripled during the First FYP.[27]

Technological changes disguised the reality of semi-artisanal practices within individual shops, as well as the proliferation of a youth subculture that challenged the authority of the foreman and the manager under the guise of "socialist competition." "Shock work" could alleviate bottlenecks, but its effectiveness was limited. Soviet planners believed that adopting U.S. methods of mass manufacture would solve the problem of how to industrialize at breakneck speed. Unfortunately they neglected issues such as storage facilities for materials and parts, and accounting practice. Where managers and technical specialists did pay attention to such issues, the production record was relatively favorable, as a comparison of the Stalingrad and Khar'kov tractor works makes clear.[28]

The most famous state-supported initiative to promote labor productivity was the Stakhanov movement. This took place during the Second FYP, when the Soviet leadership had embarked on a fresh surge of capital investment. *Pravda* (January 1, 1936) explained this in orthodox Marxist terms: "Every newly emerging social system triumphs over the old outdated mode of production because it brings about a higher productivity of labor." But most workers responded passively at best, and managers regarded the entire campaign as a pointless distraction. The outcome was chaos. By the beginning of 1937, more modest targets for labor productivity were formulated. Stakhanovism created a climate of uncertainty for Soviet managers and factory directors, who suffered the consequences of mass purges in 1937 and 1938, at a time when Stalinist ideology ostensibly celebrated managerial power.[29]

Notwithstanding all these initiatives, the prewar Soviet leadership acknowledged that the productivity gap between the USSR and its capitalist rivals remained wide. It proved much more straightforward to draft millions of additional workers than to engineer an improvement in output per person. We should also acknowledge forced labor as an integral component of the Stalinist economic system. By 1941, unfree labor contributed around 14 percent to the industrial economy, with a much larger proportion in the extraction of nonferrous metals, construction, and timber. One (in)famous example was the Belomor Canal, linking the Baltic and the White seas, whose construction cost the lives of an estimated 200,000 prisoners. The methods of construction were extremely primitive, relying largely on animate power, and the canal had to be rebuilt after World War II. Many other projects were similarly wasteful.[30]

SCIENCE, TECHNOLOGY, AND TECHNICAL SPECIALISTS

Given that the state played an important role in industrialization strategies, how far did it impinge on science and technology? Did the legal framework encourage creativity? Did the state promote or constrain scientific and technical endeavor? What part did the technical intelligentsia play in public life and economic affairs?

Czarist control over scientific and technical activity was embodied in commercial and patent law. A manifesto of 1812 stipulated that invention privileges were a special favor granted to the inventor rather than recognition of the individual's intellectual property. The underlying doctrine was service to the state. The process of application to the Manufacturing Council was cumbersome and time-consuming. One consequence was that the number of patents issued before the Revolution of 1917 was far below the corresponding figure in Germany and the United States. Thus, although late-nineteenth-century Russia boasted some remarkable individual scientists and inventors, including Tsiol'kovskii, Lodygin, Mendeleev, Timiriazev, Lebedev, Vernadskii, and Sikorskii, the legal and political framework made it difficult to capitalize on their activities.[31] The law on invention privileges formed part of a more general ethos that favored strict government control over enterprise. For example, czarist property law made it very difficult to disentangle titles to land, thus discouraging the exploitation of mineral rights. Factory owners faced other legal impediments.[32]

The czarist state devoted only 0.25 percent of budget expenditure to research activity on the eve of World War I. All the same, Russia could boast of a qualified engineering elite. The czarist technical intelligentsia—in 1913, comprising fifteen thousand engineers with higher education qualifications—embodied, like other members of the Russian professions, a self-conscious ethos of obligation toward society. (The ratio of graduate engineers to the total population or to the labor force was much greater in Germany and the United States.) A large segment was not afraid to criticize both employers and government officials, for example, over the restrictions imposed on Russian Jews. However, their voice was often drowned by officialdom. Not surprisingly, this group embraced technology as a crucial element in Russia's modernization. They received their practical and theoretical training in Russia's expanding polytechnics, such as the St. Petersburg Technical Institute, the Mining Institute, the Moscow Higher Technical School, and the Khar'kov Technical Institute. Among the leading figures in late czarist Russia were Piotr Pal'chinskii, Vasilii Grinevetskii, Nikolai Savvin, Vladimir Ipatiev, and Karl Kirsh, who affirmed the importance of professional expertise and authority. Grinevetskii, director of the prestigious Moscow Higher Technical School, brought together the Society of Technologists and the Moscow Polytechnical Society. But his attempt to create an independent national organization of engineers fell afoul of the old regime, another indication of the power of politics to constrain creative endeavor.[33]

World War I did more than anything else to advance the cause of Russian science and technology. Already there were some signs that foreign-born and foreign-trained specialists were giving way to local personnel.[34] This trend continued during the war. The new Committee for the Development of Productive Forces (KEPS) brought together Russian natural scientists and technical specialists in a project designed to establish the centrality of science in Russian public life specifically by formulating and coordinating applied research. Key problems included the expansion of hydroelectric power, the development of new mineral wealth, and scientific investigation of the properties of materials. Research institutes were planned in fields such as hydrology, petroleum, pharmaceuticals, coal, and ferrous and nonferrous metals. KEPS continued to meet and publish its findings during the civil war. War and revolution gave them a public platform and an opportunity to establish their expertise in addressing current problems, such as the need for improved technical education, developing links between polytechnics and factories, and promoting "scientific administration" at the level of the individual enterprise. These initiatives were intended to bring long-term benefits. Thus, electrical engineers hoped that the spread of electricity would transform the prospects of rural consumers, revolutionize municipal transport, and enable small-scale industry to compete with large-scale enterprise.[35]

Most of the Russian technical intelligentsia remained on Russian soil after 1917. Although they were suspicious of the doctrine of "workers' control," they nevertheless regarded the revolution as a platform denied them by czarism with improved prospects for advancement. Older specialists espoused a vision of rational and planned industrialization, which attracted them (and plenty of foreign observers too) to the Soviet modernizing project. During the 1920s, they were rewarded with a role in Soviet industry, although Lenin's espousal of these "bourgeois specialists" provoked fierce controversy within the party. Their expertise, enthusiasm, and patriotism did not protect them from the purges that eventually produced a numerous and more compliant Soviet technical intelligentsia, whose numbers stood at 290,000 on the eve of World War II, a sixfold increase since 1928.[36]

Thanks to Loren Graham, we can see something of the resulting tensions in the career of the Russian engineer Piotr Pal'chinskii, shot in 1929 (before the Great Terror) as leader of an "anti-Soviet conspiracy." Pal'chinskii consistently maintained that labor productivity could be improved only by taking account of broader issues of welfare, not just technical conditions. Unfortunately his emphasis on what he termed "humanitarian engineering" put him at odds with the Stalinist vision. Although Stalin famously stated that "cadres decide everything," this did not imply the liberation of human beings to realize their potential. More characteristic was Stalin's contemptuous depiction of people as "cogs" and "screws" (*vintiki*).[37]

Between 1917 and 1941, considerable resources were invested in Soviet science and technology.[38] For Soviet engineers and scientists who toed the party line, the rewards were considerable. Under Soviet rule, inventors were invited to contribute to the general good rather than to expect monetary reward. Moral rewards included the award of state prizes and recognition, such as being designated a Hero of Socialist Labor. (In the late 1940s, a Stalin prize was worth 100,000 rubles, around thirteen times the average wage. Awards were also given to factories and cities.) At the same time, Soviet basic science boasted people of the caliber (moral as well as scientific) of Pyotr Kapitsa, Yulii Borisovitch Khariton, Lev Davidovich Landau, Andrei Sakharov, Igor Yevgenyevich Tamm, and many others. Although cocooned in privileged scientific institutes, these physicists frequently stood up to the Soviet secret police to defend the principles of international scientific and technical cooperation against narrower conceptions of state interest. As a result, the Kremlin tended to mistrust them.[39]

How decisive was technology in transforming the Russian and Soviet economy between 1885 and 1941? Technological change was patchy. In some sectors, such as in iron and steel, chemicals, and armaments, as well as in power generation, there is strong evidence of the adoption of modern technology (1890s–1910s, 1930s). However, other industries, notably construction and, above all, agriculture, showed fewer signs of technical change. Prior to World War II, rail transport continued to be dominated by outdated technology. In metalworking, basic production was revolutionized, but auxiliary processes were not. This seems to reflect an assumption that industry could rely on abundant unskilled labor, including, where necessary, forced labor, for backbreaking tasks.

Cultures of Industrialization

Scholars have rightly drawn attention to the importance that the Stalinist system attached to spectacular technological and industrial achievements. However, this view needs to be qualified. Paul Josephson notes that large-scale technologies have been widely embraced by modern governments: "what distinguishes the USSR are differences of degree, not of essence." David Shearer also points out that "faith in the liberating power of technology wedded to the state was not peculiar to Soviet Russia." In his discussion of collectivization, James Scott suggests that Soviet initiatives resembled other projects of "authoritarian high modernism."[40]

The czarist state boasted its own grandiose schemes, chief among them the Trans-Siberian Railroad, first conceived in the 1860s. By 1900—this was typical of Witte's adroit use of modern forms of communication—the publicity material boasted that the luxury coaches were equipped with electric cigar cutters. All the same, there was in all this an element of the famous Potemkin village, because projected increases in passenger and freight volume failed to materialize.[41]

Soviet industrialization yielded impressive examples of dramatic engineering projects. Often the government embarked on them in haste, without sufficient preparation and without having resolved competing objectives. For instance, the "Arctic dominion" of Glavsevmorput (the main administration of the northern sea route) was not built on a solid base; as a consequence, capital equipment and rails suffered from corrosion or sank into the Arctic swamps.[42] The debate over the famous hydroelectric power plant at Dneprostroi revealed rival visions, with some experts supporting the view that it would revolutionize heavy industry in Ukraine while others anticipated the transformation of agriculture. Gosplan endorsed both objectives. Work began in 1927 and was completed in 1932, with assistance from foreign consultants. In terms of organization, management of resources, and labor relations, Dneprostroi was a microcosm of Soviet industrialization during the First Five-Year Plan. In practice, the enterprise relied heavily on the substitution of labor for capital. At the same time, it failed to solve problems created by rapid turnover and low labor productivity, a function in part of poor management of available cadres. Similar difficulties beset Magnitogorsk.[43]

We should also note the romance of "exploration," captured by studies of the Soviet Arctic that emphasized contemporary preoccupation with the "struggle" against an unforgiving natural world and mirrored the struggle against the human "class enemy." A czarist antecedent, the Great Northern Expedition (1733–1749), had stimulated the development of Russian biology, zoology, and geology.[44] The romance extended into aviation and the exploration of the Far North, with the heroic Soviet pilot as a "knight of old" (*bogatyr*). (In the later 1930s, the Soviet authorities appropriated this epithet to industrial managers and engineers.) The love affair with modernity was reinforced by films and novels that celebrated—sometimes ambivalently—Soviet construction and the role of the engineer. Two examples are the film *Ivan* (1932, directed by Aleksandr Dovzhenko, portraying the construction of a dam, and Fedor Gladkov's novel *Energiia* (Energy), written between 1932 and 1938. In 1939, the Soviet authorities devised an All-Union Agricultural Exhibition that became a permanent exhibition of the achievements of the national economy (VDNKh) devoting space to achievements in each of the constituent republics of the USSR.[45]

Private practice also demonstrated an endorsement of state ideology, at least in the early period of Soviet rule. Some revolutionary enthusiasts named their children "Tractor" and "Electrification" in lieu of traditional Orthodox saints. Proletarian writers invested more thought in the meaning of technological change and celebrated the factory as the "cradle of a reborn Russia" that transcended economic backwardness. In the hands of these Proletkult writers, the machine was to be embraced like an enchanting lover. However, one of its leading figures, Aleksei Gastev, anticipated a fusion of man and machine, depriving humanity of independent will and agency in an ominous foretaste of totalitarian doctrine.[46]

Not all of this cultural activity underscored the desirability of technological modernization for its own sake. In the hands of early Soviet planners, modern technology was also expected to improve the lives of households by socializing and mechanizing domestic tasks such as cooking and washing. To be sure, Soviet industrialization did not set out to—and did not—change the gendered division of labor. But it did embrace an intoxicating vision of transforming everyday life, which would enable Soviet citizens to devote more of their leisure time to cultural and educational activities. Finally, when one's life had run its course, the

Bolsheviks promised a revolution in the manner of one's death; it was a sign of modernity that the dead body was handed over to the industrialized crematorium rather than to the priest.[47]

Conclusion

What, then, of economic backwardness and the attempts to overcome it? Stalin expected the current generation to make sacrifices in order to provide the resources for investment, future prosperity, and state security. The results—curtailment of consumption and the resort to mass coercion—confirmed that expectation. Witte adopted a similar stance, although his program offered less overt coercion and more arbitrary intervention. The pressures on Russia's populace had been anticipated and roundly condemned by Alexander Herzen in the middle years of the nineteenth century: "Do you truly wish to condemn all human beings alive today to the sad role of caryatids supporting a floor for others some day to dance upon . . . or of wretched galley slaves, up to their knees in mud, dragging a barge filled with some mysterious treasure and with the humble words 'progress in the future' inscribed on its bows?" Stalin took this view to its extreme during the First Five-Year Plan and had no qualms about multiplying the number of galley slaves. But Herzen was an intellectual, not a political leader charged with steering Russia away from the stagnant waters of economic backwardness. Backwardness was real, not a figment of the imagination.[48]

In this connection, the international context of Russian and Soviet industrialization loomed large. This framework was well understood by Peter the Great, and successive military defeats and diplomatic humiliation (in 1856, 1878, and 1904) magnified the sense of Russian inferiority. International rivalry persisted after 1917, but now it had a keen ideological edge. In the 1930s, for example, the development of the Arctic North was in part a product of the race between Soviet Russia and its rivals, a full two decades before the Cold War took international rivalries into space. International relations mattered in terms of both resource allocation and sustaining a siege mentality. Czarist industrialization in the 1890s presupposed a close connection with the outside world, a far cry from the tentative and suspicious stance Soviet officials adopted toward foreign specialists.

Was Soviet industrialization distinctive in other respects? The Soviet system rested on exclusive state ownership of the means of production and formal control over the allocation of output. After 1928, enterprises were governed by the plan and subject to formal supervision. Managers retained some flexibility within this command-administrative system, exercising discretion over the acquisition of inputs, including labor, and engaging in sideline manufacturing activity. By contrast, czarist Russia offered a much greater scope for private enterprise. But its officials too intruded clumsily into business and management, thereby betraying an ambivalent attitude toward capitalist activity.

Finally, the Soviet experiment with forced industrialization had far-reaching consequences beyond Soviet borders. The Stalinist regime projected its version of industrial revolution into other societies, notably in the neighboring Baltic lands and Eastern Europe after World War II. It did so chiefly for political and strategic reasons, but also on the grounds that central economic planning offered the prospect of overcoming economic backwardness. As we revisit Russian industrialization, it is worth remembering that the Soviet model became an ideological export.[49]

NOTES

1. Alexander Gerschenkron, *Economic Backwardness in Historical Perspective* (Cambridge, MA: Harvard University Press, 1962). Compare Richard Sylla and Gianni Toniolo, eds., *Patterns of European Industrialization: The Nineteenth Century* (London: Routledge, 1991).

2. Paul Gregory, *Russian National Income, 1885–1913* (Cambridge: Cambridge University Press, 1982), 182. On Russia's vibrant artisanal culture before 1917, see Roger Portal, "The Industrialization of Russia," in *The Cambridge Economic History of Europe* vol. 6, part 2, ed. Hrothgar J. Habakkuk and Michael M. Postan (Cambridge: Cambridge University Press, 1966), 801–872, 839–843. See also Joseph Bradley, *Guns for the Tsar: American Technology and the Small Arms Industry in Nineteenth-Century Russia* (DeKalb: Northern Illinois University Press, 1990).

3. Olga Crisp, *Studies in the Russian Economy before 1914* (London: Macmillan, 1976); Peter Gatrell, *The Tsarist Economy, 1850–1917* (London: Batsford, 1986); Arcadius Kahan, *Russian Economic History: The Nineteenth Century* (Chicago: University of Chicago Press, 1989). Trotsky discussed peasants being "snatched from the plow and hurled into the factory furnace," but Robert Johnson, *Peasant and Proletarian: The Working Class of Moscow in the Late Nineteenth Century* (Leicester: Leicester University Press, 1979), demonstrates that his description was wide of the mark.

4. Robert W. Davies, Mark Harrison, and Stephen G. Wheatcroft, eds., *The Economic Transformation of the Soviet Union, 1913–1945* (Cambridge: Cambridge University Press, 1994); Paul Gregory and Valery Lazarev, eds., *The Economics of Forced Labor: The Soviet Gulag* (Stanford, CA: Hoover Institution Press, 2003); Robert C. Allen, *Farm to Factory: A Reinterpretation of the Soviet Industrial Revolution* (Princeton, NJ: Princeton University Press, 2003); David C. Engerman, *Modernization from the Other*

Shore: American Intellectuals and the Romance of Russian Development (Cambridge, MA: Harvard University Press, 2003).

5. George Yaney, *The Urge to Mobilize: Agrarian Reform in Russia, 1861–1930* (Urbana: University of Illinois Press, 1982); David R. Stone, *Hammer and Rifle: The Militarization of the Soviet Union, 1926–1933* (Lawrence: University Press of Kansas, 2000); James Heinzen, *Inventing a Soviet Countryside: State Power and the Transformation of Rural Russia, 1917–1929* (Pittsburgh: University of Pittsburgh Press, 2004).

6. John P. McKay, *Pioneers for Profit: Foreign Entrepreneurship and Russian Industrialization, 1885–1913* (Chicago: University of Chicago Press, 1970), 106–157; Portal, "The Industrialization of Russia," 829. The Urals iron industry retained traditional charcoal-iron techniques and relied on serf workers until 1861.

7. Fred V. Carstensen, *American Enterprise in Foreign Markets: Studies of Singer and International Harvester in Imperial Russia* (Chapel Hill: University of North Carolina Press, 1984); McKay, *Pioneers for Profit*, 125.

8. Martin Spechler, "The Regional Concentration of Industry in Imperial Russia, 1854–1917," *Journal of European Economic History* 9 (1980), 401–429; R. W. Davies, ed., *From Tsarism to the New Economic Policy: Continuity and Change in the Economy of the USSR* (London: Macmillan, 1990); see especially Peter Gatrell and R. W. Davies, "The Industrial Economy," 127–159.

9. Gerschenkron, *Economic Backwardness*. Elements of this story have been challenged by Olga Crisp, Arcadius Kahan, Fred Carstensen, and Paul Gregory. However, Gerschenkron's overall interpretive framework has proved remarkably durable and stimulating.

10. Alfred J. Rieber, *Merchants and Entrepreneurs in Imperial Russia* (Chapel Hill: University of North Carolina Press, 1982); James West and Iu. Petrov, eds., *Merchant Moscow: Images of Russia's Vanished Bourgeoisie* (Princeton, NJ: Princeton University Press, 1998).

11. As in Nikolai Nekrasov's famous poem, "The Railway" (1864).

12. Tim McDaniel, *Autocracy, Capitalism, and Revolution in Russia* (Princeton, NJ: Princeton University Press, 1988).

13. Thomas C. Owen, *The Corporation under Russian Law: A Study in Tsarist Economic Policy* (Cambridge: Cambridge University Press, 1991); Peter Gatrell, "Poor Russia: Environment and Government in the Long-Run Economic History of Russia" and Thomas C. Owen, "Entrepreneurship, Government and Society in Russia," in *Reinterpreting Russia*, ed. Geoffrey Hosking and Robert Service (London: Edward Arnold, 1999), 89–106 and 107–125 respectively.

14. Donald J. Raleigh, *Experiencing Russia's Civil War: Politics, Society, and Revolutionary Culture in Saratov, 1917–1921* (Princeton, NJ: Princeton University Press, 2002).

15. Julian M. Cooper and Robert A. Lewis, "Research and Technology," in Davies, ed., *From Tsarism to the New Economic Policy*, 189–211; Chris Ward, *Russia's Cotton*

Workers and the New Economic Policy: Shopfloor Culture and State Policy, 1921–1929 (Cambridge: Cambridge University Press, 1990).

16. Anthony Heywood, *Modernising Lenin's Russia: Economic Reconstruction, Foreign Trade, and the Railways, 1917–1923* (Cambridge: Cambridge University Press, 1999); Vladimir I. Grinevetskii, *Poslevoennye perspektivy russkoi promyshlennosti* (Khar'kov: Vserossiiskii Tsentral'nyi soiuz potrebitel'nykh obshchestv, 1919).

17. Edward H. Carr and Robert W. Davies, *Foundations of a Planned Economy, 1926–1929*, vol. 1 (Harmondsworth: Penguin, 1969).

18. Eugene Zaleski, *Stalinist Planning for Economic Growth, 1933–1952* (Chapel Hill: University of North Carolina Press, 1980).

19. John N. Westwood, "Transport," in Davies, Harrison, and Wheatcroft, eds., *The Economic Transformation of the Soviet Union* (Cambridge: Cambridge University Press, 1994), 158–181.

20. David R. Shearer, *Industry, State, and Society in Stalin's Russia, 1926–1941* (Ithaca, NY: Cornell University Press, 1996), 205. See also Paul Gregory, *The Political Economy of Stalinism: Evidence from the Soviet Secret Archives* (Cambridge: Cambridge University Press, 2003).

21. McKay, *Pioneers for Profit*, 135.

22. Peter Gatrell, *Government, Industry and Rearmament in Russia, 1900–1914: The Last Argument of Tsarism* (Cambridge: Cambridge University Press, 1994), 191–192.

23. Steven Marks, *The Road to Power: The Trans-Siberian Railroad and the Colonization of Asian Russia* (Ithaca, NY: Cornell University Press, 1991), 182.

24. Heather Hogan, *Forging Revolution: Metalworkers, Managers, and the State in St. Petersburg, 1890–1914* (Bloomington: Indiana University Press, 1993). Compare Gatrell, *Industry and Rearmament in Russia*, 251–252.

25. Kendall Bailes, *Technology and Society under Lenin and Stalin* (Princeton, NJ: Princeton University Press, 1978).

26. David Granick, *Soviet Metal-Fabricating and Economic Development: Practice versus Policy* (Madison: University of Wisconsin Press, 1967).

27. Shearer, *Industry, State, and Society in Stalin's Russia*, 211–219, 228.

28. Granick, *Soviet Metal-Fabricating and Economic Development*, 117–119.

29. Lewis H. Siegelbaum, *Stakhanovism and the Politics of Productivity in the USSR, 1935–1941* (Cambridge: Cambridge University Press, 1988).

30. Gregory and Lazarev, eds., *The Economics of Forced Labor;* Oleg Khlevniuk, *The History of the Gulag from Collectivization to the Great Terror* (New Haven, CT: Yale University Press, 2004).

31. Peter Lyashchenko, *History of the National Economy of Russia to the 1917 Revolution*, trans. L. M. Herman (New York: Macmillan, 1949), 425; Anneli Aer, *Patents in Imperial Russia: A History of the Russian Institution of Invention Privileges under the Old Regime* (Helsinki: Suomalainen Tiedeakatemia, 1995), 48.

32. Owen, *The Corporation under Russian Law*; Ruth A. Roosa, *Russian Industrialists in an Era of Revolution: The Association of Industry and Trade, 1906–1917* (Armonk, NY: M. E. Sharpe, 1997).

33. Harvey D. Balzer, ed., *Russia's Missing Middle Class: Professions in Russian History* (Armonk, NY: M. E. Sharpe, 1995), Cooper and Lewis, "Research and Technology," 189–211 ; and Bailes, *Technology and Society under Lenin and Stalin.* The number of students in higher technical education increased from 7,500 in 1899 to 24,800 in 1913. These figures do not account for the many self-taught engineers, or *praktiki,* who made up around one-third of the technical intelligentsia.

34. McKay, *Pioneers for Profit*, 187–189. This seems partly to do with xenophobia. Carstensen, *American Enterprise in Foreign Markets*, is more skeptical.

35. On the postwar years, see Alexei Kojevnikov, "The Great War, the Russian Civil War, and the Invention of Big Science," *Science in Context* 15 (2002), 239–275. See also Jonathan Coopersmith, *The Electrification of Russia, 1800–1926* (Ithaca, NY: Cornell University Press, 1992).

36. Bailes, *Technology and Society under Lenin and Stalin,* 218. These figures refer to engineers with higher education employed in all branches of the national economy.

37. The reference is to a victory speech in 1945. Stalin also proclaimed that "technology decides everything." Loren R. Graham, *The Ghost of the Executed Engineer: Technology and the Fall of the Soviet Union* (Cambridge, MA: Harvard University Press, 1993).

38. Robert Lewis, "Technology and the Transformation of the Soviet Economy," in Davies, Harrison, and Wheatcroft, eds., *The Economic Transformation of the Soviet Union,* 182–197. Technical aid agreements were signed with foreign firms to assist technological transfer. See Kendall Bailes, "The American Connection: Ideology and the Transfer of American Technology to the Soviet Union, 1917–1941," *Comparative Studies in Society and History* 23 (1981), 421–448.

39. David Holloway, *Stalin and the Bomb: The Soviet Union and Atomic Energy, 1939–1956* (New Haven, CT: Yale University Press, 1994).

40. Robert Conquest, Foreword to Gregory and Lazarev, eds., *The Economics of Forced Labor*, viii; Paul Josephson, "'Projects of the Century' in Soviet History: Large-Scale Technologies from Lenin to Gorbachev," *Technology and Culture* 36:3 (1995), 519–559, 559; Shearer, *Industry, State, and Society in Stalin's Russia*, 239; James C. Scott, *Seeing Like a State: How Certain Schemes to Improve the Human Condition Have Failed* (New Haven, CT: Yale University Press, 1998). Scott reaches conclusions similar to those of Pal'chinskii.

41. Portal, "The Industrialization of Russia," 836; Marks, *The Road to Power.*

42. John McCannon, *Red Arctic: Polar Exploration and the Myth of the North in the Soviet Union, 1932–1939* (New York: Oxford University Press, 1998).

43. Anne Rassweiler, *The Generation of Power: The History of Dneprostroi* (New York: Oxford University Press, 1988); Stephen Kotkin, *Magnetic Mountain: Stalinism as a Civilization* (Berkeley: University of California Press, 1995). High rates of labor turnover also applied to technical personnel.

44. Wieland Hintzsche and Thomas Nickol, eds., *Die Grosse Nordische Expedition: Georg Wilhelm Steller, Ein Lutheraner erforscht Sibirien und Alaska* (Gotha: Perthes, 1996).

45. Katerina Clark, "Engineers of Human Souls in an Age of Industrialization, 1929–1941," in *Social Dimensions of Soviet Industrialization*, ed. William G. Rosenberg and Lewis H. Siegelbaum (Bloomington: Indiana University Press, 1993), 248–264.

46. Mark D. Steinberg, *Proletarian Imagination: Self, Modernity, and the Sacred in Russia, 1910–1925* (Ithaca, NY: Cornell University Press, 2002), 189–197.

47. Wendy Goldman, *Women, the State and Revolution: Soviet Family Policy and Social Life, 1917–1936* (Cambridge: Cambridge University Press, 1993); Richard Stites, *Revolutionary Dreams: Utopian Vision and Experimental Life in the Russian Revolution* (New York: Oxford University Press, 1989), 113–114.

48. Engerman, *Modernization from the Other Shore*, 5.

49. See Ivan T. Berend, *Central and Eastern Europe, 1944–1993: Detour from the Periphery to the Periphery* (Cambridge: Cambridge University Press, 1996), for variations in the development level of societies in East-Central Europe.

12

Financing Brazil's Industrialization

Anne G. Hanley

Brazil, it can reasonably be argued, did not experience an industrial revolution. It is more proper to say that Brazil began to industrialize in the late nineteenth century in a process similar to Britain's First Industrial Revolution, but in the context of Europe's Second Industrial Revolution. While no national accounting figures assess values for the agricultural, industrial, and service sectors as a share of national income before 1939, qualitative evidence suggests that industry's share in the nineteenth-century Brazilian economy was small, highly concentrated, and characterized by the production of simple consumer goods. Brazil's most important industrial sector, cotton textiles, counted fewer than fifty factories in 1885, produced just 26 million meters of rough cloth or about 10 percent of national consumption, and was concentrated in the agriculturally prosperous southeast.[1] Urban centers housed less than 10 percent of the Brazilian population as late as 1920, keeping the market for industrial products small. It was not until the 1930s that the Brazilian state sought to stimulate industrial development directly, and it was not until the 1950s that the industrial sector moved beyond consumer nondurables to heavy industry.[2] The industrialization that scholars speak of when discussing Brazil, then, was exceedingly modest compared with the industrial revolutions of the leading industrial nations.

Still, Brazil's industrialization holds great interest for scholars of Latin American economic history because it represented one of the few instances in which a primary export product boom generated the successful and lasting transformation of an agricultural economy into an industrial economy. Most Latin American countries had some primary product that the world desperately wanted between 1870 and 1930. Rare, however, was the nation that parlayed its export fortunes into structural transformation and lasting economic development. Brazil stands apart because its modest industrial beginnings laid the foundation for a broad-based, dynamic, and uninterrupted process of industrialization that generated modern Latin America's most sophisticated industrial base.

This chapter discusses the contours of early Brazilian industrialization in the context of its most important determinant: financial capital. It argues that Brazilian industrialization was initially limited by and ultimately shaped by the nature of financial institutions available to its industrial entrepreneurs. Although the entrepreneurial impulse was fairly robust in the nineteenth century, the circumscribed development of the banking system and the periodic, critical limitations placed on the sale of equity during the Empire (1822–1889) resulted in a small industrial sector composed mainly of single proprietorships. The weight of the agricultural sector in Brazilian economic fortunes meant that industry-stimulating policies were most often an incidental effect of policies aimed at correcting some ill of the external sector.[3] It was innovation in corporate law after the fall of the Empire and the declaration of the Republic (1889) that opened up capital markets to industrialists and allowed Brazil's hallmark large-scale industrialization to develop.

Perhaps the best-known feature of Brazilian industrialization is that it occurred in the context of governmental neglect, at best, and government hostility, at worst. Since colonial times, Brazil's administrators more often limited domestic industry than stimulated it. Official Portuguese policy in the colonial period prohibited development of manufactures in its largest colony, a policy typical of the mercantilist era. Some modest colonial manufactures existed to satisfy demand in the domestic market, particularly at the poorer levels. These "mechanical trades" were practiced out of necessity, but did not constitute anything like formal industrialization.[4] Restrictions on colonial manufactures were lifted when Napoleon's invasion of Portugal forced the Bragança Court to flee to Brazil in 1808, but this new economic openness was coupled with treaties that granted England preferential access to the Brazilian market. Brazilians were finally free to form industrial ventures, but they faced direct competition from the world's leading industrial nation. By the 1830s, this new British competition had wiped out most handicraft and manufacturing production except in particularly isolated regions.[5]

The lapse of the preferential tariff agreement in 1844 marked the earliest moment when cotton textile manufactures, Brazil's predominant industrial sector, began to emerge. Its sustained development, however, came only in the 1860s when modestly favorable tariffs and the U.S. Civil War stimulated Brazilian cotton growing that provided raw material inputs to domestic textile manufactures.[6] Even then the Brazilian government was deeply ambivalent about supporting domestic economic development. Most of the government's revenues came from taxes on imports, so its policymakers did not favor domestic industries that might compete with revenue-generating imports. For the bulk of the nineteenth century, fiscal and monetary policies favored the export sector and rarely supported industrial development.[7]

Colonial and early national policies certainly hurt industrialization, but these policies probably had less to do with Brazil's limited industrialization than did the small size of the effective domestic market.[8] The market was circumscribed by at least three important problems. The first was the overwhelming predominance of slaves in Brazilian society. By definition unremunerated, by law unable to own property, one-half to three-quarters of Brazil's inhabitants were situated outside the consumer economy in the late eighteenth and early nineteenth centuries. The second major problem was the widespread poverty of the free population. Brazilian sugar production and its extractive industries created a colonial elite class that by all accounts was wealthy, but the majority of the free population existed along the margins or completely outside this elite world. The third major problem was inadequate transportation networks. Inland transport relied on mule trains, a slow and relatively inefficient and expensive means of transporting large volumes of export goods.[9] The possibilities for an integrated domestic market, then, were almost nonexistent before late-nineteenth-century attempts to improve transportation networks. Regional markets functioned on a limited scale, and most production was local and artisanal.

It took a considerable generation of wealth and attendant prosperity to rupture these conditions and to create new ones amenable to industrialization. This occurred thanks to the coffee boom. Coffee had arrived in Brazil in the eighteenth century, but was produced on a large scale for export only in the nineteenth. It surpassed sugar as Brazil's principal export in 1831 and expanded in production thereafter, fueled by seemingly insatiable European and American demand.[10] By the 1880s, extensive plantations in several provinces—most importantly in the adjacent provinces of Minas Gerais, Rio de Janeiro, and São Paulo—produced well over half of the world's coffee.[11]

This coffee boom affected every one of the major obstacles to industrialization. The dramatically increased scale of agricultural production took place in the context of a rapidly aging and shrinking slave population, given the twenty-year gap between the last slave imports (1850) and the coffee boom, making continued reliance on slave labor impossible. Using subsidized transatlantic passages, planters attracted more than 1 million immigrants to Brazil in the 1880s and 1890s to supplement and then replace slaves.[12] These wage laborers provided domestic producers with a large-scale consumer class. At the same time, planters attacked the transportation problem by financing railroad development. Government profit guarantees and private capital extended railways into the Brazilian hinterlands, connecting coffee plantations to the ports and transporting immigrants from port to plantation.[13] The coffee boom, then, increased the effective size of the market in terms of both the numbers of consumers and the geographic mobility of people, goods, and services.[14]

Coffee production created demand for goods easily produced by domestic industry. Many of these goods directly served the coffee sector, like rough cotton cloth to clothe slaves, cotton and jute sacks to bag coffee, machinery and metalworking for agricultural tools and processing machinery, and rudimentary chemical factories to produce fertilizer and insecticide destined for the coffee sector.[15] But as much or more went to satisfy the demand of wage earners in the countryside and in growing urban areas. Wheat and sugar mills, breweries and bottle makers, match factories, soap and candle manufacturers all responded to consumer demand driven by expansion in the export economy and the attendant rise in population and wealth.[16] This industrialization occurred in all three regions where coffee was grown. These same three regions, São Paulo in particular, have dominated Brazilian industrial production ever since.[17]

While the link between the export boom and industrialization is widely accepted, the institutions that allowed for this transformation of agricultural wealth into industrial development have been little studied. The major studies of Brazilian industrialization take for granted that the wealth flowed naturally, effortlessly, or inevitably to entrepreneurs. My research into the institutions that facilitated this transformation, however, clearly shows that it was neither inevitable nor effortless. I find that Brazilian industrialization was conditioned by the financial institutions at the disposal of industrialists.[18] Changes in demand conditions made industrialization attractive to entrepreneurs, but limited capital market development before 1890 kept industrial establishments small. It was institutional change after 1890, especially the use of the limited-liability joint stock corporation for business formation, that allowed the large-scale industrialization that is the hallmark of Brazil's modernization.

The most important institutional arrangements that affected business finance in the nineteenth century were laws crafted at midcentury and revised in 1890 that permitted the formation of joint-stock corporations and created the mechanisms to trade their shares. Prior to these regulatory laws, liquidity in the Brazilian economy was limited. Before the advent of a land law in 1850 that required monetary purchase, land was acquired through grants, inheritance, or squatting. Productive inputs in the agricultural sector were paid for out of export receipts. Wages were hardly a pressing concern, as fully half of Brazil's population at independence were slaves, and much of the other half worked in the countryside as tenant laborers or sharecroppers.[19] The small monetary economy that did exist in Brazil was restricted to its few urban centers.

In this environment, most Brazilian business finance rested in ties between individuals.[20] This reliance on personal financiers, which kept businesses small, persisted to the end of the Empire in 1889 in spite of legislation that permitted large-scale capital mobilization through the joint-stock format. This format, in which companies issue stock to investors, was authorized as part of a major overhaul of commercial legislation undertaken at midcentury to keep up with the quickening pace of international trade. The overhaul included laws that regulated the joint-stock companies and laws that regulated the brokers who traded equity and debt.[21] The ability to sell shares in corporations and the creation of licensed brokers to trade those shares opened up the possibility of large-scale business formation for the first time in Brazil.

In spite of the laws' potential, the central provision of the 1849 legislation permitting joint-stock companies rendered it almost unworkable: incorporation took an act of Congress. The charter requirement slowed business formation with bureaucratic steps that potentially introduced patronage into the equation.[22] An 1859 law that reinforced the charter requirement compelled joint-stock companies to produce weekly financial statements for the government's review, with severe sanctions for those caught out of compliance. An 1860 law, popularly known as the Law of Impediments, reasserted the government's authority over the formation and oversight of joint-stock companies.[23]

The imperial government eased up on domestic business in 1882 when it eliminated the requirement of a government charter for most types of companies, presumably to stimulate business development.[24] While this should have sparked business formation, the new law came with a string attached that made it unpalatable to all but the most capital-hungry businesses: unlimited liability. Shareholders were responsible for the full value of their stock for five years after purchase even if they had sold it.[25] If a firm was discovered to have fraudulently distributed dividends to shareholders, even without the knowledge of investors, shareholders were responsible for repayment of those dividends for five years from the date of the distribution. A fraud that went undetected could come back to haunt an investor who no longer owned stock in the company. Under the provisions of the 1882 law, then, it became far easier for the entrepreneur to form a company but far riskier for the investor to invest. Over and over across three decades, business regulatory legislation reflected a government that lurched between some desire to promote entrepreneurship and a seemingly greater desire to squelch it.

Because of the restrictive nature of business regulatory legislation, first requiring government charters and then introducing investor liability, few companies used the joint-stock format. The ones that did tended to be ventures whose capital requirements exceeded the kin or community resources on which most companies relied, ventures that enjoyed some government guarantee, and ventures whose practices were conservative and whose shares were closely held. These were typically banks, railroads, and utilities.[26] Through these firms, the joint-stock format had indirect benefits for industrial formation. There is no evidence that banks lent money directly to industrial firms, but they provided liquidity in the form of short-term loans and discounting services that improved the circulation of goods and services.[27] Transportation networks facilitated the importation of machinery and inputs destined for the industrial sector, thereby significantly reducing their cost. Utility companies provided power generation to factories.

The joint-stock format had no direct benefit for industrialization before 1890 because industrialists could not attract investors. Of about six hundred joint-stock companies in operation in Rio de Janeiro at some point from 1851 to 1865, just 1.5 percent were industrial firms. The amount of capital this represented was so small—less than 1 percent—as to be insignificant.[28] Minas Gerais had some joint-stock textile companies and foundries, all of them closely held, but the majority appears to have been private ventures.[29] None of the dozens of industrial firms that formed in São Paulo during the 1880s was a joint-stock company.[30]

This meant that virtually all Brazilian industrial development under the Empire was funded out of personal connections (kin and merchant finance) and retained earnings. This in turn meant that the scale of early Brazilian industrialization was small. Cotton textiles, the most important industrial sector in Brazil, were produced primarily in "one-man enterprises," kept small because the "shortage of capital in the early mills was endemic to the industry" for much of the nineteenth century.[31] The iron industry of the province of Minas Gerais, stimulated by the Paraguayan War (1865–1870) and the construction of Brazil's first railways in the

1860s, was composed of "a large number of very small firms," most of which operated a single foundry and employed fewer than twenty people.[32] Minas Gerais's textile and electricity-generating companies in the nineteenth century were similarly small by both national and international standards.[33]

The unattractiveness of industrial ventures to potential investors had to be related to the instability of their existence. The unwillingness of the Brazilian government to offer stimulus or protection to domestic industrialists left them vulnerable to major economic swings. Periods of investment in expansionary eras were followed by harsh recessions that put young businesses under. This pattern of expansion and extinction, which occurred over and over in the second half of the nineteenth century, was fed by government policy: it almost certainly dampened potential investor interest in industrial ventures and reinforced the reliance on personal sources of finance, especially with the introduction of unlimited liability in 1882.[34]

If the coffee boom broke the demand constraints on domestic industrialization, the political coup of late 1889 swept aside the institutional constraints. This coup replaced the centralized, conservative empire with a federalist, republican regime that favored domestic economic diversification and business development. Its first finance minister introduced new business regulatory legislation in January 1890 to give impetus to this priority. The central component of the 1890 reform was the introduction of limited liability of shareholders. In a stunning reversal, investors were now absolved of all responsibility for the value of their shares by simply voting to approve the company's financial statements at the annual stockholders' meetings.[35]

The response to this innovation was immediate and dramatic. In the six months after the January 1890 reform, at least 222 joint-stock companies and banks were founded in São Paulo, compared with 30 in operation in late 1887.[36] This phenomenal growth pales in comparison with the even greater growth in joint-stock corporations in Rio de Janeiro. Always the larger of the two stock markets, the Rio exchange listed 717 companies in 1891 compared to just 58 in 1889.[37] Nor were these new companies the closely held corporations of the empire. The reform of January 1890, facilitated by the Rio de Janeiro Bolsa and the São Paulo broker's association, drew thousands of investors into the market for the first time, creating so much demand for company stocks that the newspapers were filled with advertisements by anxious buyers seeking anyone who was willing to part with shares.[38]

The 1890 reforms that significantly eased business formation and eliminated shareholder liability produced a two-year speculative bubble, known as the Encilhamento, that posted fantastic numbers of new business formations and equally spectacular failures.[39] The combination of limited liability and easy access to credit in a time of expanding money supply, thanks to the 1888 bank reform that decentralized bank note issues, promoted widespread speculation along with real business formation. The government scrambled to introduce legislative modifications to curb the excesses and put new companies on more solid financial footing but could not prevent the crash.[40] The first São Paulo stock exchange, founded in 1890 to handle the trading volume generated by the spike in company formation, was a casualty of the bust. The 1888 bank reform bloated the money supply, which fed inflation and escalated foreign debt payments. A crisis of overproduction of coffee undermined its international price and threw Brazil into recession by mid-decade, compounding economic instability.[41]

Although the causes of the 1890s instability were multiple, blame was laid squarely on the Encilhamento. Policymakers after 1895 consciously sought to reorient the economy away from domestic development to Brazil's true calling: commodity exporting. Domestic producers were punished in a series of tariff policies seeking to eliminate "artificial" industry, meaning any venture reliant on imported inputs for production. In addition, the government imposed stamp taxes on most consumer products. Theoretically passed on to the consumer, these taxes raised costs to the producer because the recession had dampened demand, leaving the market unable to bear increased prices. Finally, in the late 1890s, the government induced a sharp deflation to curb the money supply, a condition of the 1898 Funding Loan from the Rothschilds destined to help Brazil meet its foreign debt obligations. In terms of sustained industrial development, the 1890s did not have much going for them.

In spite of this instability and uncertainty, the simplified incorporation procedures and the preservation of limited liability turned out to be an important catalyst for long-term Brazilian industrialization. They produced immediate if modest gains in the industrial sector during the tumultuous 1890s and positively transformed industrialization through sustained growth and diversification after Brazil's return to economic health around 1907. In the case of the cotton textile industry, the most important of Brazil's industrial sectors,[42] Albert Fishlow's study of import substitution industrialization in Brazil at the turn of the century points to a ten-fold increase in Brazilian cotton textile production from a paltry

26-million meter output in 1885, or 10 percent of domestic consumption, to an estimated domestic output of 235 million meters in 1905, representing 60 percent of domestic consumption of cotton cloth in that year.[43] Fishlow dates this expansion of cotton textile production to the boom period of the Encilhamento. New cotton textile businesses founded in the first year of the Encilhamento accounted for more than 47 million mil-réis of new capital in the industry, a figure more than double the existing total capitalization of cotton textile firms. For the joint-stock companies alone, investment in publicly owned textile firms grew from 13 million mil-réis in 1889 to 84 million mil-réis in 1892.[44] More critically, his data show that substantial numbers of firms dating from the early 1890s survived the bust and comprised an important percentage of cotton textile firms in the years just before World War I.

According to Stanley Stein, the changes in the regulations that had restricted the capital markets caused cotton textile firms to morph from single proprietorships to partnerships to joint-stock companies after 1890.[45] My research on São Paulo finance and industrialization finds that this pattern held true for textile firms in this regional market. Almost all growth in an industry dominated by single proprietorships occurred with the joint-stock format, especially after the turn of the century. Although we know of at least eighteen textile companies in existence in São Paulo before 1905, just one was quoted on the São Paulo Bolsa: it was worth just under 7 percent of all capital invested in São Paulo's cotton textile firms.[46] By 1915, just over half of São Paulo's forty-one textile companies were financed through stock issues. This means that twenty-one of the twenty-three cotton textile firms formed after 1905 used the joint-stock format to incorporate. Many of these started out as single proprietorships or limited partnerships in the 1880s and 1890s, taking advantage of the joint-stock format to fund their expansion in one year and returning to the market to issue bonded debt in the following year to further their growth. Virtually all of the subsequent growth in São Paulo's textile industries came through large joint-stock companies, propelling output from around 40 million meters per year in 1905–1907 to 121 million meters in 1915.[47] It is clear that the joint-stock format for business finance played a critical role in the expansion of the textile industry in São Paulo (See table 12.1).

Not only did the joint-stock format dominate industrial expansion, but it changed the size of Brazilian industry from small companies to large. Stein found that almost no cotton textile mills founded before 1880 were capitalized at more than 1,000 contos of réis, a cut-off mark above which

Table 12.1
Equity Finance in São Paulo's Textile Industry: Public versus Private Finance (Nominal mil-réis)

	Number of textile firms listed on Bolsa[a]	Total textile firms in São Paulo[b]	Capital of joint-stock firms[a]	Total textile capital[b]	Joint-stock K as percentage total K[b]
1905	1	18	2 million	29.6 million	6.8 percent
1910	12	24	19.5 million	46.7 million	41.6 percent
1915	22	41	46 million	81.5 million	56.3 percent

Note: Joint-stock K refers to capital.
a. Data compiled by the author from *O Estado de São Paulo Bolsa* summary page. January 1906, 1911, 1916.
b. Data from *Cano, Raízes da concentração industrial em São Paulo,* 292, Table 55.

industrial establishments were considered to be very large.[48] In 1907, however, textile mills accounted for half of all the very large industrial establishments in Brazil. These very large textile firms, worth more than 1,000 contos, or approximately U.S.$350,000, produced more than three-quarters of all textile production by value.[49] The expansion of the industry, then, was dominated by the rise of the very large industrial firm. This was made possible as textile entrepreneurs took advantage of the favorable relationship between domestic inflation, which put money in their pockets, the easy access to capital in the early 1890s, and the relatively slowly falling exchange rate to import machinery to initiate or expand production. Further decline of the exchange rate made imports expensive, thereby offering a competitive advantage to domestic producers.[50] Access to investor capital, initially through the sale of equity and then through expansionary bond finance, paid the bills.[51]

Aside from the cotton textile industry, concentrated in the southeast but reasonably widespread throughout Brazil, most industrialization of the late nineteenth century took place in São Paulo. Therefore, it is instructive to focus closely on the case of São Paulo to examine industrial finance and firm size outside the textile industry.

Located in the southeastern region of Brazil, just south of the state of Rio de Janeiro, São Paulo was emblematic of the link between agricultural wealth and industrial development in Brazil.[52] Plenty of ink has spilled to argue that São Paulo's experience should not be confounded with Brazil's experience, a valid point meant to encourage further exploration of regional cases, but the statistics from the 1920 industrial census are clear: in the thirty years following the 1890 business regulatory reform, São Paulo grew from having an insignificant industrial base to become Brazil's industrial leader.[53] Moreover, this trajectory strengthened over time. São Paulo today produces half of Brazil's total manufactures.[54] Brazilian industrialization at the turn of the twentieth century may have been modest on an international comparative scale, but São Paulo's rapid ascent was genuinely impressive.

Like industrialists throughout Brazil, before 1890, São Paulo's entrepreneurs formed small companies filling demand for rough consumer nondurable products and experimented with the joint-stock format after 1890. São Paulo business formation experienced a boom during the Encilhamento years and a bust in the years that followed. Some firms failed while others survived, but all were hurt by the recessionary policies of the government in the late 1890s. Survivors generally prospered after the crisis of overproduction in the coffee economy was remedied in 1907.[55] The turning point in São Paulo's industrial history, however, came with the introduction of limited liability in 1890. As we saw in the case of cotton textiles, when the government eased the regulations surrounding joint-stock ventures, industrialists found it easy to attract investors to new equity issues or to expand the capitalization of existing corporations. The result was a dramatic increase in the number of joint-stock companies engaged in all kinds of businesses that served urban markets, but especially industrial ventures.

Every one of Brazil's prominent industrial sectors used the joint-stock format after 1890, compared to almost none before the reform.[56] Among wheat mills, sugar refineries, breweries, and metalworks, as well as the producers of textiles, hats, shoes, and matches, only the grain mills had appeared among São Paulo's joint-stock companies in the 1880s. During the 1890s, however, at least thirty-five industrial businesses were formed as joint-stock companies and traded on the São Paulo Bolsa, representing such industries as machinery and metalworking, furniture making, textile manufacture, hat and shoe production, food processing, beer brewing, paper production, printing, rudimentary chemical production, and other manufacturing activities.[57] Every food processing firm I identified in these sectors as operating in São Paulo during the period 1890 to 1905 was traded on the Bolsa. The machinery and metalworking industries were equally well represented. Five of the six machinery and metalworking firms were organized as joint-stock companies; two of the five were the most important metalworking firms in São Paulo at the time.

Among the other manufacturing concerns organized in this period and financed by stock issues were two of three beer companies, a furniture factory, a hat producer, and the largest paper manufacturer in the country.

Not only was equity finance historically significant for funding every type of industrial venture, it funded the very large enterprises that established São Paulo's industrial leadership. Listings on the São Paulo Bolsa verify that joint-stock industrial firms were substantially larger than privately financed firms. The fourteen industrial companies listed on the São Paulo Bolsa in 1907 made up a minuscule percentage of the more than 1,000 industrial firms in São Paulo yet comprised more than 20 percent of all capital invested in São Paulo industry by 1907. By 1919, the publicly traded industrial companies still made up only 3.1 percent of all industrial companies but now represented 52 percent of all industrial capital invested in São Paulo.[58] (See table 12.2.)

That just 3 percent of industrial firms accounted for over half of industrial capital invested in São Paulo by 1919 gets to the heart of the significance of financial capital in industrial formation. The ability to tap a large pool of investors with the joint-stock format funded large industrial companies beyond the scope of traditional financing avenues. Published data on the size distribution of Brazilian firms is scarce, but one good source shows that Brazil was a nation of very small industrial firms as late as 1912, fifty years into the industrialization process. The average value of all Brazilian industrial firms in 1912 was 51 contos (about U.S.$16,000), while the median was somewhere around 1 conto (U.S.$320).[59] Publicly financed and traded companies, by contrast, were huge. São Paulo's sixty-four joint-stock companies that year averaged 1,324 contos of capital (U.S.$424,000) while the median was around 1,000 contos (U.S.$320,000). Clearly the joint-stock format played an important role in bringing to life the medium and large firms so important to São Paulo's emerging industrial base. This was possible because the central reform of 1890, the introduction of limited shareholder liability, was left untouched.

Through World War I the fortunes of the industrial sector were strongly influenced by the fortunes of the export sector, a fact of life that caused industries to stumble during the war years, but by the 1920s, industry in the southeast region had already become self-sustaining. Industrialization spread to other regions of the country by the 1930s. The growth in population and in urbanization, fed partly by continued waves of immigrants who increasingly headed directly to the cities, ensured that demand for industrial products grew. During the Great Depression, indus-

Table 12.2
Industrial Firms Listed on São Paulo Bolsa, 1907 (Nominal mil-réis)

Company (by size)	Business	Capitalization
Antârctica Paulista	Brewery	8,500,000
Mecânica e Importadora	Machinery, metalworking	5,000,000
Melhoramentos de São Paulo	Paper, ceramics, lime	3,000,000
Fabril Paulistana	Textiles	2,000,000
Industrial de São Paulo	Textiles, matches, paper	2,000,000
Moinho Santista	Sugar refinery	2,000,000
Vidraria Santa Marina	Glass (supplier to Antârctica)	1,000,000
Mac Hardy	Machinery, metalworking	978,750
Fábrica de Cimento Ítalo-Brasileiro	Cement	800,000
Tecelagem Santista	Textiles	800,000
Paulista Manufatureira de Explosivos	Chemicals	400,000
Refinadora Paulista	Sugar refinery	300,000
Tecelagem se Seda Ítalo-Brasileira	Textiles	300,000
Industrial de Kiosques	Not identified	100,000
Value of joint-stock industrial firms		27,178,750
Cano's value of all São Paulo industry in 1907		131,900,000
Joint stock capital share		20.6 percent

Sources: O Estado de São Paulo, January 23, 1907, and Cano, *Raízes da concentração industrial em São Paulo,* 163.

trial development became permanently delinked from the sagging fortunes of the export sector. By the 1950s, the Brazilian government's commitment to industrialization was unquestioned.

Compared to the industrial gains of the Great Depression era and later, Brazil's early industrialization was modest in scale and in ambition. But this rather modest industrialization demonstrates the power of institutional change in economic development. Government priorities that shape policy have a direct and profound impact on the possibilities for economic diversification. The ambivalence, and even hostility, of the Brazilian government to domestic business in general and domestic industry in particular during the empire meant that entrepreneurs worked at the margins of the economy. Demand conditions conducive to industrialization improved considerably in the last decades of the nineteenth century, but they were a necessary rather than a sufficient condition to stimulate industrial investment. The rather tenacious industrializing efforts of some entrepreneurs who saw a market opportunity were circumscribed by the availability of capital. Capital market institutions, themselves a product of business regulatory law, simply did not support long-term investment or large-scale capital mobilization before 1890.

The interest the new republican government had in altering Brazil's economic development by consciously promoting domestic business development generated new institutional arrangements that broke down the capital barrier. The introduction of limited liability for shareholders took away much of the risk of investing in a company and opened up a seemingly limitless pool of investment capital to domestic entrepreneurs. This initial euphoria did not last, but the institutional arrangement did. As macroeconomic conditions stabilized and then improved, the limited liability joint-stock corporation transformed Brazil's industrial profile. Very large companies were well funded through the equity and bond markets. Many companies formed in this era through this format lasted for decades, and some remain in business today. This 1890 innovation in business finance was the central element that transformed Brazil into Latin America's leading industrialized nation.

Notes

1. Albert Fishlow, "Origins and Consequences of Import Substitution in Brazil," in *International Economics and Development: Essays in Honor of Raul Prebisch*, ed. Luis Eugenio di Marco (New York: Academic Press, 1972), 312–313.

2. Amaury Patrick Gremaud, Flávio Azevedo Marques de Saes, and Rudinei Toneto Júnior, *Formação Econômica do Brasil* (São Paulo: Editora Atlas, 1996), 118, 135–157.

3. State-led industrial planning was absent from the Brazilian economy before the 1930s and did not take the shape of coordinated policy until the 1950s. Wilson Suzigan and Anibal Villela, *Industrial Policy in Brazil* (Campinas, São Paulo: Instituto de Economia-Unicamp, 1997), 32.

4. The largest of these, textile manufacturing, was outlawed in 1785 because Portugal feared that Brazilian textile manufactures created competition for its own limited textile industry and that economic independence through industrialization might lead to political independence. See Caio Prado, Jr., *The Colonial Background of Modern Brazil* (Berkeley: University of California Press, 1969), 256, 261–262.

5. Stanley Stein, *The Brazilian Cotton Textile Manufacture: Textile Enterprise in an Underdeveloped Area, 1850–1950* (Cambridge, MA: Harvard University Press, 1957), 4.

6. Wilson Suzigan, *Indústria Brasileira: Origem e Desenvolvimento* (São Paulo: Editora Brasiliense, 1986), 123; Stein, *The Brazilian Cotton Manufacture,* 15–17.

7. On Brazilian economic policy in the nineteenth century, see Anibal Villela and Wilson Suzigan, *Política do governo e crescimento da economia brasileira, 1889–1945* (Rio de Janeiro: IPEA/INPES, 1973), and Carlos Manuel Pelaez and Wilson Suzigan, *História monetária do Brazil: análise da política, comportamento e instituições monetárias* (Rio de Janeiro: IPEA/INPES, 1976).

8. This argument was a central point in the pioneering works of Furtado and Prado. Nathaniel Leff points out the failure of Brazilian income to grow over the course of the nineteenth century, as well as the wide disparities in income among Brazil's regions. Celso Furtado, *The Economic Growth of Brazil: A Survey from Colonial to Modern Times* (Berkeley: University of California Press, 1968); Prado, *Colonial Background*; Leff, "Economic Development and Regional Inequality: Origins of the Brazilian Case," *Quarterly Journal of Economics* 86: 2 (May 1972), 243–262.

9. Nathaniel Leff, "Economic Development in Brazil, 1822–1913," in *How Latin America Fell Behind: Essays on the Economic Histories of Brazil and Mexico, 1800–1914*, ed. Stephen Haber (Stanford, CA: Stanford University Press, 1997), 42–46.

10. Sérgio Buarque de Holanda and Pedro Moacyr Campos, *História Geral da Civilização Brasileira*, vol. 6 (São Paulo: Difusão Européia do Livro, 1971), 119; Sérgio Silva, *Expansão caféeira e origens da indústria no Brasil* (São Paulo: Editora Alfa Omega, 1978), 49.

11. Joao F. Normano, *Brazil: A Study of Economic Types* (Chapel Hill: University of North Carolina Press, 1935), 40.

12. Gremaud, Saes, and Toneto, *Formação Econômica*, 46–47. These immigrants swelled Brazil's population from under 12 million in 1880 to more than 17 million in 1900. Nathaniel Leff, *Underdevelopment and Development*, vol. 1 (London: Allen & Unwin, 1982), 241.

13. Brazil's railway network grew from 744 kilometers of track and just under 300,000 tons carried in 1870 to almost 10,000 kilometers of track and 2 million tons carried in 1890. William Summerhill, *Order Against Progress: Government, Foreign Investment, and Railroads in Brazil, 1854–1913* (Stanford, CA: Stanford University Press, 2003), tables 4.2 and 4.3, 66–69.

14. The new prosperity that created the conditions supportive of domestic industrialization was impressive but was limited to the southeastern region of Brazil. Income per capita in the sugar northeast lagged significantly behind income in the coffee-growing southeast. As a result, Brazilian industrialization was regionally concentrated. Leff, "Economic Development and Regional Inequality."

15. Suzigan, *Indústria Brasileira*, 117–118

16. These industries—textiles and foodstuffs—accounted for the majority of industrial production until the 1940s. Gremaud, Saes, and Toneto, *Formação Econômica*, 138.

17. These three states today produce 80 percent of Brazil's manufactured goods. São Paulo alone produces half of all Brazilian manufactures. Marshall C. Eakin, *Brazil: The Once and Future Country* (New York: St. Martin's Griffin, 1998), 75, 82.

18. Anne G. Hanley, *Native Capital: Financial Institutions and Economic Development in São Paulo, Brazil, 1850–1920* (Stanford, CA: Stanford University Press, 2005).

19. In 1817–1818, four years before Brazil declared independence from Portugal, the slave population was 1.9 million out of a total population of 3.8 million. Robert Conrad, *The Destruction of Brazilian Slavery, 1850–1888* (Berkeley: University of California Press, 1972), 283.

20. Sérgio de Oliveira Birchal, *Entrepreneurship in Nineteenth-Century Brazil: The Formation of a Business Environment* (New York: St. Martin's Press, 1999); Joseph Sweigart, "Financing and Marketing Brazilian Export Agriculture" (Ph.D. diss., University of Texas at Austin, 1980); Eul-Soo Pang, *In Pursuit of Honor and Power: Noblemen of the Southern Cross in Nineteenth-Century Brazil* (Tuscaloosa: University of Alabama Press, 1988); Darrell E. Levi, *The Prados of São Paulo, Brazil: An Elite Family and Social Change, 1840–1930* (Athens: University of Georgia Press, 1987).

21. Brazil, *Leis e Decretos*, Decree 575, January 10, 1849, and Law 556, June 25, 1850.

22. Entrepreneurs filed a business proposal to the provincial president. The provincial president considered whether the firm had a good chance of succeeding, especially whether the founders were well known and had the means to pay their capital installments on time. The provincial president forwarded the petitions to the Congress, where they were reviewed and voted on in the Chamber of Deputies. *Leis e Decretos*. Decree 575, January 10, 1849.

23. *Leis e Decretos*. Law 1083, August 22, 1860.

24. *Leis e Decretos*. Law 3150, November 4, 1882.

25. When corporations formed, they typically required investors to pay in the minimum capital installment of 10 percent. Additional capital calls, usually 10 percent at a time, were made periodically. According to the 1882 law, shares could be traded when 20 percent of their value was paid in. An investor's liability ended only when the full value of the share was paid. If an investor traded away a company stock with just 20 percent paid in and the subsequent holder did not pay in the balance, both would be liable for that 80 percent if the company went bankrupt. This liability expired five years after the date of investment.

26. For detailed discussion on business finance under the empire, see Hanley, *Native Capital*, chap. 3.

27. Stephen Haber, "Financial Markets and Industrial Development: A Comparative Study of Governmental Regulation, Financial Innovation, and Industrial Structure in Brazil and Mexico, 1840–1930," in Haber, ed., *How Latin America Fell Behind*, 151; Hanley, *Native Capital*, chap. 3.

28. Maria Bárbara Levy, *A Indústria do Rio de Janeiro Através das suas Sociedades Anônimas: Esboços de história empresarial* (Rio de Janeiro: Editora UFRJ, 1994), 55. Levy's data on industrial establishments in Rio jump from 1865 to the late 1880s, making it impossible to discuss the broad contours of Rio's industrialization in that interval. Stephen Haber finds only one manufacturing company listed on the Rio

stock exchange between 1850 and 1885. This is at odds with Levy's figure. Levy's assessment of the use of the joint-stock format for industrial finance should be seen as an upper-bound estimate. Haber, "Financial Markets," 151.

29. Gleaned from Birchal's discussion of entrepreneurs in *Entrepreneurship in Nineteenth-Century Brazil,* chap. 2.

30. Hanley, *Native Capital,* 72. Most of what we know about Brazilian industry in the nineteenth century comes from industrial censuses taken in 1907 and 1920. Most of the data historians use to examine and evaluate industrialization are drawn from the years 1907 and 1919. This census also captured some snapshot data for the year 1912.

31. Stein, *The Brazilian Cotton Textile Manufacture,* 25.

32. Birchal, *Entrepreneurship in Nineteenth-Century Brazil,* 70. Data on number of firms, output, and employment can be found in the pages following this quote, particularly 71–79.

33. Ibid., 84–127.

34. Suzigan, *Indústria Brasileira,* 116–121.

35. *Leis e Decretos.* Decree 164, January 17, 1890.

36. The number of business formations in 1890 is from BOVESPA, *Uma história centenária/A Centennial History* (São Paulo: BOVESPA, 1990), 15–16. The number of joint-stock companies operating in São Paulo in late 1887 is from *Correio Paulistano,* November 5, 1887.

37. For 1891, see Gail Triner, "Banks and Brazilian Economic Development: 1906–1930" (Ph.D. diss., Columbia University, 1994), 166, table 6.1. For 1889, see Levy, *História da Bolsa,* 107–108, table 7.

38. Maria Bárbara Levy, "O Encilhamento," in *Economia Brasileira: Uma Visão Histórica,* ed. Paulo Neuhaus (Rio de Janeiro: Editora Campus LTDA, 1980), 192; Hanley, *Native Capital,* 67.

39. The term *Encilhamento* comes from horse racing. Encilhamento means "saddling up" and connotes the frenzied energy in the starting gates just as the race is about to begin. While São Paulo's Encilhamento dates to the legislative reform of January 1890, Rio's boom dates to late-Empire banking reforms. These 1888 reforms eased credit and introduced liquidity into the economy, producing substantial profits for Rio banks and their shareholders. This banking boom spilled over to all stock market transactions, sparking new business formations. On the Encilhamento in Rio, see Levy, "O Encilhamento," and Stephen Topik, "Brazil's Bourgeois Revolution?" in *The Americas* 48:2 (1991), 245–271.

40. *Leis e Decretos.* Decree 850, October 13, 1890. The new decree raised the minimum level of paid-in capital necessary for operations from 10 to 30 percent and the level of paid-in capital necessary for trading to be allowed from 20 to 40 percent.

41. Political instability in the 1890s was as bad, if not worse. Over the course of the decade, the new republican regime faced coup attempts, fought a civil war in the South, and waged a military campaign against a religious separatist community in the North.

42. Cotton textiles were the most important industrial sector in the Brazilian economy up to 1939. In 1907, cotton textiles employed 34 percent of all industrial laborers and absorbed 40 percent of all capital employed in industry and all mechanical power installed in industrial firms. Suzigan, *Indústria Brasileira*, 122.

43. Fishlow, "Origins and Consequences," 313, table 1.

44. Ibid, 315.

45. Stein, *Brazilian Cotton Textile Manufacture*, 25.

46. São Paulo had eighteen cotton textile firms with a total capitalization of 29,600 contos in 1905. The single listed firm, Companhia Industrial de São Paulo, had equity capital worth 2,000 contos and a debenture bond worth 1,200 contos. A second firm, the Banco União de São Paulo, worth 5,000 contos, was almost exclusively dedicated to its Votorantim textile business at this time, but was considered as a bank in the stock exchange listings. If we include the capitalization of this second firm, the listed value of cotton textile capital relative to total cotton textile capital in 1905 rises to almost 24 percent. Wilson Cano, *Raízes da concentração industrial em São Paulo*, 2nd ed. (São Paulo: T. A. Queiroz Editora, 1981), 292, table 55, and *O Estado de São Paulo*, January 7, 1906, January 24, 1909, and January 28, 1910.

47. Cano's figures on the capital value of cotton textile industry growth from 1907 to 1915 totaled 51,900 contos. The difference in value of listed firms from 1907 to 1915 was 38,900 (excluding Votorantim in 1915), or 75 percent of all new capital formation. *Raízes da concentração industrial*, 292–293, table 55 (for capitalization) and table 56 (for output).

48. Stein, *The Brazilian Cotton Textile Manufacture*, 28.

49. Cano, *Raízes da concentração industrial*, 300, table 61.

50. Fishlow, "Origins and Consequences," 315–317.

51. On the bond market and industrial finance, see Hanley, *Native Capital*, 107–109.

52. There is a well-developed scholarly literature dedicated to examining the link between the coffee boom and Brazilian industrialization as well as the ramifications of export-led industrialization on long-term economic development. Principal among these are Cano, *Raízes da concentração industrial* and Warren Dean, *The Industrialization of São Paulo, 1880–1945* (Austin: University of Texas Press, 1969).

53. Cano, *Raízes da concentração industrial*, 296, table 59. São Paulo was responsible for just 15.9 percent of the value of total Brazilian industrial production in 1907 compared to Rio de Janeiro and Guanabara's combined 37.8 percent. By 1919, São Paulo surpassed Rio/Guanabara (28.2 percent) producing 31.5 percent of total industrial output by value. In 1949, it produced close to half of all Brazilian industrial output by value.

54. Eakin, *Brazil,* 82.

55. For more detail on the recovery and its impact on economic growth, see Hanley, *Native Capital,* 99–100.

56. The leading industries of the nineteenth century are analyzed in Suzigan, *Indústria Brasileir,* chap. 3.

57. For sources and listings of all identified companies, see Appendix A in Anne G. Hanley, "Capital Markets in the Coffee Economy: Financial Institutions and Economic Change in São Paulo, Brazil, 1850–1905," (Ph.D. diss., Stanford University, 1995).

58. Using the 1907 industrial census, Cano estimates that there were 1,114 industrial establishments in the state of São Paulo, valued at a total of 131,900 contos. For 1919, Cano separates out investment in industry by joint-stock companies, sole proprietor firms, and "other." The total universe of industrial companies in the 1920 census was 4,145 firms. The 3.1 percent that were joint-stock companies roughly equaled 128 joint-stock industrial companies. Cano, *Raízes da concentração industrial,* 163, 221–222, 225, 303.

59. Cano, *Raízes da concentração industrial,* 306, table 68. The exchange rate for this year was U.S.$0.32 per mil-réis. Rates ranged between U.S.$0.31 and U.S.$0.33 during the period 1905 to 1913. Exchange rates come from Antônio Emílio Muniz Barreto, "Relações econômicas e o novo alinhamento internacional do Brasil (1870–1930)," (Tese de Livre-Docência, Faculdade de Economia e Administração, Departamento de Economia, University of São Paulo, 1977).

13

Trade and Industry in the Indian Subcontinent, 1750–1913

Prasannan Parthasarathi

At the end of the nineteenth century, industrial capacity in India was limited. In the decade 1900–1910, industry accounted for only 11 percent of national income and a little more than 10 percent of total employment. By contrast, in the same period, industrial production contributed 22 percent of gross domestic product (GDP) in the United States and 40 percent in Britain. In 1913, per capita levels of industrialization in India were less than 2 percent of those in Britain and less than one-third of those for Brazil and Mexico. The limited contribution of industrial production to GDP in the subcontinent in the early twentieth century was a stark contrast to the situation some 150 years earlier, when the region accounted for a quarter of the world's manufacturing production. By 1900, this figure had plummeted to less than 2 percent.[1]

The reasons for the limited industrial development of India have been hotly debated. The opposing camps may be divided into internalist versus externalist positions. Externalists place primary responsibility for the lack of industrialization on British colonialism in the subcontinent. In the words of Romesh Dutt, the great critic of British rule: "All the old industries, for which India had been noted from ancient times, had declined under the jealous commercial policy of the East India Company; and when Queen Victoria ascended the throne in 1837, agriculture was left the only national industry of the people. Little was done to foster new industries after the Crown assumed the administration of India in 1858; and the last decades of the century still found the Indian manufacturer and artisan in a state of poverty and decline."[2] The internalists give primacy to social, cultural, or economic conditions within India. Vera Anstey has argued: "We have seen that certain religious ideas and conventions, and the rigid social stratification and conservatism based upon those ideas and conventions, still pervade every sphere of life, and limit economic development at every step. The resulting weakness of the 'economic motive,' and lack of economic enterprise, have prevented full advantage from being taken of existing knowledge, and have prevented adequate use from being made of India's great natural resources."[3]

For much of the twentieth century, the externalists dominated the debate, especially within India, but in recent years, the internalist position has been on the ascendant. Tirthankar Roy has argued that Indian resource constraints were the most important determinants of economic development in the nineteenth and twentieth centuries: "A first step toward a useful alternative approach would be to replace imperialism with economic structure—that is, the constraints and opportunities that took shape under resource endowment patterns that took time to change. In a labor surplus economy facing persistent high risks, conditions of manual labor and behavior toward risks must be the principal links between the past and the present. Indeed, economic change in colonial India can be seen as a process wherein resource constraints were temporarily overcome by reallocation and increasingly industrious labor."[4] Imperial revisionism has gone so far, however, that Niall Ferguson has declared: "Victorian India . . . was booming. Immense sums of British capital were being invested in a range of new industries: cotton and jute spinning, coal mining and steel production."[5]

This chapter argues that colonialism cannot be left out of discussions of Indian industrial development in the nineteenth century. Most critically, colonial trade policies had a powerful impact on the activities of manufacturing enterprises throughout the period. These policies were not only destructive, in that they checked and constrained industrial activity, but also creative, as they channeled manufacturing work along certain lines. The colonial state shaped industrialization not only by its actions, but also by its inactions: its failure to invest in public goods such as education and the expansion of technical knowledge hampered the industrial potential of the subcontinent.

This chapter begins in the eighteenth century, but not from a belief that the Indian subcontinent was on the verge of a breakthrough to modern industry in that period. The social and economic pressures for such a path were lacking in the subcontinent at that time. Milder pressures, however, built up over the course of the eighteenth century to produce significant changes in the relationship between the state and the economy in several areas of South Asia that included a push for new production technologies and a growing interest in technical knowledge among educated classes, especially those connected to the levers of political power. In the closing decades of the eighteenth century, this combination of forces was pushing for important changes in systems of production.

The establishment of British rule in the subcontinent dampened and then, over the span of a few decades, destroyed this eighteenth-century dynamism and redirected economic life in the Indian subcontinent in vastly different directions. This is a familiar claim, but this chapter moves away from a strict focus on political economy that centers on the imperial regime of free trade. Free trade is certainly a central part of the nineteenth-century Indian story, but it must be supplemented by both knowledge and the mastery of techniques of production, which were both reworked in dramatic fashion by the establishment of colonial rule from the late eighteenth century.

THE EIGHTEENTH-CENTURY BACKGROUND

The eighteenth century was a period of rapid social, political, and economic change in several parts of the Indian subcontinent. A number of military innovations placed greater fiscal burdens on states, which has been labeled an Indian-style military fiscalism. The growth of British power in eastern and southern India further worsened fiscal problems as the ever-expanding armies of the British posed a growing threat to the political autonomy of indigenous states.

These fiscal imperatives produced far-reaching changes in the organization of state administrative and revenue systems. In many cases, these changes were extensions of seventeenth-century developments, including standardization of measurement and the growth of bureaucracy. This restructuring of state administrations has received some attention from historians, but less studied are state-sponsored improvements in production technologies. These technological changes served either to increase the revenue base of a territory or to manufacture armaments more effectively. To accomplish these aims, there was a growing interest in knowledge and its codification and diffusion.

Some of these changes have been most abundantly documented for agricultural production. Iqbal Ghani Khan found a significant shift in the content of agricultural manuals from the seventeenth to the eighteenth centuries. In Mughal northern India, an enormous number of manuals for managing elite households were produced. The eighteenth-century versions included far more detailed technical information on yields, soil types, irrigation, and the labor requirements for various crops and tasks facilitating the more careful supervision of agricultural work, substantially raising productivity and output, and thereby the taxable base. The eighteenth-century texts were also more focused on the agriculture of specific regions.[6]

States also directly undertook research to improve techniques of production in the eighteenth century. An outstanding example comes from the northern Indian state of Kota, where state-sponsored work led to the development of new seed strains, novel techniques of grafting, improved breeds of cattle, and a new plow that made it easier to work heavy soils. These discoveries were disseminated throughout the kingdom by means of specially organized agricultural fairs.[7]

The codification of knowledge and the search for new production technologies also extended to manufacturing. In the seventeenth century, knowledge of production methods was collected and recorded in a variety of works, including a series of texts from the Mughal empire known as the imperial book of regulations, which compiled useful knowledge about iron, gold, and other metalworking; sewing, weaving, and cotton carding; and pottery, dyeing, cooking, and alchemy. There were also works devoted to a single topic, especially for metalworking and dyeing.[8]

These types of works continued to be produced in the eighteenth century. In the state of Mysore, Tipu Sultan commissioned compilations of knowledge on diverse topics, including natural history, medicine, metallurgy, and dyeing. Texts were also produced on the manufacture of guns and artillery. In Mysore and elsewhere in the subcontinent in the eighteenth century, interest was also growing in European works that contained technical material. In northern India, Persian and Urdu translations were made of English handbooks on cannons and guns. In Mysore, a large number of European scientific works were translated into Persian under Tipu's patronage. Indian rulers also attempted to gain access to European technology through the Europeans who came to India as soldiers, traders, and adventurers. Some were trained in crafts and came to serve Indian states in a variety of technical capacities. It became routine for Indian rulers to interrogate passing Europeans about their technical knowledge.[9]

Indian state interest in European knowledge continued into the nineteenth century. In the Punjab, which came under British rule only in the 1840s, there was a large contingent of French who assisted the state, mainly with military matters.[10] Even in areas that had come under British domination, nominally independent rulers continued to pursue European knowledge, much as their predecessors in the eighteenth century had. In 1817, the nawab of Awadh was keenly interested in European science and set up a printing press and translation project. Similarly, at the other end of the subcontinent, in the southern kingdom of Tanjore, Raja Serfoji inherited a great library. In the early nineteenth century, he expanded it and supplemented it with scientific equipment. He also sponsored translations of scientific writings from European languages.[11]

Outside the Punjab, these nineteenth-century technical projects took place in a vastly different political, economic, and military context, however. They were more like the leisure-time pursuits of noble dilettantes and were not linked to political projects or attempts at economic transformation. Perhaps even more important, they became the exception rather than the rule in the nineteenth century as the British colonial state took little interest in patronage for technical change or in compiling and diffusing technical knowledge.

Colonial Rule and the Nineteenth Century

Since the late nineteenth century, the deleterious impact of British rule on Indian industry has been central to the critique of colonialism. A long list of writers, from Romesh Dutt to Amiya Bagchi, has criticized the British for permitting the destruction of the Indian cotton textile industry, a historic transformation that has been immortalized in Marx's evocative description of the bones of the handloom weavers bleaching the plains of India. Unlike Europe, however, India experienced no process of creative destruction in which artisanal and hand manufacturing gave way to "modern" industry. This failure has been the other major plank of critique: the absence of British policies to encourage the industrialization of India.

British policy in India served the interests of British industry. Long before Britain itself turned fully to free trade in the 1840s, a regime of free trade was imposed on India, and Indian manufacturers received no protection from British imports. The hostility to Indian manufacturing interests was evident as early as the 1780s, when British cotton manufacturers formed an association and agitated for protection from the East India Company's muslin imports from India. In discussions, the manufacturers suggested that the company give up the muslin trade and instead import raw cotton from the subcontinent. Such a trade, the muslin manufacturers declared, would benefit both Britain and India![12]

The purchasing policies of the British Indian state, which gave preference to British-made goods, further undermined the development of local industry. This policy was perhaps most glaringly evident in the construction of the railways.[13] Colonial trade and purchasing policies have been at the center of the analysis of Indian industrialization under British rule. Less attention has been given to issues of technical knowledge and its generation and diffusion, and there has been little analysis of the fate under British rule of the institutions that had been built in the eighteenth century to perform these functions.

The British established a highly centralized state with unprecedented geographical reach in the subcontinent, which led to a profound transformation in the operation of state power. Colonial rule put an end to the intense competition that had flourished in the late precolonial period between the many kingdoms that dotted the Indian landscape. In addition, the unitary British Indian state dissolved the negotiations and conflicts between different levels of political authority that were armed and ready to protect their claims to a share of sovereignty and a piece of political power. The dissolution of political competition dampened economic life by reducing competition between states, which led to a decline in the technical dynamism of the subcontinent because so much support for the development and diffusion of technical knowledge in the eighteenth century derived from a desire to strengthen state power.

Like its Indian predecessors, the British Indian state was driven to maximize its revenues. The methods by which it achieved this aim departed radically from those of precolonial polities. While the military fiscal pressures of the eighteenth century compelled Indian rulers to push forward improvements in production, both agricultural and industrial, company policy shared this concern only to a very limited extent.

Under British rule in the early nineteenth century, some very limited resources were put into the creation of experimental farms, chiefly devoted to cotton, which had become a valuable export to China. This was coupled with expanding the cultivation of long-stapled varieties that would suit the requirements of the spinning machines of Lancashire. Such British efforts over the whole of the subcontinent were on a limited scale; they were perhaps equivalent to the eighteenth-century efforts of a small or medium-size state such as Kota, Awadh, or Mysore. Nor did the British support other elements of eighteenth-century policies for agricultural improvement, including revenue incentives for the cultivation of high-quality land or high-value crops. And under the British, state support for investment in agriculture through advances and loans dwindled to insignificant levels.[14] Nevertheless, the British were able to collect unprecedented levels of agricultural revenue from their holdings in India. They did this through a ruthless and efficient system of taxation, which was enforced through military supremacy. Instead of higher revenues being achieved through agricultural improvement, they were gained through coercion.

The story of manufacturing under British rule is also one of state disinterest and, in some cases, outright hostility to technical and economic improvement. For eighteenth-century states, there were enormous incentives to adopt new techniques, most critically in the metal industries, in order to produce better armaments. With the establishment of British rule, Indian society was largely demilitarized as the pluralistic political system, with many political centers and armies, was replaced with a unitary system and a single military force, that of the English East India Company. With the elimination of military competition, the need to propel advances in manufacturing no longer pressed on political authorities in the subcontinent.

The army that the British built in nineteenth-century India also came to rely heavily on imported military equipment. In the case of small arms, despite the high quality of Indian-made guns in the late eighteenth century, the East India Company's army relied almost exclusively on weapons imported from Britain. The only exceptions were guns captured in battle, which were sometimes refurbished locally and then issued to company troops. After an absence of about a century, small arms manufacturing was established in British India only in the early twentieth century when a factory was built in Ishapore in 1905. The first Indian rifle was completed in 1907.[15]

British rule did not mean the end of the manufacture of big guns, but the scale on which this activity was conducted in the subcontinent shrank sharply by the early nineteenth century. The English East India Company maintained only one foundry in Bengal for casting brass cannon and howitzers. Iron guns continued to be imported from Britain. In the 1820s, new boring lathes powered by steam engines were introduced at the foundry, and in the opinion of some, the guns manufactured in Bengal were superior to those imported from Britain. From the 1830s, after the foundry had been shifted from Fort William to Cossipore, iron shot and shell were also cast. In the 1850s, a substantial modernization of the foundry was authorized, with new steam engines, a steam hammer, a Ryders patent forge, a Whitworth universal shaping machine, and two new boilers. But the East India Company foundry started to fall behind European big gun production, which began to center on muzzle-loading guns fabricated from wrought iron and steel. By the 1880s, Britain had adopted the breech-loading system, for which steel was the "only possible material." It took several decades for these new armament technologies to reach India. In the meantime, from the 1860s to the early twentieth century, India was reliant on guns imported from Britain as manufacturing activity at Cossipore ground to a halt. Only in 1905 was the manufacture of ordnance resumed.[16]

In the twentieth century, armaments manufacturing exploded in India, especially during the massive buildups of the two world wars. But in the nineteenth century, the production of arms, a major manufacturing activity in the eighteenth century, shrank in size enormously, with profound consequences for Indian industrialization. Under British rule, it was not only production that was scaled back. The creation and diffusion of knowledge was also reduced from eighteenth-century levels. In part, this was because the British ruled India on the cheap: its interests lay in extracting resources for the present, not in investing for the future.

Typical of this approach was the government's response in 1853 to a request from the Cossipore foundry for "some scientific books and periodicals." A majority of the Military Board thought this to be a "quite unnecessary expense and presumed that Superintendents usually bought those they desired at their own expense." The governor-general of India himself declared that there was no need to subscribe to any professional periodicals.[17]

In the early days of their rule, the British were also hesitant to transmit knowledge to Indians. In 1770, the East India Company informed Bombay "that the natives must be kept as ignorant as possible both of the theory and practise of artillery." More than forty years later, in 1813, the Court of Directors of the company feared to disseminate the knowledge of casting ordnance to Indians. British secrecy was not confined to military technology. Well into the nineteenth century, the British in India sought to maintain a monopoly on the art of surveying. As the surveyor-general of India wrote, "The Government ha[s] notified to me that they wish to throw cold water on all natives being taught, or employed in making Geographical discoveries." The active eighteenth-century state patronage for the expansion of knowledge was replaced in the nineteenth century by active state discouragement.[18]

Accompanying the decline in state support was the destruction of major eighteenth-century libraries. The massive library in Lucknow, capital of the state of Awadh, was looted in the late eighteenth century and its books and manuscripts dispersed over the subcontinent and Europe. Similarly, victorious British forces carried off the collections of the Mysore library, and our knowledge of its holdings comes from a list compiled by Charles Stewart, a Persian scholar who worked for the company.

Despite the lack of state support for economic improvement and the inhospitable climate for the growth of manufacturing, there were several impressive industrial ventures in nineteenth-century India, of which the best known is the Bombay cotton industry. Less celebrated and less well known are major undertakings in the metal industries, especially in the first half of the nineteenth century. These ventures, from iron, to cotton, to steel, demonstrate the technical potential that existed in nineteenth-century India and the entrepreneurial skills that were found in the subcontinent. Nevertheless, these abilities had only a limited industrial impact because they were hampered by state policies.

IRON SMELTING IN SOUTH INDIA

In the late eighteenth and early nineteenth centuries, the quality of Indian iron was widely acclaimed in Europe. Descriptions of the iron smelted in the subcontinent were published in the *Philosophical Transactions,* the *Journal of the Royal Asiatic Society,* the *Journal of the Asiatic Society of Bengal,* the *Quarterly Journal of Science, Literature, and the Arts,* and several travelers' accounts. Even Michael Faraday conducted experiments on wootz, or Indian steel.

In 1825, the Madras Council received a proposal from J. M. Heath to construct an iron-smelting works along European principles.[19] Before embarking on the project, he requested an exclusive privilege to produce iron with those methods for the duration of the East India Company's charter. Heath repeated the request in 1829 and reiterated his demand for "provision which should defend me from competition for a reasonable period." He supported this request with the examples of England, where "such protection was given in the reign of King James, I think, to the persons who first attempted to make iron with pit coal," and Russia, where "such protection was afforded to the first persons who established iron works." When the Madras Council considered Heath's first petition in 1825, it noted that "the policy, as well as justice, of affording temporary protection from competition to the authors of any new inventions, or to those who encounter the difficulty, expense, and risk of endeavouring to establish any new manufacture, or to open any new channel of commerce, is, we believe generally admitted" and that "without an example set by him it will not be engaged in by any other person within the period for which this privilege is proposed to be granted." Heath's exclusive privilege was granted after the second request, but progress on the construction of the ironworks was not made until the 1830s.

In the early 1830s, Heath constructed his ironworks in the town of Porto Novo, on the coast of Tamil country, about two hundred miles south of Madras. Iron ore was brought from the interior, in the vicinity of Salem, and was sailed down the Cauvery River, which had sufficient water for this purpose six months of the year. The ore was of very high quality, and its supply, according to Heath, was "inexhaustible." He proceeded to construct two furnaces, which were larger than those customarily used in South India. The yield of each was estimated to be 20 tons of pig iron per week. In addition, Heath built two cupola or blast furnaces and two reverberatory furnaces. The machinery for the works consisted of a 4-horsepower steam engine, a blower driven by bullocks, and a rolling mill. Using wood and charcoal fuel, Heath designed a smelting works to produce wrought iron using the puddling and rolling method.

An East India Company inspection of the facility in late December 1837 reported that the works consisted of four smelting furnaces with two steam engines of 12 and 16 horsepower and a blowing apparatus attached to each engine. There was also a casting house with a powerful crane and other equipment, as well as a forge with two heavy hammers and a tilt worked by a 25-horsepower steam engine. There were five puddling furnaces and eight refineries for converting the pig iron to wrought along with blowing apparatus. Most of the machinery for a rolling mill was on the spot, including a 35-horsepower steam engine. A building had even been erected to house it, but the rolling mill had not yet been put into operation. In addition, there were several smaller furnaces, godowns for storing charcoal, and blacksmith shops. The value of the works was estimated to be over 260,000 rupees.

From the mid-1830s, the works exported substantial quantities of iron to Britain. South Indian iron received rave reviews from several British iron experts, who declared that the quality of the Porto Novo product was equal to that of Sweden. In 1839, it was reported that Porto Novo iron was being used in the Royal Arsenal of Woolwich in place of Swedish iron. Along with its high quality, the Porto Novo iron was extraordinarily cheap. Including the cost of transport, the wrought iron sold for 20 pounds sterling per ton in Britain. This price compared very favorably with the prices for Swedish iron, the best grades of which sold for 40 pounds sterling per ton. Even the best British-made charcoal sheet iron was priced at 35 pounds sterling per ton.

In establishing his manufacturing enterprise, Heath had great success with Indian workers. He brought two expert workmen from Britain to set up and operate the works. After only a year, Heath wrote, "When we commenced the manufacture of malleable iron we had no hope that we should be able to dispense with European labourers in several branches of the business; during the year which has since elapsed, the natives have made such progress in all departments of the manufacture that we think it probable that when the current engagements with our European workman shall have expired, we shall not find it necessary to procure any new men from England." The Indians, therefore, were quick learners and adept at puddling iron. It would also appear that Indian workers were skilled at maintaining and repairing the imported machinery. This is not surprising given other early-nineteenth-century accounts of Indian workers repairing British instruments and machines or adopting them for manufacturing processes.

Technical problems were, in some respects, the least of the difficulties that the Porto Novo ironworks faced. Throughout the 1830s, Heath complained that the enterprise was undercapitalized because the demands of both fixed and working capital were very high in the establishment and operation of the works. A substantial manufacturing center had been built from scratch by sinking funds into machinery. The machinery then had to be transported to India, a six-month voyage, and set to work, which took many months. There was a long stretch of time, therefore, before the investment began to reap a return. Much working capital was also sunk into the final wrought iron, which did not yield a return until it had been transported to Britain and sold. In the early years of the enterprise, markets for Indian iron had to be established, which further lengthened the time before realization of the investment.

Heath was able to raise some capital from Europeans living in Madras and also worked tirelessly to obtain additional capital in Britain. His financial savior, however, came in the form of the East India Company, which extended substantial loans to him, eventually even canceling interest payments. For the company to make such loans to a private manufacturing enterprise was extraordinary, to say the least. And it was unusual circumstances that led to company support for the ironworks.

In the mid-1820s, South India began to descend into a great depression that lasted until the late 1840s. From the early 1830s, East India Company officials began to worry about the bleak commercial situation in Madras. The export of piece goods from the region had virtually disappeared, save for a small trade in blue goods to the Guinea coast. Exports of raw cotton to China had also plummeted, and trade in that item to

Europe was minor. Finally, indigo exports had dwindled to insignificance. As a consequence, there was a steady export of specie to Europe to pay for imports from Britain. If an export trade in iron could be established, Madras would have a lucrative commodity that could be exchanged for goods from abroad.

Indian iron was attractive to East India Company officials because it did not compete with a British manufacture. When exported to Britain, it displaced imports of Swedish iron, not a British-made good. British manufacturers relied on Sweden for high-quality iron used in the production of steel. The possibility of substituting Indian for Swedish metal was raised in 1814 by the naturalist Benjamin Heyne, who wrote: "A quantity sufficient to substitute that which is derived from Sweden might be allowed to be introduced from India without detriment to any but the foreign manufacturer of the article and the merchant engaged in the Swedish trade."[20]

The condition that South Indian iron was not to compete with British metalworks, limited markets. These markets tended to be located outside the subcontinent, largely in Europe, which greatly increased the working capital requirements for the venture. The Porto Novo ironworks found it difficult to sell their product locally because, by the 1830s, substantial quantities of British pig iron were imported to the subcontinent. British iron, although of lower quality, sold for about half the price of Porto Novo iron, and it was adequate for many local uses. Enlarging the Indian market for Porto Novo by taxing imported iron was completely ruled out: the East India Company would not implement a policy that would reduce the market for British manufacturers. At this time, Swedish iron imports into Britain paid a tariff of over 6 pounds sterling per ton.

The Porto Novo ironworks operated within these constraints for a few decades until it quietly closed down in the 1860s. Its story brings to the fore several of the major elements that shaped Indian industrial development between the early nineteenth century and World War I. First, Indian workers were skilled and quickly became highly competent technically. Second, the market for Indian industrial products was circumscribed because Indian manufacturers, subordinate to British interests, received little or no protection. Finally, to succeed in this economic context required enormous entrepreneurial talents. The ways in which these forces shaped industrial development are well illustrated in the cotton textile industry of western India, which began to emerge from the 1850s.

COTTON MANUFACTURING

Machines for spinning cotton were imported to the subcontinent from the second decade of the nineteenth century. A few were brought into Bengal and set to work, but that enterprise had a short life. In the 1830s, a modern cotton spinning enterprise was established in Pondicherry, which was under French rule, for the production of indigo-dyed cloth for export to Senegambia. This cloth was exchanged for Senegalese gum, an essential ingredient for dyeing and printing cloth. With massive support from the French state, including funding to ship the machinery, low-cost loans, and production subsidies, a private spinning factory was established with French machinery. The enterprise operated and even expanded, for about a decade, but it failed with the collapse of the gum trade in 1842.[21]

Although the Pondicherry cotton industry was quite successful, it was unique in the extent of state support and the niche market that it supplied. There were no further attempts to build a local cotton industry with imported machinery until the 1850s, when merchants in western India invested in Bombay and then later in Ahmedabad. The capital that was invested in these undertakings was accumulated over several decades from trade profits with Asia. In particular, this included trade in raw cotton and opium to China, activities in which Indian merchants from Bombay participated on a comparable footing with European trading houses until the mid-nineteenth century. Many of these merchants were also involved in marketing subcontinent cotton yarn and piece goods imported from Britain. Therefore, the establishment of a cotton industry in the city of Bombay brought together two arms of trade for these merchants. In the conduct of these trades, several merchants also came to be acquainted with imported machinery, including steam engines for the operation of cotton presses.

Domestic investment in cotton manufacturing was also propelled by the growing subordination of Indian traders to Europeans from the 1850s. As European markets began to be a more important destination for Indian raw cotton, Indian traders were displaced from the export trade, which began to be dominated by large European managing agencies. Cotton spinning was an alternative economic activity that held some promise of profit. To gain flexibility in their business operations and to hedge their risks due to fluctuations in price and demand for cotton, Indian traders began to invest in cotton machinery. This allowed them to select from a larger portfolio of business options depending on prices and markets. A number of Britons and other non-Indians also started mills in Bombay.[22]

By the end of 1860, ten companies had been organized in Bombay, but the climate for these undertakings deteriorated for several years because of the outbreak of the American Civil War, which produced a cotton export boom followed by a deep bust. From the 1870s, however, textile mills began to be established at a steady clip in western India, and by 1914, there were eighty-five in operation in Bombay and forty-nine in Ahmedabad. Cotton textile factories were also being established in northern and southern India as well as in Bengal.

As is well known, textile manufacturers in India received no protection from the competition of British imports. In the late nineteenth century, when the government of British India enacted tariffs on imported cotton goods for revenue reasons, an equivalent countervailing excise was placed on locally made cottons. As a result of this commercial policy, textile mills in Bombay and elsewhere concentrated their production on coarser yarn, where they had an advantage due to lower labor costs. In the 1890s, 6 percent of total Indian output consisted of better-quality yarn, whereas only 18 percent of British yarn imported into Bombay fell into the coarse ranks. Therefore, little direct competition emerged in the subcontinent between the British and Indian cotton spinning industries during the late nineteenth century. Piece-good imports from Britain also tended to be of higher-quality cloth, generally woven at least in part from yarn of at least middling quality.[23]

Although there was little competition in India itself between local and imported yarns, Indian goods began to displace British yarn in other markets from the 1870s, most important in China, where Bombay rapidly established a commanding position in the market. In 1893–1894, of the 373 million pounds of yarn spun in India, 170 million pounds—nearly 50 percent—were exported. Virtually all of these exports were coarse yarns for Chinese handloom weavers. As in local markets, Indian producers were able to exploit lower labor costs to outcompete British coarse goods in China. Indian exports of yarn to China peaked in 1905–1906, reaching nearly 300 million pounds, and then declined as Chinese and Japanese manufacturers undercut them.[24]

The machinery for Indian producers came largely from Lancashire. Before electrification, it was powered by imported steam engines fueled by a combination of wood and coal. Indian technical staffs, along with British experts, were actively involved in the modification of the machinery to suit conditions in Bombay, Ahmedabad, and elsewhere. In 1895, 30 of the 51 spinning masters, 31 of the 51 carding masters, and 38 of the 61 engineers in the factories of Bombay were Indians. Taking mill overseers as a whole, 141 of the 245 in Bombay city were Indians. By 1940, the proportion of Indians had increased to 347 out of 415 in Bombay city. The Indian technical staff tended to be concentrated, especially in the early days of the industry, in Indian-owned firms.[25]

From the 1920s, there were also some efforts to build machinery in India.[26] There was little British encouragement for textile machinery manufacturing, and the competitive niche that Indian producers had developed to withstand competition from Lancashire limited the possibilities for an indigenous machine-building industry. Beginning in the late nineteenth century, Indian cotton manufacturers took advantage of cheap and flexible supplies of labor, not technologically advanced systems of production. As the industry migrated from Bombay and Ahmedabad to up-country areas that were closer to cotton-growing districts where labor was cheaper, these tendencies were further accentuated as new mill owners set up shop with discarded and second-hand machines. Many of these machines came from mills in Bombay that had gone bankrupt. Only a few producers invested in new equipment. According to Rajnarayan Chandavarkar, "By the 1930s, the Sassoon group of mills preferred to destroy rather than sell their discarded machinery for scrap lest the upstart millowners of Coimbatore bought them and then used these to jostle Bombay out of the domestic market."[27]

Because of free trade, the Indian cotton industry concentrated on the production of coarser weaves with lower thread counts to take advantage of cheap labor costs. Nevertheless, Indian entrepreneurs were able to develop technical expertise and compete effectively with British goods both at home and abroad. Although constraints from British competition were relaxed after World War I, economic stagnation and low wages in India continued to push the cotton industry in labor-intensive directions. As a consequence, technological change in textile manufacturing—and in the economy as a whole—was held back. There was little demand for modernization of machinery or processes. Although this may have been the story in cotton, it does not hold to the same extent for iron and steel, where there was less scope for labor to replace modern machinery.[28]

THE TATA IRON AND STEEL COMPANY

The Tata Iron and Steel Company (TISCO) is one of the great entrepreneurial success stories of twentieth-century India. For the first third of the twentieth century, TISCO *was* the Indian steel industry. The founder of the firm, Jamsetji Tata, made his fortune in cotton manufacturing in the central Indian city of Nagpur. In the 1880s, he learned of the vast deposits of high-quality iron ore in India and decided to manufacture iron and steel in the subcontinent. Based on his experience in cotton manufacturing, where he came to value state-of-the-art technology, he decided to draw on top experts to construct an advanced steel plant. For this, he called on the experience and knowledge of Kennedy, Sahlin and Company of Pittsburgh. After a painstaking search, Tata settled on a site for his works in central India where abundant supplies of water, iron, coal, and dolomite were readily available.[29]

Once the site had been selected, Tata had to raise approximately 2 million pounds sterling. Indian investors were at first reluctant to put money into a risky and unfamiliar project. Tata was unwilling to borrow the money in London because of the control that British financiers would demand over the undertaking. In any case, there may not have been sufficient takers because many Britons doubted that high-quality steel could be manufactured in the subcontinent. Tata's financial problems were resolved as a result of an upsurge of anti-British sentiment in the subcontinent in the first decade of the twentieth century. From 1905, the Swadeshi movement was in full swing, with its push for home production of goods and the boycott of British imports. The Tata family issued a prospectus for shares in their steel works in August 1907, and more than 1.5 million pounds sterling worth of shares were sold swiftly. The subscribers ranged from local princes to the middle classes.[30]

TISCO began production in late 1911 with a technical support staff of Germans and Americans. During World War I, the Germans were interned, and Indians stepped in to perform technical duties with support from American experts. Early in the war, two-thirds of the thirty-one workers in the blast furnace crews were Indians. World War I not only accelerated the technical training of Indians but also created major demand for Tata steel.[31]

Steel production in British India received no protection from competition from imports until 1924, when tariffs were enacted on a wide variety of steel products. During the war, however, steel production was effectively insulated from European competition. European steel was needed for the war effort, and trade routes were disrupted by the operation of submarines. As a consequence, TISCO began to supply the government of India with rails, shells, carriage wheels, and even ferromanganese for a brief time in 1916 because of a global shortage of the alloy. In the second half of 1918, the government purchased 90 percent of Tata's production. The steel had to be up to British standards, and the government set prices at a quarter to a half of those prevailing in the Indian open market. TISCO therefore gave up substantial profits but received in exchange a steady and substantial buyer for its output while earning the goodwill of the government.[32]

With the conclusion of the war, TISCO faced several tough years. During the postwar slump, European steel, especially from Belgium, was dumped in Indian markets, and TISCO was forced to sell below cost on occasion. The British Indian government responded in 1924 with protective tariffs, which were structured so that steel imports from Britain paid a lower duty than goods from continental Europe. This was, of course, the famous system of Imperial Preference. Despite this two-tiered system, the protection from imports was invaluable for the expansion of steel production in the subcontinent. From the mid-1920s, TISCO ambitiously expanded capacity and enlarged two blast furnaces, installed four blowers of very high capacity, and added assorted new equipment. In these years, there was also an Indianization of the technical staff as Europeans and Americans were replaced by newly qualified Indians, many of whom were trained at the Jamshedpur Technical Institute.[33]

In 1934, tariffs on steel were lowered, but demand for steel was rising. This led to the founding of several new firms to manufacture steel in both British and princely India, including the Mysore Iron and Steel Works and the Steel Corporation of Bengal. The expansion of steel capacity was a great benefit for the government of India during World War II, but this transformation moves the story into a new historical phase.

Conclusion

Industrialization in India began to take off in the interwar period when Indian industry was insulated from foreign, chiefly British, competition. Tariffs on imports lent some measure of protection to local manufacturing. The decline in world trade with the onset of the Great Depression meant that fewer European goods made their way to India. In addition, the Raj's precarious finances led to a shift in government purchasing policy toward locally made goods since they were cheaper than British goods.

The industrial activity that began in these decades could not, however, make up for the limited growth in manufacturing in nineteenth-century India. Twentieth-century India paid a heavy price for these nineteenth-century failures. From the 1920s, population growth accelerated. Wages stagnated, and even fell during the Depression; per capita incomes also languished in the first half of the twentieth century. Therefore, the expansion of industrial and manufacturing work in the 1920s and 1930s was as likely to be based on sweated labor as on advanced technology. This form of industrial development has been best explored for the textile trades, which experienced a veritable explosion of dispersed production in weaving during the Depression. It was found in other areas as well, especially in the metal trades with a profusion of small-scale ventures.[34]

Economic life in nineteenth-century India represented a break with the eighteenth century. Nineteenth-century India lacked neither technical skills nor entrepreneurship. What it lacked was the power to make economic policy. As a consequence, Indian manufacturing interests were subordinated to those of Britain, which dealt a deep and enduring blow to the industrial future of the subcontinent. Limited industrial development in the nineteenth century also meant that by the twentieth century there was insufficient capital for industrialization when the task had become far more costly and complex.

Notes

1. This paragraph is based on the calculations of Paul Bairoch, "International Industrialization Levels from 1750 to 1980," *Journal of European Economic History* 11 (1982), 269–333.

2. Romesh Dutt, *The Economic History of India,* vol. 2, *In the Victorian Age, 1837–1900* (Delhi: Low Price Publications, 1990 [1906]), 388.

3. Vera Anstey, *The Economic Development of India*, 3rd ed. (London: Longmans, Green, 1936), 471.

4. Tirthankar Roy, "Economic History and Modern India: Redefining the Link," *Journal of Economic Perspectives* 16 (2002), 128.

5. Niall Ferguson, *Empire: The Rise and Demise of the British World Order and the Lessons for Global Power* (New York: Basic Books, 2003), 164.

6. Iqbal Ghani Khan, "Revenue, Agriculture and Warfare in North India: Technical Knowledge and the Post-Mughal Elites, from the mid-18th to the Early 19th Century" (PhD diss., London University, 1990).

7. Norbert Peabody, *Hindu Kingship and Polity in Pre-Colonial India* (Cambridge: Cambridge University Press, 2003), 131–132.

8. Khan, "Revenue, Agriculture and Warfare."

9. Irfan Habib, ed., *State and Diplomacy under Tipu Sultan: Documents and Essays* (New Delhi: Tulika, 2001); Charles Stewart, *A Descriptive Catalogue of the Oriental Library of the Late Tippoo Sultan of Mysore: To Which Are Prefixed Memoirs of Hyder Aly Khan and His Son Tippoo Sultan* (Cambridge: Cambridge University Press, 1809); Khan, "Revenue, Agriculture and Warfare."

10. Jean-Marie Lafont, "The French in the Sikh Kingdom of the Punjab, 1822–1849," in his *Indika: Essays in Indo-French Relations, 1630–1976* (New Delhi: Manohar, 2000).

11. Iqbal Ghani Khan, "The Awadh Scientific Renaissance and the Role of the French: c. 1750–1820," *Indian Journal of History of Science* 38 (2003), 273–301; Raja Jayaraman, *Sarasvati Mahal: A Short History and Guide* (Thanjavur: Tanjore Maharaja Serfoji's Sarasvati Mahal Library, 1981).

12. Great Britain Board of Trade, BT/6/140, ff. 53–54, Public Record Office, London.

13. Daniel Thorner, "The Pattern of Railway Development in India," *Far Eastern Quarterly* 14 (1955), 201–206.

14. Christopher A. Bayly, *Rulers, Townsmen and Bazaars* (Cambridge: Cambridge University Press, 1983).

15. Henry A. Young, *The East India Company's Arsenals and Manufactories* (Oxford: Clarendon Press, 1937), 223–224.

16. Ibid., chap. 13.

17. Ibid., 147.

18. Ibid., 133, 138.

19. The papers on the Porto Novo works, which are the basis for this section, are found in the correspondence of the Board of Control, Oriental and India Office Collections, British Library, London.

20. Benjamin Heyne, "On the Establishment of Copper and Iron Works in India," 1814, Home Miscellaneous Series, H/258, 562, Oriental and India Office Collections, British Library, London.

21. Richard Roberts, "West Africa and the Pondicherry Textile Industry," in Tirthankar Roy, ed., *Cloth and Commerce: Textiles in Colonial India* (New Delhi: Sage Publications, 1996).

22. S. D. Mehta, *The Indian Cotton Textile Industry: An Economic Analysis* (Bombay: G. K. Ved, 1953), 1–2.

23. Amiya Kumar Bagchi, *Private Investment in India, 1900–1939* (Cambridge: Cambridge University Press, 1972), 229–230.

24. Ibid., 29.

25. Mehta, *Indian Cotton Textile Industry*, 58.

26. Colin Simmons, Helen Clay, and Robert Kirk, "Machinery Manufacture in a Colonial Economy: The Pioneering Role of George Hattersley and Sons, Lt., in India, 1919–43," *Indian Economic and Social History Review* 20 (1983), 277–315.

27. Rajnarayan Chandavarkar, *Imperial Power and Popular Politics: Class, Resistance and the State in India, c. 1850–1950* (Cambridge: Cambridge University Press, 1998), 60.

28. This is why the first modern steel mills in China and Japan were state enterprises.

29. U. K. Jha, "Iron and Steel," in K. V. Mital, ed., *History of Technology in India,* vol. 3, *From 1801 to 1947 A. D.* (New Delhi: Indian National Science Academy, 2001), 490–498.

30. Jha, "Iron and Steel," 498–499.

31. Bagchi, *Private Investment,* chap. 9.

32. Ibid.

33. Ibid.; Dileep M Wagle, "Imperial Preference and the Indian Steel Industry, 1924–39," *Economic History Review* 34 (1981), 121–131.

34. Chandavarkar, *Imperial Power and Popular Politics.*

14

Cultural Engineering and the Industrialization of Japan, circa 1868–1912

Ian Inkster

Any initial conception of Japanese industrialization should perhaps begin with some acceptable benchmarks.[1] Until the political changes of 1868, the autarchic political economy of Tokugawa Japan did not effectively cocoon the economic system of over 30 million people in a dark age, but it did severely limit the free movement of the factors of production and diluted any injection of useful and reliable knowledge from outside. It also taxed land heavily and reduced the commercial motivations of innovative groups.[2] Recent reconceptualizations of Tokugawa performance have stressed demographic change, institutional evolution, and agricultural growth over a very long term, and this has in turn (e.g., through the implications for the rate of growth in agriculture) dampened Meiji economic performance.[3] Debate remains on such matters, and at this stage, we might conclude that a Meiji watershed remains, but there was some continuity of key institutional and cultural forms. A significant by-employments sector thrived during industrialization and thus carried over pre-1868 commercial practices and work skills into the industrialization process.[4] Third, Meiji industrialization is associated with some interesting simple ratios. In terms of approximate growth performance, the agricultural sector grew at around 2 percent, manufacturing at around 4 percent, and infrastructure (including transport) at around 10 percent per annum. Even in the 1890s, the proportion of factory output from mechanized establishments (principally textiles) to net national product was around 1 percent.[5] Initially the economy was saddled with a potentially unproductive, expensive, and resistant aristocratic caste numbering over 7 percent of the population (over 2 million people) that represented more than half the number of all those employed in productive artisanal or commercial occupations. The ratio of foreign trade to national income during Meiji stood at around 10 percent, and by the 1890s, some half of all exports were of cotton yarn and raw silk. Government expenditure as a percentage of national income during Meiji stood at around 10 percent. The initial ratio of the urban population to the total population was approximately 20 percent.

How has Japanese industrialization been conceptualized until recently?[6] The available statistics and a modified Rostovianism[7] produced a view that Japanese industrialization, though possessing many of the features of a late developer, could be interpreted in broadly conventional economic terms and could be seen to a large extent as a model of development for the underdeveloped nations of the Third World.[8] So relatively mature markets for products and factors seem to have been assumed, just as if there had been no Meiji project for sovereign industrial catch-up and modernization.[9]

Conventional questions relate to the character of such markets, and in the Japanese case, this leads to the nature of so-called industrial dualism.[10] A population of some 40 million by the end of the Meiji era had for some time experienced increased money-income differentiation, agricultural improvement, and enhanced production of commercial crops. It was complemented in Meiji by a rejuvenated, mostly nonmetallic by-employment sector–cement making, sugar, glass, pottery, brewing, sake, paper, chemicals, and fertilizer production. A mechanized textile sector supplied 50 percent of Japanese exports in the 1890s. Modernized heavy industries served infrastructure (especially railroads, docks, arsenals, and shipyards), the demands of the military, and the state until at least the Russo-Japanese War of 1904–1905.[11] Indeed, transport and communications strike a first major contrast with China, the vastness of which absorbed without impact all the expenditures on railways and tramway systems in the late nineteenth century.[12] In Meiji Japan, improvements were both effective and low cost because of historical demographic factors. On average, Europe was about half as urban as Japan in 1700, where about 10 percent of the population of 30 million lived in urban areas of over ten thousand inhabitants. Population density combined with *sankin kōtai* (from 1635) and the associated Tokugawa control mechanisms to create–in comparative terms—an especially well-connected civil society long before 1868.[13]

This civil society was a skilled society, and again this must testify to the central importance of by-employments, thriving in transition and linked to new external export markets. Around 1868, literacy was equivalent or superior to that found in Western industrial systems, and almost 1 million youngsters attended some twelve thousand *terakoya* primary schools, many of them from families at least surviving, if not always prospering, in the handicraft system.[14] It should be emphasized that prior to 1868 Japan was living in an essentially wooden age. Most skills were honed by craftsmen in a non-metal-dependent commercial economy, and thus comparison with the power transmission, cogwheel, and gearing techniques of Europe must remain limited—even admitting that until the eighteenth century, many such features developed in Europe using hardwoods, as with wooden wedges for cogs in nineteenth-century corn mills.[15] In Japan, the huge new urbanism of the seventeenth and eighteenth centuries did, however, induce the significant growth of differentiated samurai groupings, the higher and middling levels of whom took over large-scale administrative and commercial tasks as they abandoned both their swords and took up literacy, interurban cultural forms, and new philosophies of work-for-the-national-polity as expressions of codes of noble conduct.[16] As the seventeenth-century writer Saikaku Ihara put it, "It is training rather than birth which counts."[17] Most social mobility took place within the pre-1868 estates, but some allowance for adoption and intermarriage across the estates widened the pools of talent and increased social motives for profit making and commercial success in the face of a more restrictive, traditional moral economy of status and obligation.[18]

Meiji risk taking[19] was clearly and favorably affected by the income redistribution effects of the 1869–1872 reforms that reclassified the samurai as commoners and led to the conversion of their stipends to taxed bonds in 1873–1874, followed by the government's selling-up policy that centered on the Matsukata deflation of the early 1880s. The first of these activist state policies reduced the income stream flowing to former samurai by anywhere from 10 to 75 percent while increasing government income.[20] As this was associated with lump-sum payments and greater freedom of movement and investment, it has been argued that many former samurai and commoners fared very well, and "in particular the landowners, moneylenders, and petty manufacturers at the upper levels of rural society, thrived in the more open social order of the Meiji era."[21] The second state effort, named after Finance Minister Masayoshi Matsukata (1835–1924), diffused technologies and augmented capital at low cost. Combined with social control activity, model factories inducing imitation,[22] and infrastructure development, it seems clear that the environment for entrepreneurial activity improved considerably between 1868 and 1885 under the singular, small-budget auspices of the Meiji state.[23]

Amid much debate concerning the role of the Meiji state, one conclusion is sound: this was a very cheap state. Despite its slogan of *shokusan kōgyō* (increase production and encourage industry), public expenditure as a proportion of national income was around 10 to 12 percent in the mid-1880s, rising to over 20 percent by 1910. This rise reflected the success of two major wars and the resultant credit-worthiness of the Japanese government at home and abroad.[24] A cultural explanation for this lies in the process of cultural engineering: the cost of hiring foreigners and establishing model factories and experimental stations, rails, and telegraphs was kept to a minimum through the inexpensive effectiveness of cultural suasion and momentum.

I might add that the social distinctions between samurai (many becoming entrepreneurs and capitalists) and townsfolk and commoners were formally legalistic and socially fairly clear, and that increased mobilities and interactions were not sufficient during Meiji to bring equivalence with Western patterns of early industrial revolution. That is, despite selective adoption into noble families and cultural expenditures by government, interactions between many entrepreneurs and craftsmen or townsfolk would have been limited in comparison to eighteenth-century Europe. Against this feature of distinctive social locations, spatial locations within large urban environs might have been more favorable to information dispersals early in the Meiji period.[25] In this respect, and with comparison once more in mind, it might be noted that the transition of merchants to industrial entrepreneurship was a weak feature of Japanese industrialization.[26] Many wholesale or ton-ya merchants often resisted Meiji industrialization as a disturbance to the system of social obligations that for so long had provided them with usurious debt-servicing incomes.[27] Again, we should not neglect the manner in which many Meiji craftsmen, well employed in the small-shop part of the economy, combined the functions of innovator, entrepreneur, and capitalist.[28]

We might also briefly consider several of the less conventional characteristics of industrializing Japan, the first of which concerns the nature of technological change. The need of the state was for the absorption and adaptation of best techniques from the West within the constraint of the relative absence of foreign private capital borrowing, low government expenditure, and high rates of labor underemployment.[29] Concentration was thus on low-cost revamping and spreading of existing best techniques in agriculture, handicrafts, and domestic industries, which, after the turn of the century, could employ such highly modern imported technologies as electrical power and machinery in small factories.[30] Western technologies were principally acquired through trade and transfer mechanisms and developed through a variety of agencies situated in key places, against the background of thriving traditional industries feeding off both retained values and new foreign demand. Here, especially in the first years of Meiji, government initiation was crucial, for such transfers of industrial technologies demanded "the spread of information, contact with foreign countries, formation of skills, and change in factory structure,"[31] and the cost and time of all this was primarily borne by the public sector.

A standard thesis is that effective absorption of the "best from the West" took place without cultural transformation or disturbance and alongside the retention of such famous features as group identity, respect for ancestors, a strong work ethos (in which work gives meaning to life, in contrast to Genesis, where man is condemned to work), and a particular appreciation of human relations as well as of traditional artistic expressions of underlying values. In short, Japan boasted the social resolution rather than the legal resolution of disputes.[32] Included among the cultural forces for modernity were an established religious plurality and a relative lack of even a pantheistic metaphysics. A high level of formal education dating from the sixteenth century was massively boosted by the 1872 legislation and encouraged by the greater social mobility associated with the abolition of the noble classes after 1868, which served to increase incentives further. Notions of the "group before the individual" were based on values of social harmony derived from neo-Confucianism, perhaps the basis of Gustav Ranis's "community-spirited entrepreneur,"[33] and daily reflected in complex honorific forms in Japanese grammar and by the seeming vagueness of Japanese expressions available to express refusals, rejections, and negatives, as well as the seemingly nebulous character of the Japanese smile and the commonality of gift giving. This social homogeneity was expressed in the popular aphorism that "the stake that sticks up gets hit."

A related claim is that because cultural borrowing was an early tradition in Japan, technological borrowing was eased during Meiji. Shiratori stresses such cases as Japanese borrowing from (and perfecting) the staple production of tea from China from the ninth century, manufacture of cotton from China and Korea in the sixteenth century, transfers of sugarcane via China, Taiwan, and the Ryūkyū Islands, and silk from China. [34] According to this sort of argument, such a tradition of borrowing set Japan apart from China and India, epitomized perhaps by the absorption of Chinese Buddhism from the sixth century and its mergers with Shintoism, which through shamanism (spirits dwell in all things, including ancestors) created an ancestor-worshiping Buddhism that was an entirely new cultural element. Exogenous cultural impacts created accretions, not revolutions: hence the modern Japanese emphasis on flexibility in business or fusion in science and technology.

During the Tokugawa period, the state as controller did create trajectories of information and influence, particularly with regard to *sankin kōtai* and the monitoring of *Rangaku*, the limited trickle of Dutch-based Western learning centered on Deshima after the 1640s.[35] But a major function of the Meiji state was to convert the essential tacit knowledge embodied in Western technique into something approaching useful and reliable codified knowledge in Japan, and it did this through translations and cheap editions, model enterprises, employment of teachers and skilled workers, and sending students to key locations overseas.[36] In the absence of direct foreign capital importation, no other agency was in a position to cover such a complex transformative role.[37]

The debate is yet open on the contribution to technological progress of a longer-term backdrop of critical inquiry in the natural and physical sciences, a common claim for industrial revolutions elsewhere.[38] Japanese historians generally argue for both a very early period of Western "technique" adoption from the sixteenth century (small firearms, cannon, shipbuilding, navigation, mining, metallurgy, printing, paper manufacture) and a measurable but somewhat separate transition from traditional to modern "science" in Japan from around 1720 to 1854 that did not especially embrace technique. Any group identifiable as a scientific establishment was bureaucratic or composed of employees of the state and the court, whose influence, if at all important, was exerted through such regime interventions as *sankin kōtai* or the Deshima-*rangaku* framework of monitored connection with the outside world.[39]

Historians do not, then, generally argue for a linearity of knowledge-technique development that would be pre- or corequisite to the more general process of industrial modernization, at least not in any manner that conforms easily to the Eurocentric dualism of "science and technology." Sugimoto and Swain at one point judge that during the intellectual breakthroughs of the seventeenth century, there was in fact little development or interest in "techniques strategic to modernizing processes" and that earlier armament development fell away precisely in these years. Archimedean screw pumps that forever broke down were not impressive when labor was underemployed. Until well into the eighteenth century, then, there appears to have been something of a disjuncture between formal knowledge and technique in Japan.[40] More recently, Bartholomew has suggested that certain lacunae were forged in the social and cultural incommensurability of mathematicians and natural scientists in Japan.[41]

A better attack on the question of useful knowledge might be to pursue the shorter term idea of a scientific-technical information frontier that was soon converted into a system of useful and reliable knowledge using especially nurtured sites of interaction between the West and

Japan. The latter ranged from shipyards such as the French-dominated Yokosuka arsenal established in the 1870s to engineering colleges such as the Tokyo College of Engineering dominated by the British or the Sapporo Agricultural School strongly influenced by American experts and advisors. The main function of modernized national capabilities during a period of late development may be not so much that of transfer and settlement as that of translation, selection, and specification. Which bits or parcels of the "best from the West" are potentially reliable and useful? How are these to be recognized and owned? Something of an answer was found in the institution building of the Tempō-Bakumatsu years (1830s–1868)[42] and the post-1868 years, and the creation of an instant scientific establishment.[43] From these years, we glimpse the tortuous efforts of men such as the physician Yoan Udagawa (1798–1846) to translate Western chemical terms and principles into Japanese meanings in his *Seimi Kaisō* (*Foundation of Chemistry 1837–47*).[44] In effect, then, doubts may be cast on any formative role for an earlier scientific establishment, while the Meiji scientific establishment was itself an offshoot of military-industrial modernization. A subject such as chemistry was effectively institutionalized in early Meiji within twenty to thirty years by the Dutch chemist K. W. Gratama's chemical lectures in Osaka in 1869, the publication of textbooks and the foundation of a chemistry department in Tokyo University (1866–1877), and the inauguration of the Chemical Society of Japan in 1878.[45]

We may begin developing the novel elements in a reconceptualization with the claim by Joel Mokyr that within Europe in the late eighteenth century, "the relative contribution of technological progress to economic growth compared to other elements began to increase, and the institutional basis supporting this progress was transformed. The result was the Industrial Revolution." Thus, in a blow, Mokyr dislodges capital and labor from center stage in the conceptualization of the Industrial Revolution in Europe. Technological change succeeds Smithian growth processes and can be understood only in broadly institutional terms, and the material success of the West "must be explained through developments in the intellectual realm concerning . . . useful knowledge."[46] We are thereby introduced to notions of institutions providing knowledge and skill as more than ancillary to a process of technological change that is at the heart of an industrial revolution. If we add to this process approaches that differentiate early industrializers from later ones, we may introduce the notion of a non-profit-maximizing state as a provider of useful and reliable knowledge and its institutions as part and parcel of its search for modern technologies and military industries.

For both early and late industrializers may we abandon the term *science* in favor of the notion of useful and reliable knowledge, which stresses the locations of knowledge both physically and socially, and inhibits our dwelling on the erroneous notion that "applications of knowledge" somehow await "investments in education" and easily measurable increases in human capital formation? Here I am thinking of something like the scholarly publications of the Tokugawa technologist Okura Nagatsune in the early nineteenth century that generalized the experiential knowledge that he had developed as an agriculturalist and cultivator, shifted methods toward better practice through the itinerant persuasion of village *osataru hito* (leading men), and attempted to explicate tacit knowledge, while diffusing tools such as the *eburi* or light raking hoe and the *kusakezuri* weeder.[47]

When we interpose the state, we may introduce motives and measures other than those derivable from more formal economic history. The Japanese state was motivated by both fear and resentment before it became much of an economic animal, but survived this transition without breakdown. The impact of the arrival of Commodore Matthew Perry's fleet in 1854 and his technological demonstration at all levels was broadened with the unequal treaties and the resulting outflow of gold.[48] With motives deeper than the avaricious, measures other than profit become available. Under such conditions, the state will spend disproportionately on means of persuasion and education in order to keep the social costs of industrial revolution down, will take pecuniary losses on strategic projects well beyond those of private enterprise, and may thus act as a loss leader to subsequent entrepreneurial-based industrialization. Such forces of backwardness may have been heightened in the Japanese case through the strong Confucian tradition that denied the invisible hand and emphasized service to the state. Thus the great champion of competitive enterprise, Shibusawa Eiichi, described his aim as being "to build modern enterprise with the abacus and the Analects of Confucius."[49] Successful entrepreneurship might thus go hand in hand with dependency on state contracts, subsidies, or other market interventions. As Crawcour has summarized, such dependency actually validated the civil motives of the businessman, and any failure to obtain such validation "could be both socially and financially disastrous, a situation that gave the government great power vis-à-vis the top ranks of modern business leadership."[50] It certainly might help explain the juxtaposition of an energetically intervening government and relatively small public expenditures in economic affairs.

In Japan and Britain and perhaps elsewhere, basic reconceptualizations may rest on the notion of a strong carry-over of human capital formation from the pre–Industrial Revolution economy into the Industrial Revolution economy.[51] In the case of Britain, this process was itself strongly associated with the earlier formation of a civil society and a complexity of sites and agencies through which useful and reliable knowledge was harnessed to the processes of innovation, entrepreneurship, and capitalist investment, without political breakdown.[52] This allowed a continued exploitation of tacit and codified knowledge using the established skill base. In Japan, in contrast, this process was associated with a watershed regime change that protected cultural forms, invested further and massively in human capital formation, and married this to a novel and inexpensive system of cultural engineering, an intent of which was to show that the past was revocable without revolution.[53] This transition allowed a process of in-transfer of technique and knowledge as principal inputs into the Industrial Revolution without political breakdown despite the dual influence of exogenous ideology and internal pressure on the cultural and consumption standards of small farmers and lower aristocratic groups.[54] In turn, such internal pressures were reduced by the ability of the handicrafts and by-employments sector to benefit from both cultural engineering (through demand) and foreign techniques (through supply).[55] My thesis here is that cultural engineering in Japan further reduced the costs of internal disruption (or, to an equivalent effect, increased the efficiency of expenditures on human capital formation and associated social overheads). This view demands that we reconceptualize the Industrial Revolution as a socioindustrial phenomenon, the trajectory of which was determined by the early institutional innovations of the political regime. It should also be emphasized that successful cultural engineering did not merely act as an inexpensive information system. The latter term tends to be reduced to mechanisms of information dispersal, as if information was a neutral asset. Cultural engineering embraces the emotions, passions, commitments, and ideologies that are engendered by a great deal of new information, especially information that appears to threaten the existing assets of its recipients. In Japan, state-based cultural engineering carried new information about institutions, technologies, and practices within a positive emotive framework.[56]

Something of this may be measured from the statistics of government expenditure. Expenditure on human capital formation and social overhead capital amounted to some 26 percent of government expenditure overall, but averaged 33 percent for the years of nonwar expenditures between 1868 and 1892.[57] This was far ahead of government expenditure on capital formation, which lay at around 13 percent, but reached even this figure only because of very large expenditures after 1898. Most of government human capital formation expenditure was in the form of education and transfer payments to former feudal elements, for example, as with the commutation payments described earlier. Such expenditures financed institutions: something more was required to imbue those institutions with an ideology of civilization, progress, and civic concern.[58] Our conceptualization of Japanese industrialization is forged from the historical relations between such expenditures and the process of cultural engineering. The potency of the latter reduced the cost of the former in producing the transferred technology and information system of the Japanese Industrial Revolution. Public education was clearly the closest institutional link between human capital formation, cultural engineering, and technology transfer,[59] but the power of transfer payments should not be undervalued: as Koichi Emi has commented, they dampened dissent and its costs and were fundamental to the construction of "the new and coming capitalist society to support the initial political revolution."[60] In this formulation, it becomes difficult to distinguish among the possible explanations of the industrial conjuncture.

What we may say is that cultural engineering was at the core of the Meiji industrializing program, explicitly and from the beginning. It focused on an overall cultural program mounted by bureaucrats, scholars, and educators of *bunmeikaika*, a very particular notion of "Westernized civilization" for Japan, which among other features attempted to define the ethical or moral principles common to China, Japan, and the West.[61] Such a program for Japan linked the concepts of rationality, utility, scientism, education, enlightenment, and "the promise of national unification provided by common language habits and textbooks."[62] In such a manner, cultural engineering served to both increase and select from the cultural stock.[63]

Persuasion centered on education, aided and abetted by judicious emphases on cultural traditions by the College of Historiography, reorganized during the political crisis of 1881 when an authoritative version of Japanese history was demanded by the conservative forces in government.[64] Here the emphasis was on *kokutai* (national polity) and its embodiment in the *tennōsei* (emperor system) as established in the Meiji Constitution of 1889.[65] Such Meiji institutions represented a reformulation of earlier traditions of cultural engineering. In how many European states of the sixteenth century can we quote sovereign leaders to the effect that the "printing and diffusion of books is the most important task of a benevolent government"?[66] In a new context, a new system of national learning (*kokugaku*) was appropriate.[67]

It was during the disruptive Tempō reform period (1830–1843)[68] that the regime sponsored the new Confucianism, *bushidō,* and other cultural forms as means of control and social integration, beginning that general articulation of samurai values that became the cultural backbone of industrialization after 1868.[69] Soft rule was inexpensive; it relied on media laws, confinement of assembly, a huge number of administrative formalities,[70] arrests of key dissidents, and official appointments for radical critics. Some increase in democracy at the level of local assemblies and a promise of a full liberal constitution at a later date curtailed the anti-industrialization movements of the 1880s.[71] State distribution of key texts dominated the 1880s and followed the 1879 dictum of the official "Key Concepts of Education" to the effect that the goal of learning and culture was the "drawing into the brain before all else a sense of the moral obligation of filial piety." The Education Decree of 1880–1881 elevated moral training as the top cultural priority, and the textbooks of the following year were dominated by notions of the family-state.[72] As Eifu Motoda, Confucian tutor to the emperor, had proclaimed a little earlier, civilization and enlightenment were more than a mere matter of "knowledge and skills . . . high-sounding ideas and empty theories"; they involved retention of ancestral precepts of benevolence and piety.[73]

All such cultural engineering induced the institutionalization of technological progress, for it encouraged a groupist or communal learning process, action and response by association, a social psychology of acceptance of change for the sake of others, and specific mechanisms of technological diffusion. Thus, agricultural production could be made measurably more productive by the importation of new fertilizers and varieties alongside the institution of a system of itinerant agricultural instructors, agricultural experimental stations and exhibitions, and locally organized *Nodankai* (agricultural discussion societies) and *Hinshukokankai* (societies for the exchange of seeds).[74] From 1867 to 1910, Japanese delegates participated in thirty-eight major international exhibitions. Internal exhibitions of technology began around 1877, and by the 1890s, they attracted over 1 million visitors. The *kyōshinkai* or cooperative-competitive exhibitions were designed for the diffusion of existing techniques and during their peak in the 1880s, they lasted thirty to one hundred days.[75] The establishment of hundreds of trade associations under government auspices from 1883 was designed initially to control product quality, but they soon emerged as regional technical extension centers. If technological change across entire systems is something of a matter of information and numbers, then in Japan, its success depended on prior and contemporaneous processes of cultural engineering.[76]

Because Japan is the only recognizable case of sustained Industrial Revolution outside the Euro-American system prior to 1970, it has become a point of huge interest and a large bone of contention among historians. I have argued that there is a case for some reconceptualization of Japanese industrialization along lines that move it away from many conventional analyses of Europe and the United States. There are specific advantages attached to this approach. It seems to allow a strong differentiation of the Japanese from the Chinese case that does not invoke a simple cultural interpretation. It permits a strong statement to be made about the timing of the Industrial Revolution, because it explicitly distinguishes between the measures of industrial growth and the possible causes of such growth. The approach encourages a perspective on the state or the public sector that goes well beyond investment or labor management or even education in estimating its real function, and allows the notion of the Industrial Revolution as a core phenomenon of an even more complex whole. This approach also encourages the development of a comparative method that might focus on the historical relations of useful and reliable knowledge, cultural engineering, and human capital formation. Finally, the emphasis on a variety of agents and on institution building is in sympathy with any approach to cultural economy that is not too elusive or ill defined.

Notes

1. Japanese economic history was transformed by the Hitotsubashi group, which eventually produced Kazushi Ohkawa, Miyohei Shinohara, and Mataji Umemura, eds., *Estimates of Long-Term Economic Statistics since 1868* (vols. 1–14) (Tokyo: Toyokeizai Shinposha, 1965–); Kazushi Ohkawa, *Differential Structure and Agriculture: Essays in Dualistic Growth* (Tokyo: Kinokuniya, 1972); Kazushi Ohkawa and Miyohei Shinohara, *Patterns of Japanese Economic Development* (New Haven, CT: Yale University Press, 1979).

2. Peter Duus, *Feudalism in Japan* (New York: Knopf, 1969); James I. Nakamura and Matao Miyamoto, "Social Structure and Population Change: A Comparative Study of Tokugawa Japan and Ch'ing China," *Economic Development and Cultural Change* 30 (1982), 229–269; James I. Nakamura, *Agricultural Production and the Economic Development of Japan, 1873–1922* (Princeton, NJ: Princeton University Press, 1966). See also Philip C. Brown, "The Mismeasurement of Land: Land Surveying in the Tokugawa

Period," *Monumenta Nipponica* 42 (1987), 115–155; William W. Kelly, *Deference and Defiance in Nineteenth-Century Japan* (Princeton, NJ: Princeton University Press, 1985).

3. As varied examples, see E. Sydney Crawcour, "The Development of a Credit System in Seventeenth-Century Japan," *Journal of Economic History* 20 (1961), 342–360, and Kōzō Yamamura, "The Agricultural and Commercial Revolution in Japan, 1550–1650," *Research in Economic History* 5 (1980), 85–107.

4. In light of my later argument, it might also be noted that the thriving production of traditional industries meant that the bulk of the populace during Meiji were not much disturbed by new consumption standards. See, in particular, chap. 7 of Susan B. Hanley, *Everyday Things in Premodern Japan: The Hidden Legacy of Material Culture* (Berkeley: University of California Press, 1997).

5. Calculated as 30 percent of all factory output, which in turn was around 3 percent of all national output.

6. It should be mentioned that this must be an idiosyncratic exercise. I depend primarily on the analyses of Ryōshin Minami, *The Economic Development of Japan: A Quantitative Study* (London: Macmillan, 1986), and the twenty-one "classic" essays republished in William J. Macpherson, ed., *The Industrial Revolutions*, vol. 7, *The Industrialization of Japan,* ed. Roy A. Church and E. Anthony Wrigley (Oxford: Economic History Society and Blackwell, 1994).

7. This was modified principally because the new writers did not necessarily adhere to Walt W. Rostow's dictum that the net savings rate had to rise from 5 percent to 10 or 12 percent at the onset of industrial revolution.

8. For different views see Smith, Inkster, Kelley and Williamson, Rosovsky, Dore, and Blumenthal in Macpherson, ed., *Industrialization.*

9. Yet there is every reason to stress that the dissolution of the old guild restrictions around 1873–1874 was not accompanied by market formation but by flooding by inferior products, fraud, cheating, and bankruptcy.

10. This is the existence over time of two industrial sectors: one based on modern technology and a high capital-to-labor ratio and the other with mostly by-employment with opposite characteristics. See Seymour Broadbridge, *Industrial Dualism in Japan: A Problem of Economic Growth and Structural Change* (London: Frank Cass, 1966).

11. Hyman Kublin "The Modern Army of Early Meiji Japan," *Far Eastern Quarterly* 9 (1949), 20–41; Kōzō Yamamura, "Success Illgotten? The Role of Meiji Militarism in Japan's Technological Progress," *Journal of Economic History* 37 (1977), 113–135.

12. It is precisely in the field of transport that loosely formulated contrasts between China and Japan are too often constructed. For counterpoint, see Ting-yee Kuo and Kwang-ching Liu, "Self-Strengthening: The Pursuit of Western Technology," in *The Cambridge History of China,* vol. 10, part 1, ed. John K. Fairbank (Cambridge: Cambridge University Press, 1978), 491–542; Nathan Sivin and Z. John Zhang, "Steam Power and Networks in China, 1860–98," *History of Technology* 25 (2004), 203–210.

13. *Sankin-kōtai* or dual residence was imposed in the early seventeenth century as a regime control system and demanded that nobility provide hostages to central authorities. See especially Thomas C. Smith, *Native Sources of Japanese Industrialization, 1750–1920* (Berkeley: University of California Press, 1988); Toshio G. Tsukahira, *Feudal Control in Tokugawa Japan: The Sankin-Kōtai System* (Cambridge, MA: Harvard University Press, 1970); Nobu Nakai and James L. McClain, "Commercial Change and Urban Growth in Early Modern Japan," in John W. Hall, ed., *The Cambridge History of Japan*, vol. 4, *Early Modern Japan* (Cambridge: Cambridge University Press, 1991), 519–595.

14. Kōichi Emi, "Economic Development and Educational Investment in the Meiji Era," in Mary J. Bowman et al., eds., *Readings in the Economics of Education* (Paris: UNESCO, 1968), 146–167.

15. For eighteenth-century Japanese technology in the world context, see Ian Inkster, "Technological and Industrial Change: A Comparative Essay," in *The Cambridge History of Science*, vol. 4, *Eighteenth-Century Science*, ed. Roy Porter (Cambridge: Cambridge University Press, 2003), 846–882.

16. This is especially true of the *hirazamurai* or middling *samurai* in urban areas. It should be remembered in building any comparative approach that the upper estate in Japan was enormous. How they were cheaply harnessed to industrialization remains a crucial question for all historians of Japan.

17. Saikaku Ihara, *The Japanese Family Storehouse*, trans. George W. Sargent (Cambridge: Cambridge University Press, 1959), 24.

18. Charles P. Sheldon, *The Rise of the Merchant Class in Tokugawa Japan, 1600–1868* (New York: Russell, 1958); R. A. Moore, "Adoption and Samurai Mobility in Tokugawa Japan," *Journal of Japanese Studies* 29 (1970), 617–632; Shigeru Yamamoto, "The Capitalist Logic of the Samurai," *Entrepreneurship* (Tokyo) 7 (December 1983), 1–11; Smith, *Native Sources*.

19. If we wish to consider this comparatively, we should consider two levels: What were the levels of risk calculated by the relevant agents, and what was the likelihood of such calculations being at all accurate post hoc?

20. See William G. Beasley, *The Meiji Restoration* (Stanford, CA: Stanford University Press, 1972).

21. Gordon, *A Modern History of Japan*, 65.

22. See Thomas C. Smith, *Political Change and Industrial Development in Japan: Government Enterprise, 1868–1880* (Stanford, CA: Stanford University Press, 1955).

23. Kōzō Yamamura, "The Role of the Samurai in the Development of Modern Banking in Japan," *Journal of Economic History* 27 (1967), 198–220.

24. Obviously war saw temporary rises financed by extraordinary measures–the Russo-Japanese War, 1904–1905, led to 75 percent of government expenditure being military.

25. I am suggesting here that questions of state-entrepreneur interaction might usefully distinguish social, ideational, and spatial relations. In the early *jōkamachi,* or castle towns, townsfolk might occupy only 20 percent of the land area, the rest given up to daimyo and samurai and religious establishments and their commercial agents. Distinctions between such spatial areas weakened over time.

26. George C. Allen, *A Short Economic History of Modern Japan* (London: Allen & Unwin, 1968), 31–32.

27. Sheldon, *Rise of the Merchant Class in Tokugawa Japan.*

28. Kōzō Yamamura has argued that entrepreneurship may have been more socially widespread. See his "A Re-Examination of Entrepreneurship in Meiji Japan, 1868–1912," *Economic History Review* 21 (1968), 144–158.

29. For the notion of a technological shelf from which the Japanese could choose techniques, see Gustav Ranis, "Industrial Sector Labour Absorption," *Economic Development and Cultural Change* 21 (1973), 387–408; Ian Inkster, *The Japanese Industrial Economy: Late Development and Cultural Causation* (London: Routledge, 2001), chap. 4.

30. Ryōshin Minami, *Power Revolution in the Industrialization of Japan, 1885–1940* (Tokyo: Kinokuniya, 1987).

31. Blumenthal in Macpherson, ed., *Industrialization,* 436.

32. For a particular interpretation of this, see Takie S. Lebra, *The Japanese Self in Cultural Logic* (Honolulu: University of Hawaii Press, 2004), chap. 2.

33. Gustav Ranis, "The Community-Centered Entrepreneur in Japanese Development," *Explorations in Entrepreneurial History* 8 (1955), 80–98.

34. Fumiko Shiratori, *The Cultural Background of Japanese Economic Development* (Manila: De La Salle University Press, 1995).

35. A major source on Japan-West interaction is the marvelous series of reprints by Japan-Netherland Institute, Tokyo and CHEE, Leiden, replete with wonderful illustrations of the Deshima Dagregisters from 1639 forward in lavish volumes.

36. It should be understood that there was little even in the Japanese language that could stand for reasonable representations of Western scientific concepts; thus there was a "translatability of culture" problem that no private agency could take on as a commercial proposition.

37. For the analysis in an entirely different context, see David J. Teece, *Managing Intellectual Capital: Organizational, Strategic and Policy Dimensions* (Oxford: Oxford University Press, 2000), chaps. 1, 4, 6.

38. Ian Inkster, "Potentially Global: A Story of Useful and Reliable Knowledge and Material Progress in Europe circa 1474–1914," *International History Review* 18 (2006), 237–286.

39. It is appropriate to emphasize that the Dutch-Deshima connection was tenuous and perhaps has been overemphasized, see Inkster, *The Japanese Industrial Economy*, 100–101, 136.

40. See chapter 5 of Masayoshi Sugimoto and David Swain, *Science and Culture in Traditional Japan, 600–1854* (Cambridge, MA: MIT Press, 1978), 231.

41. See chapter 2 of James Bartholomew, *The Formation of Science in Japan: Building a Research Tradition* (New Haven, CT: Yale University Press, 1989).

42. In this period, the prime motive of Han authorities in gaining Western knowledge was *i no jutsu wo motte I wo seisu,* to "suppress the barbarians with the arts of barbarians."

43. Bartholomew estimated that there were some fourteen hundred holders of the *hakushi* doctoral degree 1888–1920, some 53 percent drawn from *shizoku* or the former samurai stratum.

44. Togo Tsukahara, *Affinity and Shinwa Ryōku: Introduction of Western Chemical Concepts in Early Nineteenth-Century Japan* (Amsterdam: J. C. Gieben, 1993).

45. Mori Tanaka, *Seimi Kaiso Kenkyū* (Tokyo: Kodansha, 1975).

46. Joel Mokyr, "The Intellectual Origins of Modern Economic Growth," *Journal of Economic History* 65 (June 2005), 285–287.

47. For a fine and characteristically nuanced account, see the essay on Nagatsune as chap. 8 of Smith, *Native Sources*.

48. Through the exploitative treaties of Commerce and Amity, by trading Japanese silver coins for gold coins, foreigners gained gold at around three and a half times the value of the silver metal exchanged. See Tara Ohkura and Hichio Shimbo, "The Tokugawa Monetary Policy in the Eighteenth and Nineteenth Centuries," *Explorations in Economic History* 15 (1978), 101–124.

49. As quoted in Johannes Hirschmeier, "Shibusawa Eiichi: Industrial Pioneer," in William W. Lockwood, ed., *The State and Economic Enterprise in Japan* (Princeton, NJ: Princeton University Press, 1965), 243.

50. E. Sydney Crawcour, "Industrialization and Technological Change, 1885–1920," in *The Economic Emergence of Modern Japan*, ed. Kōzō Yamamura (Cambridge: Cambridge University Press, 1997), 50–116, 113.

51. All of the recent work that emphasizes slow industrial change over the years from around 1660 in Britain or around 1550 in Japan lends itself to this type of approach.

52. A major argument has been that political disruption from below was severely limited by the divisions among industrial workers.

53. For the general quandary, see Gary S. Morson, *Narrative and Freedom: The Shadows of Time* (New Haven, CT: Yale University Press, 1994), 203–209.

54. All those on stipends or in more traditional occupations suffered from the inflation of the 1860s to 1880s. Much dissatisfaction was indeed politicized.

55. Note that even mechanized textile production drew on traditional skills from by-employments and remained relatively labor intensive.

56. For something on information mechanisms after 1884, see Kaoru Sugihara, "The Development of an Informational Infrastructure in Meiji Japan," in *Information Acumen: The Understanding and Use of Knowledge in Modern Business*, ed. Lisa Bud-Frierman (London: Routledge, 1994), 75–97.

57. For sources and more detail, see Kōichi Emi, *Government Fiscal Activity and Economic Growth in Japan, 1868–1960* (Tokyo: Kunokuniya Bookstore Co., 1963); Inkster, *Japanese Industrial Economy*, 89–95.

58. Carol Gluck, *Japan's Modern Myths: Ideology in the Late Meiji Period*, (Princeton, NJ: Princeton University Press, 1985).

59. There remains little doubt that the principal focus of education expenditures was on cultural engineering rather than technical training.

60. Emi, *Government Fiscal Activity*, 109.

61. For detailed writing on *bunmeikaikam*, see S. Hattori in Toshimichi Ōkubo, ed., *Meiji keimō shisō shū* (Tokyo: Chikuma shobō, 1930), 417–424; Carmen Blacker, *The Japanese Enlightenment: A Study of the Writings of Fukuzawa Yukichi* (Cambridge: Cambridge University Press, 1964); Tatsusaburō Hayashiya, ed., *Bunmeikaika no kenkyū* (Tokyo: Iwanami shoten, 1979).

62. Douglas R. Howland, *Translating the West: Language and Political Reason in Nineteenth-Century Japan* (Honolulu: University of Hawaii Press, 2002), 17.

63. It is thus possible to argue that a long-existing Japanese "cultural stock" asset lay dormant until Western interaction triggered growth-promoting reactions after 1868.

64. Margaret Mehl, *History and the State in Nineteenth-Century Japan* (Houndsmills: Macmillan, 1998), 26–34.

65. *Kokutai* lay well beyond *seitai* (the political system), for it bonded national morality and rational consciousness into a state-centered spiritual force.

66. The best example is Tokugawa Ieyasu (1542–1616), who had as much right to the title of warrior and strategist as to that of intellectual. See Warren W. Smith, *Confucianism in Modern Japan*, 2nd ed. (Tokyo: Hokuseido Press, 1973), 10.

67. The renowned academic historian Ukichi Taguchi (1855–1905), in his *Short History of Japan's Civilization* (Tokyo: Taguchi Ukichi, 1877–1882) asserted the status of Japan not by emphasizing its uniqueness but by identifying its similarities with the West.

68. It should be emphasized that peasant and regional rebellions had been increasing for some long time. See Herbert Bix, *Peasant Protest in Japan, 1590–1884* (New Haven,

CT: Yale University Press, 1986); Anne Walthall, *Social Protest and Popular Culture in Eighteenth-Century Japan* (Tucson: University of Arizona Press, 1986).

69. See chapter 3 of Ian Inkster, *Technology and Industrialisation: Historical Case Studies and International Perspectives* (London: Variorum, 1998), 256–265.

70. In the short pre-Satsuma years (1873–1877), 4,555 notifications addressing matters of social and civil control were issued. Calculated from *Kinji Hyōron,* January 3, 1878, 2.

71. Inkster, *Technology and Industrialisation*, 114–123.

72. Munesuke Mita, *Social Psychology of Modern Japan*, trans. Stephen Sutoway (London: Kegan Paul International, 1992), 225–229.

73. Quoted in Gluck, *Japan's Modern Myths*, 105.

74. For the impact of biological improvement in agriculture, see Yujiro Hayami and Vernon W. Ruttan, "Factor Prices and Technical Change in Agricultural Development," *Journal of Political Economy* 78 (1970), 1115–1141.

75. In 1887 alone, 317 exhibitions were opened, involving products from 180,000 exhibitors displaying 430,000 articles before over 2 million people.

76. The role in information diffusion and cultural suasion of the new system of meritocracy in the civil service that developed during 1885–1888 may have been significant.

15

WHAT PRICE EMPIRE? THE INDUSTRIAL REVOLUTION AND THE CASE OF CHINA

Peter C. Perdue

Two questions dominate the study of industrialization in China: Was imperial China a contender to be the first country to industrialize, and why was China so slow to industrialize once the Industrial Revolution had begun? Twenty years ago, no one had ever considered the first question remotely plausible, but the great burst of Chinese economic growth since the reforms of the 1980s and further research on the vigor of the imperial Chinese economy have shown how advanced China was on the eve of industrialization. Recognizing China as substantially equal to Europe in economic terms around the year 1800 also questions conventional explanations of the Industrial Revolution in Europe.

Even if we accept the argument that China was nearly equal to Europe, and that the "Great Divergence" between East and West began only in the early decades of the nineteenth century, we still need to explain why China was so slow to catch up. Despite efforts at industrialization in the late nineteenth century, China made much less progress than Western Europe, the United States, Russia, or Japan, a fact that became obvious with its loss to Japan in 1895. What happened to the potential for economic growth seen in 1800? Here again, many of the conventional explanations seem less convincing in light of what we now know.

Some new research on China's economic history calls into question inherited explanations of European and Chinese industrialization. I have no final answers, only further questions for research. But I would insist that the industrial revolution now has to be examined in a comparative, global framework. Yet local ecologies—particular combinations of natural resources, topography, demography, political structures, and trading institutions—all decisively shaped the timing and character of industrialization.

Kenneth Pomeranz's *The Great Divergence* impressively synthesized research on late imperial China's economy and challenged conventional assumptions about European economic and technological superiority. He showed that many measures of economic progress, like standards of living, agricultural productivity, the spread of markets, urbanization, demographic structures, property rights, and environmental pressures, put imperial China and Europe at approximately equal levels in the late eighteenth century.[1]

Pomeranz, unlike many other scholars, tried to compare similar units. China was a vast continental empire; Europe was a collection of states. Since most economic data are organized by nation-state, most comparative analyses set the entire Chinese empire against England, for example.[2] Simply because China was so diverse, its average state of development must be less than the advanced parts of England. It is much more appropriate to compare the lower Yangzi valley (or Jiangnan), the most economically advanced area of China, with England or other advanced regions of Europe. Well provided with water transport and mercantile capital, a textile center, and a key site of trade with all parts of the Empire, it had an area of 43,000 square kilometers, only one-third the size of England, and about 16 percent larger than the Netherlands. Its population was roughly 20 million in 1620, compared to England's 4 million, rising to 35 million in 1850, when England had 18 million.[3]

More difficult comparisons would measure the entire Chinese empire in the Qing period against European Eurasia (up to the Ural Mountains). Qing China at its maximum was somewhat larger and more densely populated than Europe, including Russia. The Qing Empire had a land area of 12 million square kilometers and a population of 150 to 200 million in the early eighteenth century, while Europe had 9.6 million square kilometers and 120 to 150 million people.[4] Closer comparisons would set "China proper" (the eighteen core provinces of Qing China, omitting Manchuria, Mongolia, Xinjiang, and Tibet) against Europe without Russia, both at roughly 4 million square kilometers. Because data sources varied widely across Europe and reliability varied widely across the Chinese empire, these comparisons would be even rougher than the Lower Yangzi and England. On the whole, however, one would have difficulty showing that the total production of Europe was significantly larger than imperial China by 1800, or that Europe as a whole grew faster than China in the eighteenth century. Angus Maddison estimates that from 1700 to 1820, China's gross domestic product (GDP) grew at an annual rate of 0.85 percent, while Europe (excluding Russia) grew at 0.68 percent per year. He finds that China's GDP was 17 percent larger than Europe's in 1820, but its per capita GDP was only 60 percent of Europe's.[5]

Life expectancy at birth in England ranged from around 32 in 1650 to 40 in 1750. In north China, a poorer area than the lower Yangzi, it was 35.7 for males around 1800.[6] Average caloric intake was similar—

about 2,000 to 2,500 calories per day. The European capital stock was not obviously larger than China's, although Europeans and Chinese accumulated different things. Europeans held more livestock, allowing them to eat more meat and to plow northern wet soils; Chinese had higher yields per acre, however, because of the efficient recycling of human manure from cities and substantial investment in water conservancy. The positive effects of dikes and the digging of channels on agricultural production in China mimicked those of animals in Europe.

We may undermine assertions of European uniqueness either by showing that China had similar features or that the unique features of Europe had little direct consequence for economic growth. Here I will sketch only the most common arguments about demographic structure, market structure and property rights, agrarian productivity, mercantile capital, and the European state system.

Many historians have emphasized that the Western European demographic system, with a late age of marriage, a relatively high percentage of unmarried, and primogeniture inheritance, kept population growth below that of agrarian production. The resulting surplus could be invested in urban, commercial, and industrial growth. Asian populations, by contrast, which had early and relatively universal marriage, as well as a frequent equal division of property among sons, were thought to have no incentives to control fertility. They bred up to the limits of agrarian production, keeping the entire population at bare subsistence. This argument constantly recurs to explain the poverty of Asia compared to Europe.

James Z. Lee and his collaborators have decisively refuted these Malthusian explanations by using detailed demographic records to show that Chinese peasant populations did respond to food scarcity by controlling fertility within marriage through abstention, herbal abortifacients, and infanticide. Married women in China had fertility rates markedly lower than their counterparts in Europe. Chinese women had at most 6 children over their reproductive lifetime, while married European women averaged 7.5 to 9.[7] The differing demographic regimes still had comparable effects on the ratio of population to production. The lower Yangzi valley grew at the slow rate of 0.3 percent per year during the eighteenth century, much more slowly than agricultural production, allowing the creation of a significant surplus.[8] This is why living standards and life expectancies were comparable between China and Europe. We cannot divide China and Europe on the basis of demography.

Chinese markets, on the whole, constrained capital movements less than most of Europe, and China's labor was relatively free to move. The empire had considerable migration, competitive markets linked in a hierarchy of marketing centers, frequent land sales by contract, and property rights enforced in magistrates' courts.[9] Much of Europe controlled labor movement through serfdom or enforced apprenticeship, limited land sales by entailment, and controlled market prices in the interest of mercantilism or to protect urban subsistence. On the whole, imperial China's land and commercial systems looked more like the most advanced regions of early modern Europe, the Netherlands and southern England, than like Eastern Europe or Russia.[10]

Other scholars have argued that the competitive European state system stimulated early-modern economic growth. Unlike Smithian arguments for the benefits of free trade, they rely on Werner Sombart's and Joseph Schumpeter's stress on the benefits of war and monopoly. Indeed, the aggressive expansion of European colonial powers after 1500 generated pressures to rationalize political and economic systems in the interest of state power and wealth. Peer Vries, however, in his review of this literature, points out the neglected costs of war and the benefits of Asian empire.[11] He doubts that the net effect of these destructive wars helped Europe as a whole. He notes also that the lengthy peace created by Asian empires and their relatively favorable or hands-off policies toward the economy encouraged long-term capital investment.

Wars, of course, kill people, destroy productive capital, and ravage economies, while diverting resources into economically unproductive activities. But they can also stimulate spin-off innovations that lead in the long run to technical progress.[12] Only certain wars, however, bring more progress than destruction. Most obviously, low-cost wars yielding quick victories fought on other people's territory can be profitable and innovative for the victors. Fortunate England fought its wars at sea; most of continental Europe did not have this luxury. European wars were not the secret of industrial growth.

Where does this leave us? We might look for other long-term processes and structural differences between Europe and China that should have significant economic effects. The new emphasis on "applied knowledge" seems to follow this well-worn track. Yet most economic historians still agree that the particular discoveries of the seventeenth-century scientific revolution in Europe had no direct effect on the development of productive technology. Some argue that only Enlightenment Europe created the productive synergy between theory and practice that generated useful and reliable knowledge for technological change. This view recognizes that Chinese scholars used many of the same intellectual techniques as scientific researchers in Europe—clarification of hypotheses, empirical research, open critique and disputation of theories—but claims that they applied them only to philology. In fact, the *jingshi*, or "statecraft," school of scholars in seventeenth- and eighteenth-century China devoted themselves precisely to linking theory to practical knowledge. They stressed that officials and local elites must understand the realities of farming, trade, and water conservancy in order to practice good governance. They wrote agricultural manuals to encourage farming improvement and tried out their theories in demonstration plots.[13] To give only a few examples, Xu Guangqi, a leading seventeenth-century scholar, published the *Nongzheng Quanshu* (*Comprehensive Treatise of Agricultural Management*) in 1625–1628, following a long tradition of scholarly works on agricultural technology.[14] Song Yingxing's *Tiangong Kaiwu* (*Works and Materials of Heaven*), published in 1637, contained detailed descriptions and illustrations of production techniques.[15] The *Huangchao Jingshi Wenbian* (*Imperial Collection of Essays on Statecraft*), first published by He Changling in 1826 in 120 volumes, with many sequels during the nineteenth century, discussed such topics as land clearance, grain trade, famine relief, currency control, water conservancy, and the improvement of literacy.[16] These practical officials distributed their writings widely to the literate public. (Even imperially commissioned encyclopedias attracted private imitators who published cheaper editions.) William T. Rowe's brilliant biography of one such official, Chen Hongmou, shows how he used this knowledge to improve the lives of the people he governed.[17]

This is not to claim that China, any more than Europe, had perfectly joined the world of the scholar and the world of the artisan or farmer. Still, Chinese social norms allowed close contact between literati elites and the common people. Most literati returned to their rural homes after serving in office or during three-year leaves from office for the funerals of their parents. Their communities expected them to contribute to local welfare. They sponsored projects like bridges, roads, waterways, and orphanages to improve their reputations. Officials posted around the Empire worked closely with these knowledgeable elites. Both groups shared interests in improving agriculture and promoting economic growth. Furthermore, agriculture, commerce, and scholarship were not sharply separate occupations: unlike in Japan or France, elites who engaged in farming or trade were not despised.[18] Sons of many elite families went into business, while the sharpest minds studied for examinations. Since merchants bought low-level degrees in the examination system and financed successful scholars, they shared an interest in its perpetuation.

Official contracts gave merchants opportunities to trade in grain, salt, textiles, metals, tea, and other goods in daily use. No wonder the liberal physiocrats idealized China as a model society. It was an idealization, but one not entirely removed from reality.

It is more difficult to study the connection of officials and elites to artisans. Any comparative discussion of innovation in the technical arts and the role of the European scientific revolution would have to cover the shared technical culture of East Asia. We can at least say that East Asian artisans demonstrated spectacular creativity in areas such as porcelain, lacquer, and sword making. It was the brilliant quality of tea trays and porcelain that attracted Europeans to Asia in the first place and stimulated them to produce substitutes.

A common misperception holds that technical innovation ceased in China after the year 1400. The cancellation of the great imperially sponsored voyages into Southeast Asia and the Indian Ocean in the 1430s, however, ended only one imperial project. In the next century, Ming China completed two of the world's greatest large-scale engineering works: the Grand Canal and the Great Wall. Both had substantial economic and technological effects. Other institutional changes in the Ming and Qing periods, like the single whip tax reform, which converted nearly all tax payments from labor service and agricultural goods to silver, greatly stimulated the commercialization of agriculture. New World crops like maize, sweet potatoes, tobacco, and indigo, once adapted to Chinese soils, generated new cash-crop opportunities while improving the nutrition of farmers on hill lands. Li Bozhong argues that there were "extremely important" innovations in the technology of spinning and the specialization of labor in the Jiangnan textile industry in the seventeenth and eighteenth centuries.[19] Once farmers learned how to hand spin cotton thread in moist underground chambers, the textile industry also flourished in dry northern China. These institutional and economic innovations were not as spectacular as, for example, the Portuguese ability to place guns on ships, but they arguably had a more positive average economic impact.

So I am generally skeptical of arguments that claim greater innovative activity for Europeans than East Asians in the centuries preceding the Industrial Revolution. Not all innovations have positive effects on economic growth; much military technology, for example, has a mainly negative impact. Historians of technology know well that technologies need to be examined as systems, not as isolated new artifacts.[20] Different societies innovate in different ways, and comparable measures are difficult to find.

If, for the sake of argument, we posit the rough technological equality of China and Europe, what follows for explanations of the Industrial Revolution? Rather than a long-term development of many centuries based in uniquely European characteristics, it looks more like a fairly sudden, unexpected event beginning in Britain in the late eighteenth century and spreading to much of Europe in the nineteenth century.

The "late and sudden" thesis does not require us to call the Industrial Revolution an "accident." It means only that long-term structural processes do not inevitably determine outcomes and that we need to specify more carefully what aspects of institutions have definite economic effects. Many layers of mediation separate general cultural or epistemological characteristics from productive output.

STRUCTURE VERSUS POLITICS IN ECONOMIC GROWTH: QING CHINA IN THE EIGHTEENTH CENTURY

Most of the studies in this book do not invoke long-term structural causes for industrialization. Instead, they focus on the relatively short-term activities of state officials and enterprising individuals. They show how politics shapes economic decisions. For example, British-branded products sought global markets on the basis of quality, modernity, and national origin, and they sold especially well among colonists of British ancestry abroad. Britain's first government trade mission, the Macartney embassy, went to China in 1793; China's insulting refusal of open trade led to later efforts to open China by force.

Several Chinese imperial decisions significantly affected economic and technical change. During times of expansion, military policies for provisioning stimulated concern for civilian welfare and had positive effects on economic growth. After the empire reached its peak in the mid-eighteenth century, however, other forces began to undermine its ability to stimulate the economy and secure its borders.

The Qing dynasty (1636–1911), particularly during its flourishing age (c. 1670–1760), stands out as a time of military victory, territorial expansion, government consolidation, and economic growth. Qing expansion fostered institutional and economic innovations, which made possible the high standard of living of the eighteenth century.[21] One key institutional innovation, the Grand Council, was created to facilitate faster military decision making and communication. This small group of high-ranking officials discussed secret reports that went directly from provincial officials and field generals to the emperor.[22] Originally devised by the

Yongzheng emperor (r. 1723–1735) to support his wars against the Zunghar Mongols, the Grand Council under the Qianlong emperor (r. 1736–1796) became the empire's most important decision-making apparatus. The Qianlong emperor expanded the functions of the Grand Council well beyond military mobilization. He demanded monthly reports of grain prices throughout all of China's more than 150 prefectures and insisted on careful audits of the empire-wide granary system that distributed grain to local markets in order to level price fluctuations. The price reports were not public, but officials used them to make grain marketing predictable and to stabilize prices. They used granaries not to substitute for private markets, but to prime market flows so that merchants delivered grain where it was needed. For example, if shortages tempted merchants who were expecting further price rises to stockpile grain, officials could lower prices by releasing supplies to the market, forcing speculators to sell instead of hoard. Officials, through their granary distributions, thus transmitted information about grain prices to merchants over the entire empire. In the eighteenth century, the empire-wide granary system could directly distribute 7 percent of the annually marketed grain in the empire. In famine years, provincial granaries could feed up to 5 percent of the population for at least two months. Because of the inelastic demand for grain, such changes in supply could significantly affect market prices. This vast system faced many obstacles, including rotting grain, official peculation, and conflicts over where and how to distribute relief. Still, when they worked, granaries facilitated market exchange and improved the welfare of the average subject.[23]

After conquering Xinjiang in 1760, the Qing promoted large-scale immigration and agrarian settlement by Han Chinese peasants. Military governors strictly supervised cultivation by soldiers and new settlers. Officials investigated supplies of water, soil quality, and animal power, relying on their experience with military colonies during the expansionary period. The data on land yields from these military and civilian farms are the most accurate we have for any region of China.[24] Here, since rich landowners could not conceal surpluses and shift taxes to poorer cultivators, accurate data on yields made for more equitable taxation.

Thus, the secret memorial system, a rapid communication system originating in military demands, ensured stable grain prices for the entire civilian population of China and fair taxes for those under military rule. Since it rapidly disseminated useful, reliable knowledge about market conditions and circumvented routine procedures, it offset incentives to conceal harvest difficulties, peculate funds, or neglect granary upkeep. This institution depended on military expansion. In the flourishing age, the emperors had to guarantee both military supply and tax collection. They had incentives and capabilities to stimulate market exchange and higher peasant living standards. The emperors and their officials policed the bureaucracy through auditing, and they rewarded local elites who helped to relieve famines. The high standard of living of the eighteenth century was the result not just of favorable economic conditions but of an activist state.

Official-merchant cooperation in frontier provisioning offers another example of militarily induced innovation. After winning over Mongol allies, the Qing settled them in defined pastures. The Qing now had to feed settled or confined Mongols in remote, relatively unproductive agrarian regions, in addition to the frontier Han settlers. And scarce herds had to supply meat and militarily capable horses. The answer was to promote frontier trade with both surrendered Mongols and others like the Kazakhs who had abundant herds and needed food and other goods from the interior. At first, Qing officials tried to control this trade under the guise of the "tribute system." The tribute system confined trade to specific frontier towns and years, limiting the size of trade missions and the goods to be exchanged, but canny Central Eurasian merchants soon found ways around these restrictions. They brought excess goods to the frontier markets and pleaded for generosity, threatening to dump the goods or touch off Mongol revolts if they were refused. Exasperated officials often ended up disposing of the unwanted goods at fire-sale prices.[25] It soon became apparent that subcontracting trade to merchants was far more effective. Merchant families who specialized in frontier trade extended their networks from textile-producing regions in the lower Yangzi to the Mongolian frontier. These merchants delivered Qing products to the frontiers, licensed under the policy of "merchant management under official supervision." In the nineteenth century, this subcontracting arrangement supported China's industrialization drive, but the idea originated in the frontier wars of the eighteenth century.[26]

Promoting merchant contractors helped the Qing to integrate the newly conquered regions into the core of the empire. As merchants penetrated Mongolia and Xinjiang, they tied these regions into networks that soon went beyond officially licensed trade. Nomadic tribal leaders settled in towns and developed a taste for Chinese goods from the interior. In Ürümchi, the capital of Xinjiang, exiled literati discovered the delights of entertainment quarters and exotic customs. Military intervention in southwest China likewise stimulated the regional economy.[27] As with the granary system, military demands positively affected the civilian economy of the frontier.

These two examples indicate how the Qing state actively fostered both integrative market exchange and military expansion. Military victories brought new allies who joined larger market networks. Until the late eighteenth century, the cost of these frontier wars did not break the budget. Emperors boasted about their abilities to spare their subjects the burden of large military expenditures. Some Jiangnan literati criticized the expensive frontier wars: they were repressed, but their critiques kept the emperor and his officials on their toes. An emperor who purported to follow Confucian ideals could not openly support excessive military expenditure.

In sum, the eighteenth-century Qing state promoted economic growth directly and indirectly. A number of officials who specialized in frontier governance actively supported agrarian, commercial, and educational development.[28] Expanded territory offered new opportunities for farmers to clear land, and Qing officials offered them seeds, tools, animals, and tax-free loans to help them get started. Frontier trade with government support offered merchants new opportunities for profits. Greater territory helped to relieve demographic pressure in the core of China, and the defeat of the Mongols brought more security on the northwest border.

Why, then, was China not the first country to industrialize? As R. Bin Wong points out, the factors causing Smithian growth—extension of markets, division of labor, commercialization—are distinct from and not necessarily connected to the technological innovation that started industrialization.[29] Jiangnan in China, Britain, and other places like the Netherlands were at roughly similar stages in developing the preconditions for industrial growth. The actual breakthrough was not predictable from these conditions alone. As Pomeranz argues, China followed the "normal" path, and Britain could well have been a failed Jiangnan, rather than vice versa.

European economic historians confirm that radical technological changes do not derive directly from general indicators of commercialization and market expansion. Joel Mokyr has compared the cluster of macroinventions producing the First Industrial Revolution to the large random mutations that create radical shifts in the direction of biological evolutionary change. Gregory Clark proposes the counterintuitive hypothesis that inventors of radically new technologies do not act from economic self-interest, because their innovations are so different that they do not offer immediate material returns: "If technical progress occurs it must be because the society rewards innovators in other than material ways, through fame and admiration. Alternatively, technical progress can only occur if innovators themselves value it for other than the material rewards it conveys."[30]

China had just as many creative, innovative people as any other society, but the Chinese put their efforts mainly into areas that promised material rewards. These included shipping strategic goods to the frontier, relieving food crises, studying for examinations, or repairing dikes. They did not include steam engines, mechanized paper mills, or the exploitation of maritime colonies. Where creative Chinese put their technical skills depended at least in part on incentives offered to them by the state, which set the boundaries of possibilities for technical change.

We can propose a simple model connecting Chinese military mobilization to economic change. It only applies to wars and economies under specific conditions. In Qing China, military mobilization for expansion to the northwest stimulated institutional changes to increase the speed of communications, state knowledge of the economy, and cooperation with merchants. These changes had direct and indirect effects that stimulated economic growth. The state directly invested in increased agrarian production to support its peasants and soldiers and in the commercial exchange of grain and other products, like copper. Indirectly, territorial expansion stabilized peasant welfare and extended the reach of market capital into new frontiers.

Failures of Industrialization in the Nineteenth Century

Beginning in the late eighteenth century, however, these positive feedback effects reversed themselves. Many of the institutions created in the flourishing age proved inadequate to keep domestic stability or meet the challenge of the industrializing West. We still do not have very specific reasons for this institutional inadequacy. Attributing it to "corruption" is too vague: there were spectacular corruption cases in this period, but why did they flourish just at this time? Population pressure is another common explanation, but again, why did it have such deleterious effects at this time and not earlier?

I suggest that we can address the difficulties of the Qing in the nineteenth century more adequately if we understand the difficulty it had in maintaining the achievements of the previous century. The effort to hang on to the newly conquered territories seriously hampered policymaking to deal with the new threat from the southern coast. This particular form of "imperial overreach" limited the ability of even the savviest officials to put resources into industrialization. Conventional explanations blaming unequal treaties, indemnities, and opium imports or pointing to differences between China and Japan tend to neglect these constraints.

China, unlike France or India, did not have to concern itself with the threat of British manufactured goods for quite some time. Until 1890, the dominant Chinese import was not a manufactured product but opium, a product of Britain's agrarian colony, India.[31] Even after 1890, cotton cloth and yarn were only about one-third of Chinese imports by value, and Chinese imports and exports were roughly in balance. Handicraft spinning did decline sharply by the early twentieth century, but handicraft weaving nearly held its own, continuing to supply three-quarters of cloth production in 1910. British manufacturers found themselves quite frustrated by Chinese resistance to textile imports. In 1859, when England had still failed to increase its textile exports to China significantly, W. H. Mitchell noted that the British "were about to start in competition with the greatest manufacturing people in the world, with a people who manufactured cloth for themselves when the nations of the West wore sheepskins, and that any development of our manufactures in this country must necessarily be very slow," because the "beautiful and simple economy" of peasant household production "renders the system literally impregnable against all the assaults of foreign competition."[32]

Astute traders like Mitchell saw the Chinese domestic economy as highly competitive, but ideologues blamed the closure of the Chinese market on obstructive traditional "mandarins" and urged forcible measures to open markets. In the Second Opium War (or Arrow War) of 1856–1860, the British opened ten more ports to trade and reduced Chinese tariffs on imported goods. Chinese historians have placed the blame on these "unequal treaties" as the main cause of China's failure to industrialize in the nineteenth century. But thanks to these treaties, China increased its tea exports to Britain, while the British did not succeed in taking over the large domestic Chinese market in woolen and cotton textiles, as they had expected.[33]

The unequal treaties did limit Chinese capabilities to respond to foreign encroachment, but their impact has been exaggerated. Opening ports to trade by itself did not lead to a flood of Western goods into the country. The indemnities, before the Sino-Japanese War of 1895, amounted to a relatively small part of the state budget. The takeover of the Maritime Customs by an international staff in order to pay the indemnities improved the efficiency of revenue collection. Certain Chinese merchants took advantage of the new global trading system to amass huge amounts of capital.[34] The Cantonese merchant Wu Bingjian, worth U.S.$56 million in 1834, was one of the wealthiest men in the world, exceeding his European contemporary, Nathan Rothschild. So it was not shortage of capital or imperialist pressure on the treasury that limited Chinese investment in industrial technology.

Comparisons between China's failure and Japan's success at late-nineteenth-century industrialization often focus on differences between the two countries in the premodern era, but the two societies were similar in many ways. Both China and Japan were predominantly agrarian societies with highly developed marketing networks, dense populations, large cities, and shared cultural traditions. Both used the same slogan, "rich country strong military," to guide their drive for wealth and power. But one clear difference between China and Japan stands out: China was a true empire, and Japan, until 1895, was not. Although both states had emperors who made universal claims to dominion based on heavenly support, only Qing China after the eighteenth century ruled peoples with radically different cultural traditions and ecologies. Although Japan contained diversity among its more than two hundred domains, and Hokkaido and the Ryūkyūs had significant contacts with other peoples, Tokugawa Japan did not attempt to encompass the large open spaces and large cultural gaps of the Qing empire. The Qing effort to maintain the entire territorial space conquered by the eighteenth-century emperors significantly affected its response to Western imperial pressure in the nineteenth century.

An imperial debate in the 1870s over land versus sea defense shows how imperial legacies constrained the Qing drive to improve its military, fiscal, and economic strength. By this time, the empire had lost several wars to Western powers and had suffered a series of cataclysmic internal uprisings. Yet the dynasty survived, with foreign support, and launched its "self-strengthening movement." Leading provincial governors like Li Hongzhang, Zeng Guofan, and Zuo Zongtang focused on military modernization as the means of rebuilding national strength. Like Meiji Japan, they founded centers of new technical education, relying on Western advisors. The Fuzhou dockyard and Naval Academy, begun in 1866 with French support, and the Jiangnan arsenal, created in 1867, were two key sites.[35]

Li Hongzhang (1823–1901), governor-general of Zhili province, where the capital Beijing was located, strongly advocated maritime defense. His rival, Zuo Zongtang (1812–1885), an energetic military man and great proponent of military and economic modernization, had been sent to Xinjiang in 1867 to put down a major uprising led by the Turkic Muslim leader Yakub Beg. During the 1870s, Western and Japanese imperial powers put strong pressure on the Qing's defenses. The Russians moved into the Ili valley of Xinjiang to safeguard their interests. In 1870, a Chinese riot that killed French missionaries in Tianjin caused the French to mobilize warships, forcing China to issue an apology and to pay reparations of 250,000 taels. Japan temporarily seized Okinawa in 1874, withdrawing only after China paid reparations of 500,000 taels. A few years later, one of the worst droughts in China's history struck northern China from 1876 to 1879.

In the environment of crisis in 1874, Li Hongzhang and Zuo Zongtang vigorously debated where to allocate the state's resources. Li, mainly fearing Japanese aggression, argued strongly for reinforcing maritime defenses and building a navy. Because it lacked a strong navy, China had been forced to buy off Japan instead of fighting to recover Okinawa. Zuo Zongtang, on the other hand, having defeated Muslim rebels in Shaanxi and Gansu in 1873, argued for strengthening land defenses in the northwest to ward off Russian advances and to put down Yakub Beg.

Both sides had plausible arguments. Advocates of coastal defense linked the promotion of a navy to the building of dockyards and munitions factories. They pointed to the advantages Japan had already achieved from industrial and military development. Li Hongzhang argued forcefully for "chang[ing] our [old] methods and employing [new] talents." He estimated the total costs of naval defense at 10 million taels per year. He regarded Xinjiang as a drain on the empire's resources that was "not worthwhile," and he did not think Xinjiang could be held for long. He proposed "temporary renunciation" of control of the region, leaving local Muslim chieftains autonomous while Han Chinese remained in military colonies. The money saved should be shifted to coastal development and defense.[36]

The coastal advocates' arguments mirrored those that had been raised in the eighteenth century against the emperors' campaigns: Xinjiang was mainly barren, too far away, and not worth defending. In the past, expansion might have aided both security and prosperity, but now the empire had to conserve its resources to meet the larger threat from the sea.

Zuo and his supporters, however, argued that the expanded Qing borders had to be defended. If they allowed the Russians to advance in the west, the British and other powers would place further demands on the coastal defenses. Ding Baozhen compared Russia's threat to a "sickness of the heart and stomach, which is near and serious," while the maritime threat was "like the sickness of the limbs, which is distant and light."[37] Zuo himself knew the importance of coastal defense: he had directed the Fuzhou dockyard on the southeast coast. Yet he argued that Western powers wanted only harbors and ports, not large amounts of territory. He also urged the lessons of history: the Qianlong emperor had rejected doubts about the frontier campaigns and achieved great success. He invoked a long-standing imperial tradition of mobilizing the most important strategic assets against any force that threatened to take Mongolia, and then Xinjiang.

In the end, the court supported Zuo's arguments, giving him the funds he needed to recover Xinjiang. From 1875 to 1881, Zuo spent 51 million taels to suppress the Xinjiang rebellion and restore the economy of the region. The navy was allocated only 4 million taels per year, of which only 1 million taels per year were actually appropriated.

Certainly in hindsight, the strategic priorities of the court seem to have been dictated by an "obsolescent" obsession with the threat of nomadic warriors and blindness to the new challenge from the sea, but at the time, the Russian threat was serious. Furthermore, the debate over land versus sea defense was not just about military strategy; it also affected the direction of efforts to industrialize. Coastal defense appropriations supported related industrial enterprises, while funds sent to Xinjiang went to the old formations of horsemen, spears, infantry, and land settlement. China's naval defeat by Japan in 1894 provokes speculation about what would have happened if the navy had received the funds it needed. The court made a fateful strategic choice that had significant economic implications.

Shortly after this debate, Qing officials faced a third demand for resources. The famine of 1876–1879 struck all the provinces of northern China, especially Shanxi, inflicting death by starvation on 10 to 13 million Chinese peasants. These years were a time of global climatic disturbance, during one of the most severe El Niño–Southern Oscillation phenomena of recent centuries. Mike Davis has described in gripping detail the suffering created during these years in the colonized world and in China.[38] He blames much of the death toll on the cruel indifference of colonial regimes to the plight of their non-European subjects. China, however, was not a colony, and it still had freedom of choice. Li Hongzhang led relief campaigns against major flooding that hit Zhili province in the 1870s. Relief efforts in the late 1870s were more limited but still significant. Shanxi alone received 700,000 taels in direct relief and 270,000 taels in retained land taxes and customs revenue. Other provinces loaned Shanxi 1.5 million taels, and private contributions from local gentry and merchants were estimated at 12 million taels. About 3.4 million of Shanxi's total population of 15 million received some relief.[39] Foreign missionaries raised 200,000 taels for distribution to famine victims in the first international relief campaigns directed at China. Cash relief was preferable to grain distribution because it was easier to carry, and those who had enough money were able to buy grain. The use of cash relief indicates that China's grain markets still functioned in the midst of disaster. Those who starved suffered more from "entitlement failure," in Amartya Sen's terms, than an absolute lack of grain. On the other hand, much of the relief did not reach those who suffered the most. Poor transport blocked many grain routes, and pervasive corruption diverted large sums of money.[40]

Yet the Qing relief campaign pales by comparison with the effort put into recovering Xinjiang. Zuo Zongtang even bought 5.5 million kilograms of grain from the Russians to relieve the peasantry of the far northwest, and took out a loan of 10 million taels to suppress the rebellion. The great debate over land versus maritime defense did not address the poverty of the peasantry of northern China.

This neglect provokes further speculation: What if the Qing had put most of its funds into relieving and reviving the northern Chinese peasantry instead of into frontier defense at either border? The key defect in famine relief delivery was the poor roads of northern China. As one missionary reported: "Grain has been bought in abundance at Tientsin and elsewhere by the Governor's agents, but all the beasts of burden in [Shanxi] and the adjoining provinces are not sufficient to carry it. When matters have come to such a pass, it is a small thing to say that the roads are so narrow in the mountains that half of the carriers are obliged to travel by night, whilst the other half travel by day to prevent delay in waiting at the defiles."[41]

Why did the Qing state not invest in better roads, not to mention rail transportation, in this region? We might expect millions of lives to have been saved, not to mention the improvement of agriculture and the further development of the cotton textile industry that had grown in the eighteenth century. The Boxer Rebellion of 1900, a famine-induced rebellion in northern China that dealt a fatal blow to the dynasty's chances for survival, might not have occurred. The Chinese economy depended far more crucially on the welfare of the peasantry in its core northern region than on distant frontier trade. By the early twentieth century, neglect of the infrastructure of the rural economy of northern China had allowed many once-prosperous areas to become abandoned, insecure, poverty-stricken peripheries.[42] Yet in the eighteenth century, the Qing military had built major roads into remote Mongolia and Xinjiang through the deserts and grasslands to support its campaigns.

These reflections suggest that the territorial expansion of the eighteenth century and the Qing determination to hold onto the fruits of empire seriously constrained its ability to industrialize in the nineteenth century. Some conservatives defended tradition simply for its own sake, but leading statesmen agreed on the need to strengthen the country through an energetic program of military and economic modernization. They could not, however, resolve the dilemmas of defending an overexpanded empire in the face of an onslaught by a multitude of powers. Their inability to agree with each other weakened their ability to overcome conservative resistance. They ended up neglecting the agrarian heartland of the empire: these were the people who brought the empire down.

The fate of Qing China shows that long-term structural changes do not adequately predict or determine economic outcomes, and successful industrialization depends on delicate timing and appropriate policies that respond to changing global geopolitical and economic conditions. In the late nineteenth century, both China and Japan tried to learn from the West, but China's response was regionally diverse and strongly inflected by its imperial past. Not until China became a united nation after 1949 could it launch a sustained industrialization drive that benefited all of its people while still preserving most of the imperial legacy.

Notes

1. Kenneth Pomeranz, *The Great Divergence: China, Europe, and the Making of the Modern World Economy* (Princeton, NJ: Princeton University Press, 2000). Reviewed by Peter C. Perdue in H-World listserv (www.h-net.msu.edu/~world), August 2000, with extensive discussion on H-World and H-Econ listservs. Pomeranz rebuts Philip Huang's criticism in the following exchanges: Philip C. C. Huang, "Development or Involution in Eighteenth-Century Britain and China? A Review of Kenneth Pomeranz's *The Great Divergence: China, Europe, and the Making of the Modern World Economy*," *Journal of Asian Studies* 61–62 (2002), 157–167, and Kenneth Pomeranz, "Beyond the East-West Binary: Resituating Development Paths in the Eighteenth-Century World," *Journal of Asian Studies* 61–62 (May 2002), 167–181. Also see Robert Brenner and Christopher Isett, "England's Divergence from China's Yangzi Delta: Property Relations, Microeconomics, and Patterns of Development," *Journal of Asian Studies* 61 (2002) and Roy Bin Wong's comments at http://www.aasianst.org/catalog/wong.pdf.

2. Peer Vries, surveying this literature, compares England with China for "pragmatic" reasons, since so many others do, not to justify it analytically. Peer H. H. Vries, *Via Peking back to Manchester: Britain, the Industrial Revolution, and China* (Leiden: Research School CNWS Leiden University, 2003).

3. Li Bozhong, *Agricultural Development in Jiangnan, 1620–1850* (New York: St. Martin's Press, 1998), 3, and *Jiangnan di zaoqi gongyehua* [Early Industrialization in Jiangnan] (Beijing: Shehui KexueWenxian Chubanshe, 2000), 19. Review in English by Roy Bin Wong in *International Journal of Asian Studies* 1:1 (2004), 185–187. Li Bozhong uses a minimal definition of the region, restricted to eight prefectures located in modern Jiangsu, Shanghai, and Zhejiang. Other authors include wider areas in this region. The province of Jiangsu as a whole, for example, has an area of 100,000 square kilometers.

4. Colin McEvedy and Richard Jones, *Atlas of World Population History* (New York: Penguin, 1978).

5. Angus Maddison and Organisation for Economic Co-operation and Development. Development Centre, *Chinese Economic Performance in the Long Run* (Paris: OECD, 1998), 40. Maddison, however, assumes a constant per capita GDP in China of $600 from 1280 to 1820, which seems implausible, and he does not include Russia in Europe, while using the maximal Qing area (12 million square kilometers) for China. Including Russia would give Europe nearly equal GDP to China in 1820, but lower its per capita GDP to 1,042 instead of 1,129.

6. James Z. Lee and Cameron Campbell, *Fate and Fortune in Rural China: Social Organization and Population Behavior in Liaoning, 1774–1873* (Cambridge: Cambridge University Press, 1997), 60, 76–81; Pomeranz, *Great Divergence*, 37.

7. James Z. Lee and Wang Feng, *One Quarter of Humanity: Malthusian Mythology and Chinese Realities* (Cambridge, MA: Harvard University Press, 1999), 85–87.

8. Bozhong, *Agricultural Development in Jiangnan*, 20.

9. Peter C. Perdue, "Constructing Chinese Property Rights, East and West," in *Constituting Modernity: Private Property in the East and West*, ed. Huri Islamoglu (New York: I. B. Tauris, 2004).

10. Jan de Vries and Ad van der Woude, *The First Modern Economy: Success, Failure, and Perseverance of the Dutch Economy, 1500–1815* (Cambridge: Cambridge University Press, 1997).

11. Vries, *Via Peking back to Manchester.*

12. Merritt Roe Smith, *Harpers Ferry Armory and the New Technology* (Ithaca, NY: Cornell University Press, 1977).

13. Timothy Brook, "The Spread of Rice Cultivation and Rice Technology into the Hebei Region in the Ming and Qing," in *Explorations in the History of Science and Technology in China*, ed. Li Guohao, Zhang Mengwen, and Cao Tianqin (Shanghai: Chinese Classics Publishing House, 1982); Peter C. Perdue, *Exhausting the Earth: State and Peasant in Hunan, 1500–1850* (Cambridge, MA: Harvard University Press, 1987).

14. Arthur W. Hummel, ed., *Eminent Chinese of the Ch'ing Period, 1644–1912* (Washington, D.C.: U.S. Government Printing Office, 1943–1944), 318.

15. Song Yingxing, *Tiangong Kaiwu* (Beijing: Zhonghua Shuju, 1978); Sung Yingxing, *T'ien-kung k'ai-wu: Chinese Technology in the Seventeenth Century*, trans. Tu Zen Sun and Shiou-chuan Sun (University Park: Pennsylvania State University Press, 1966). Further evidence is in Francesca Bray, "Instructive and Nourishing Landscapes: Natural Resources, People, and the State in Late Imperial China," in Greg Bankoff and Peter Boomgaard, eds., *A History of Natural Resources in Asia: The Wealth of Nature, 1500–2000* (New York: Palgrave Macmillan, 2007).

16. Selected translations into English of these essays are found in Helen Dunstan, *Conflicting Counsels to Confuse the Age: A Documentary Study of Political Economy in Qing China, 1644–1840* (Ann Arbor: University of Michigan Press, 1996).

17. William T. Rowe, *Saving the World: Chen Hongmou and Elite Consciousness in Eighteenth-Century China* (Stanford, CA: Stanford University Press, 2001).

18. For a biography of one man who engaged in all these occupations during his lifetime, see Henrietta Harrison, *The Man Awakened from Dreams: One Man's Life in a North China Village, 1857–1942* (Stanford, CA: Stanford University Press, 2005).

19. Bozhong, *Early Industrialization in Jiangnan*, 37–38.

20. Thomas P. Hughes, *Networks of Power: Electrification in Western Society, 1880–1930* (Baltimore, MD: Johns Hopkins University Press, 1983); Merritt Roe Smith and Leo Marx, eds., *Does Technology Drive History? The Dilemma of Technological Determinism* (Cambridge, MA: MIT Press, 1994).

21. Peter C. Perdue, *China Marches West: The Qing Conquest of Central Eurasia* (Cambridge, MA: Harvard University Press, 2005), 547–565.

22. Beatrice S. Bartlett, *Monarchs and Ministers: The Grand Council in Mid-Ch'ing China, 1723–1820* (Berkeley: University of California Press, 1991).

23. Pierre-Etienne Will and Roy Bin Wong, eds., *Nourish the People: The State Civilian Granary System in China, 1650–1850* (Ann Arbor: University of Michigan Press, 1991), 481, 484.

24. See data in Perdue, *China Marches West*, 583–587.

25. Sources translated in Ibid., 575–581.

26. Ibid., 263.

27. Dai Yingcong, *A Key Strategic Area: The Southwest Under the Qing* (Seattle: University of Washington Press, 2005).

28. For an excellent biography of one of these officials, see Rowe, *Saving the World*.

29. Roy Bin Wong, *China Transformed: Historical Change and the Limits of European Experience* (Ithaca, NY: Cornell University Press, 1997), 13–70; Pomeranz, *The Great Divergence*.

30. Gregory Clark, "Economic Growth in History and in Theory," *Theory and Society* 22 (1993), 871–886; Joel Mokyr, *The Lever of Riches: Technological Creativity and Economic Progress* (New York: Oxford University Press, 1990).

31. Albert Feuerwerker, "Economic Trends in the Late Ching Empire, 1870–1911," in *The Cambridge History of China,* vol. 11, part 2, *Late Ch'ing, 1800–1911,* ed. John K. Fairbank and Kwang-ching Liu (Cambridge, MA: Harvard University Press, 1980), 25, 48.

32. W. H. Mitchell, "Correspondence Relative to the Earl of Elgin's Special Missions to China and Japan, 1857–1859," in *Chûgoku Kindai Keizaishi Kenkyû Josetsu,* ed. Masatoshi Tanaka (Tokyo: Tokyo Daigaku Shuppankai, 1973).

33. Jo Yung Wong, *Deadly Dreams: Opium and the Arrow War in China* (Cambridge: Cambridge University Press, 1998), 451.

34. Sucheta Mazumdar, "Chinese Hong Merchants and American Partners: International Networks in a New Age of Global Commerce, ca. 1750–1850," *Journal of World History.* Forthcoming.

35. Hummel, ed., *Eminent Chinese,* 643, 721.

36. Immanuel C. Y. Hsu, "The Great Policy Debate in China 1874: Maritime Defense vs. Frontier Defense," *Harvard Journal of Asiatic Studies* 25 (1964–1965), 215.

37. Ibid., 219.

38. Mike Davis, *Late Victorian Holocausts: El Niño Famines and the Making of the Third World* (New York: Verso, 2001).

39. Paul Richard Bohr, *Famine in China and the Missionary: Timothy Richard as Relief Administrator and Advocate of National Reform, 1876–1884* (Cambridge, MA: Harvard University Press, 1972), 54–55.

40. Basic studies of the famine include Bohr, *Famine*; China Famine Relief Committee, *The Great Famine* (Shanghai: American Presbyterian Mission Press, 1879); He Hanwei, *Guangxu chunian Huabei di da hanzai (1876–79)* (Hong Kong: Zhongwen Daxue Chubanshe, 1980).

41. China Famine Relief Committee, *The Great Famine,* 49.

42. Kenneth Pomeranz, *The Making of a Hinterland: State, Society, and Economy in Inland North China, 1853–1937* (Berkeley: University of California Press, 1993).

Index